Antenna Applications Reference Guide

The McGraw-Hill Engineering Reference Guide Series

This series makes available to professionals and students a wide variety of engineering information and data available in McGraw-Hill's library of highly acclaimed books and publications. The books in the Series are drawn directly from this vast resource of titles. Each one is either a condensation of a single title or a collection of sections culled from several titles. The Project Editors responsible for the books in the Series are highly respected professionals in the engineering areas covered. Each Editor selected only the most relevant and current information available in the McGraw-Hill library, adding further details and commentary where necessary.

Hicks • PLUMBING DESIGN AND INSTALLATION REFERENCE GUIDE

Covers the fundamentals of plumbing design and installation for a wide variety of industrial buildings and structures. Culled by Tyler G. Hicks from several McGraw-Hill books.

Hicks • POWER PLANT EVALUATION AND DESIGN REFERENCE GUIDE

Provides concise evaluation and design information for power plants serving many different needs—utility, industrial, and commercial. Culled by Tyler G. Hicks from several McGraw-Hill books and magazine articles.

Johnson & Jasik • ANTENNA APPLICATIONS REFERENCE GUIDE

Includes practical information and guidelines to antenna applications in all areas of communication. Comprised of one full section of Johnson and Jasik's Antenna Engineering Handbook, *Second Edition. Prepared by Richard C. Johnson.*

Markus and Weston • CLASSIC CIRCUITS REFERENCE GUIDE

Collects in one source hundreds of electronic circuits immediately useful in a wide variety of applications. Culled by Charles D. Weston from Markus's Sourcebook of Electronic Circuits, Electronics Circuits Manual, *and* Guidebook of Electronics Circuits.

Merritt • CIVIL ENGINEERING REFERENCE GUIDE

Offers quick reference to major civil engineering fields: structural design, surveying; geotechnical, environmental, and water engineering. A condensation by Max Kurtz of Merritt's Standard Handbook for Civil Engineers, *Third Edition.*

Woodson • HUMAN FACTORS REFERENCE GUIDE FOR ELECTRONICS AND COMPUTER PROFESSIONALS

Presents all essential data on human factors (ergonomics) relevant to the electronics and computer fields. Compiled by Wesley E. Woodson from his Human Factors Design Handbook.

Woodson • HUMAN FACTORS REFERENCE GUIDE FOR PROCESS PLANTS

Makes available to engineers and specialists all essential data on human factors (ergonomics) relevant to the process industries. Compiled by Nicholas P. Chopey from Woodson's Human Factors Design Handbook.

Antenna Applications Reference Guide

Editors

Richard C. Johnson

Georgia Institute of Technology,
Atlanta, Georgia

Henry Jasik

(Deceased)
Formerly Vice President,
AIL Division of Eaton Corporation
Deer Park, Long Island, New York

McGraw-Hill Book Company

New York   St. Louis   San Francisco   Auckland   Bogotá   Hamburg
Johannesburg   London   Madrid   Mexico   Milan   Montreal   New Delhi
Panama   Paris   São Paulo   Singapore   Sydney   Tokyo   Toronto

Library of Congress Cataloging-in-Publication Data

Antenna applications reference guide.

  (The McGraw-Hill engineering reference guide series)
  Originally published as part 3 of: Antenna
engineering handbook. 2nd ed. 1984.
  Includes index.
  1. Antennas (Electronics)—Handbooks, manuals, etc.
I. Johnson, Richard C. (Richard Clayton), 1930-
II. Antenna engineering handbook, 2nd ed.   III. Series.
TK7871.6.A49   1987   621.38′028′3   86-20807
ISBN 0-07-032284-8

_1)_
_621·38028'3_
JOH

1234567890   DOC/DOC   8932109876

ISBN 0-07-032284-8

Antenna Applications Reference Guide reproduces Part 3 of the
Antenna Engineering Handbook, 2d ed., McGraw-Hill, New York, 1984.

Printed and bound by R. R. Donnelley & Sons Company.

Contents

Contributors

William P. Allen, Jr. *Staff Engineer, Lockheed-Georgia Company, Marietta, Georgia.* (CHAP. 14)

M. C. Bailey *Senior Research Engineer, NASA Langley Research Center, Hampton, Virginia.* (CHAP. 8)

Brian S. Collins *Technical Director, C & S Antennas, Ltd., Rochester, England.* (CHAP. 4)

James H. Cook, Jr. *Principal engineer, Scientific-Atlanta, Inc., Atlanta, Georgia.* (CHAP. 13)

William F. Croswell *Head of RF Design Section, Harris Corporation, Melbourne, Florida.* (CHAP. 8)

Raymond H. DuHamel *Antenna Consultant, Los Altos Hills, California.* (CHAP. 5)

Boynton G. Hagaman *Senior Engineer, Kershner & Wright Consulting Engineers, Alexandria, Virginia.* (CHAP. 1)

Maurice M. Hallum III *Chief of Systems Evaluation Branch, Systems Simulation and Development Directorate, U.S. Army Missile Command, Redstone Arsenal, Alabama.* (CHAP. 15)

Howard T. Head *Consulting Radio Engineer, A. D. Ring & Associates, Washington, D.C.* (CHAP. 2)

Edward B. Joy *Professor, School of Electrical Engineering, Georgia Institute of Technology, Atlanta, Georgia.* (CHAP. 6)

Hugh D. Kennedy *Director of Marketing, Technology for Communications International, Mountain View, California.* (CHAP. 16)

Charles M. Knop *Director, Antenna Research, Andrew Corporation, Orland Park, Illinois.* (CHAP. 7)

Dennis J. Kozakoff *President, Millimeter Wave Technology, Inc., Atlanta, Georgia.* (CHAP. 15)

John D. Kraus *Director, Ohio State University Radio Observatory, Columbus, Ohio.* (CHAP. 18)

John A. Lundin *Consulting Radio Engineer, A. D. Ring & Associates, Washington, D.C.* (CHAP. 2)

Josh T. Nessmith *Principal Research Engineer, Georgia Tech Research Institute, Georgia Institute of Technology, Atlanta, Georgia.* (CHAP. 11)

Willard T. Patton *Manager, Advanced Antenna and Microwave Technology, Missile and Surface Radar, RCA Corporation, Moorestown, New Jersey.* (CHAP. 11)

Paul E. Rawlinson   *Manager, Microwave & Antenna Department, Equipment Division, Raytheon Company, Wayland, Massachusetts.* (CHAP. 9)

Leon J. Ricardi   *Head, MILSTAR Technical Advisory Office, Lincoln Laboratory, Massachusetts Institute of Technology, Los Angeles, California.* (CHAP. 12)

Charles E. Ryan, Jr.   *Principal Research Engineer, Georgia Tech Research Institute, Georgia Institute of Technology, Atlanta, Georgia.* (CHAP. 14)

James M. Schuchardt   *Vice President, Millimeter Wave Technology, Inc., Atlanta, Georgia.* (CHAP. 15)

Vernon C. Sundberg   *Section Head, GTE Systems, Mountain View, California.* (CHAP. 17)

Jean-Claude Sureau   *Technical Director, Radant Systems, Inc., Stow, Massachusetts.* (CHAP. 10)

Harold R. Ward   *Consulting Scientist, Equipment Division, Raytheon Company, Wayland, Massachusetts.* (CHAP. 9)

William Wharton   *Technical Consultant, Technology for Communications International, London, England.* (CHAPS. 3 and 16)

Ronald Wilensky   *Manager, High-Frequency Products, Technology for Communications International, Mountain View, California.* (CHAP. 3)

Daniel F. Yaw   *Advisory Engineer, Electronic Warfare Systems Engineering, Westinghouse Defense and Electronic Systems, Baltimore, Maryland.* (CHAP. 17)

Preface

Antenna technology is a very broad and complex subject. The second edition of the *Antenna Engineering Handbook,* recently published by McGraw-Hill Book Company, covers the fundamentals, types, design techniques, and applications of antennas, as well as some closely related topics. The handbook is designed for antenna engineers who need a reference book that covers the subject in its entirety.

It now is apparent that many engineers and managers share a particular interest in antenna applications. The *Antenna Applications Reference Guide* was designed for such people. Extracted from the handbook, the material in this guide covers the major applications of antenna technology with emphasis on how antennas are employed to meet the requirements of electronic systems. The design methods which are peculiar to the applications are presented as well.

Thanks are due to the many publishers who have granted permission to use material from their publications. We have conscientiously tried to credit all sources of information by references and think we have been thorough. We apologize for any omissions or mistakes. Certainly, they were inadvertent.

The work of many outstanding engineers who prepared the individual chapters has made this guide possible. The fruits of their efforts now are available to the reader. It would be impossible to list here everyone else who contributed time and work to this book, so I would like to take this opportunity to express our sincere gratitude to them collectively.

Richard C. Johnson

Antenna Applications Reference Guide

Chapter 1

Low-Frequency Antennas

Boynton G. Hagaman

Kershner & Wright Consulting Engineers

1-1 GENERAL APPLICATION*

The low-frequency (LF) portion of the radio-frequency (RF) spectrum is allocated to a number of special services to which the low attenuation rate and relatively stable propagation characteristics at low frequencies are of particular importance. Aeronautical and maritime navigational and communication services extend from approximately 10 to 500 kHz. Pulsed hyperbolic navigation systems operating near 100 kHz provide medium-range navigational aids. A phase-stable hyperbolic navigation system (Omega) transmitting within the 10.2- to 13.6-kHz region provides global navigation coverage. Other allocations include radio-location and maritime communications systems and LF fixed public broadcast. These services are of particular importance in the polar regions, which are subject to frequent and severe ionospheric disturbances.

Numerous strategic communication systems are currently operating within the region below 100 kHz, including a number of very-low-frequency (VLF) facilities capable of radiating hundreds of kilowatts within the 14.5- to 30-kHz band. These systems provide highly reliable service to airborne, surface, and subsurface transport.

VLF-LF transmitting systems involve large and expensive radiation systems and require extensive real estate and support facilities. Despite these drawbacks, activity within this portion of the spectrum is increasing, and many new systems have become operational within the past two decades.

1-2 BASIC CONFIGURATIONS

Transmitting Antennas

Early radiators for low-frequency transmitting application generally consisted of flat-top T or inverted-L arrangements using multiwire panels supported between two masts. As the operating frequency of interest moved further into the VLF region and available transmitter power increased, additional top loading was required. Additional masts and support catenaries were provided to suspend larger top-hat panels. These *triatic* configurations were employed on the majority of VLF radiators, although a variation termed the *umbrella type* was also successful. The top hat of the umbrella antennas consisted of several multiple-wire panels supported by a central mast and a number of peripheral masts. The top-hat panels were suspended by insulators, and the support masts were grounded. In some cases, the center support mast was insulated from ground at its base to reduce its effect on antenna performance.

Many examples of these configurations are still in use, although the T and inverted-L variations are used primarily for relatively low-power applications within the upper portion of the low-frequency band.

Base-insulated, freestanding, or guyed towers have come into general use as radiators for public broadcast service in the medium-frequency (MF) band (535 to 1605 kHz). At these frequencies, tower heights of one-sixth to five-eighths wavelength are practicable. In some cases, various top-loading arrangements were added to the guyed towers in an attempt to improve their performance. These experiments were quite successful.[1]

---

*10 to 300 kHz in this instance.

The most widely used variation of the base-insulated antennas for use at low frequencies is the top-loaded monopole. In this antenna, top loading is provided by additional "active" radial guys attached to the top of the tower and broken up by insulators at some point down from the top. In some cases, the top-loading radials also function as support guys.

Receiving Antennas

LF antennas designed for general receiving applications have very little requirement for efficiency. The receiving system is generally atmospheric-noise-limited, and air or ferrite loops, whips, or active probes are usually adequate. The design of low-frequency receiving antennas is not included within the scope of this chapter.

1-3   CHARACTERISTICS

Wavelengths within the low-frequency band range from 1 to 30 km. The physical size of low-frequency radiators is generally limited by structural or economic considerations to a small fraction of a wavelength. These antennas therefore fall into the *electrically small category* and are subject to certain fundamental limitations, clearly defined by Wheeler.[2] These limitations become increasingly apparent at very low frequencies and are a primary consideration in the design of practical VLF-LF radiation systems.

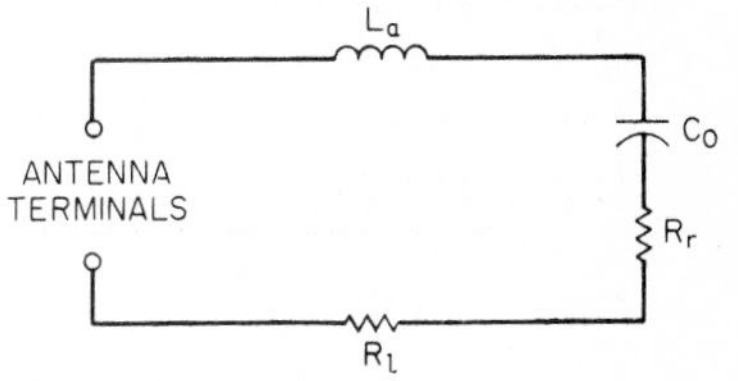

FIG. 1-1   Equivalent-circuit low-frequency antennas.

Since the low-frequency antenna is very small in terms of wavelength, its equivalent electrical circuit may be quite accurately represented by using lumped constants of inductance, capacitance, and resistance, as shown by Fig. 1-1,

where L_a = antenna inductance

C_0 = antenna capacitance

R_r = antenna radiation resistance

R_ℓ = radiation-system loss resistance

The antenna radiation resistance R_r accounts for the radiation of useful power. It is a function of the effective height H_e of the antenna:

$$R_r = 160\pi^2(H_e/\lambda)^2 \quad \Omega$$

The antenna effective height is equivalent to the average height of the electrical charge on the antenna. The effective height of an electrically short, thin monopole of uniform cross section is approximately equal to one-half of its physical height. For other, more practical antenna systems, the value of effective height is not simply related to the physical dimensions of the antenna structure. It is difficult to compute accurately the electrical height of complex multiple-panel antenna systems requiring numerous support masts and guy systems. Computer codes which will eventually make such analysis feasible are being developed. At present, the effective height and other basic design parameters of such an antenna are more effectively determined by scale-model measurement (see Sec. 1-8).

Radiation-system losses are often expressed in terms of equivalent resistance. The total loss resistance R_ℓ is the sum of all individual circuit losses:

$$R_\ell = R_g + R_t + R_c + R_d + R_{\mathrm{misc}}$$

where R_g = ground loss
  R_t = tuning loss
  R_c = conductor loss
  R_d = equivalent-series dielectric loss
  R_{misc} = loss in structural-support system

The principal loss in most systems occurs in the ground system and in the tuning-network inductors. Other losses occur in antenna conductors, insulators, and miscellaneous items of the structural-support system. In some locations, antenna icing, snow cover, and frozen soil may also contribute to antenna losses.

Antenna efficiency η is determined by the relative values of radiation resistance and antenna circuit loss. It is expressed by

$$\eta = R_r/(R_r + R_\ell)$$

To achieve a given efficiency, there is a maximum value of radiation-system loss resistance R_ℓ that can be allowed. The individual-circuit losses, particularly those contributed by the tuning network and the ground system, can often be allocated during the detailed-design phase so as to minimize overall cost.

The lossless, or intrinsic, antenna bandwidth (to 3-dB points) is derived from the ratio f/Q, where Q is equal to the circuit reactance-to-resistance ratio. Therefore,

$$\mathrm{BW\ (intrinsic)} = 1.10245 \times 10^{-7} C_0 H_e^2 f^4$$

where H_e = effective height, m
  C_0 = static capacitance, μf
  f = operating frequency, kHz

The intrinsic bandwidth is useful as a convenient means of antenna comparison.

The bandwidth of the radiation system (antenna, ground system, and tuning network) is inversely proportional to antenna efficiency:

$$\mathrm{BW\ (radiation\ system)} = \mathrm{BW\ (intrinsic)}/\eta$$

This expression defines the bandwidth of the radiation system. It neglects, however, the reflected resistance of the transmitter output amplifier R_t, which can increase the overall bandwidth substantially, depending upon the particular amplifier design and coefficient of antenna coupling. In practice, the actual operating bandwidth of a typical transmitting system driven from a vacuum-tube amplifier is 40 to 80 percent greater than that of the radiation system alone. This may not be the case when the power is generated by a solid-state amplifier of high efficiency.

VLF antennas are normally operated at a frequency below self-resonance, and the principal tuning component is a high-Q inductor, usually referred to as the *helix*. It is possible to increase the operating bandwidth of the radiation system appreciably by dynamically varying the effective inductance of the helix in a manner to accommodate the bandwidth requirements of data transmission. This method of synchronous tuning is accomplished with a magnetic-core *saturable reactor* shunted across a portion of the helix.

The inductance of the reactor is a function of the relative permeability of its magnetic core. The relative permeability is changed by regulating the amplitude of a bias current flowing through a separate winding on the magnetic core. A control signal derived from the data stream is processed and used to regulate the bias current and to synchronize the antenna tuning in accordance with the instantaneous requirements of the modulation system.

1-4 PRACTICAL DESIGN CONSIDERATIONS

The first task in the design of a low-frequency radiation system is the selection of the basic antenna configuration most suitable for the application. There are, at present, few choices capable of effective performance within the 10- to 300-kHz frequency range. The most common types are identified and illustrated by Fig. 1-2. Except for the conventional base-insulated monopole, they are all structural variations of a short, top-loaded current element. The principal radiated field is vertically polarized and essentially omnidirectional. The ultimate performance of each configuration is closely related to its effective height and volume. Therefore, antenna selection depends primarily on the frequency, power-radiating capability, and bandwidth requirement of the proposed system.

Vertical Radiators

The conventional base-insulated tower of Fig. 1-2a may be adequate for low-power application within the upper portion of the low-frequency spectrum. The theoretical radiation resistance for simple vertical antennas of this type, if linear and sinusoidal current distribution is assumed and metal-structure towers, guy wires, etc., are ignored, is shown by Figs. 1-3 and 1-4.

The input reactance is represented by the equation

$$X_a = -Z_0 \cot (2\pi \ell / \lambda) \qquad \Omega$$

which, for the case of a vertical cylindrical radiator of height h and radius $(a \ll h)$, is usually evaluated by letting $\ell = h$ and

$$Z_0 = 60[\ln (2h/a) - 1] \qquad \Omega$$

If the tower is very short, greater accuracy is obtained by using[4]

$$Z_0 = 60[\ln (h/a) - 1] \qquad \Omega$$

Top-Loaded Monopoles

The base-insulated tower (Fig. 1-2b) in which top loading is provided by active radials or sections of the upper support guys is likely to be more cost-effective for low frequencies than the simple vertical radiator or the T or inverted-L types discussed later. A comprehensive parametric study of this antenna was completed by the Naval Electronics Laboratory in 1966. The study report contains numerous design curves that permit accurate evaluation of the performance of top-loaded monopoles of various configurations.[5] The data were acquired by scale-model measurement. The top-loading radials were held taut with little sag. A reduction of approximately 4 percent

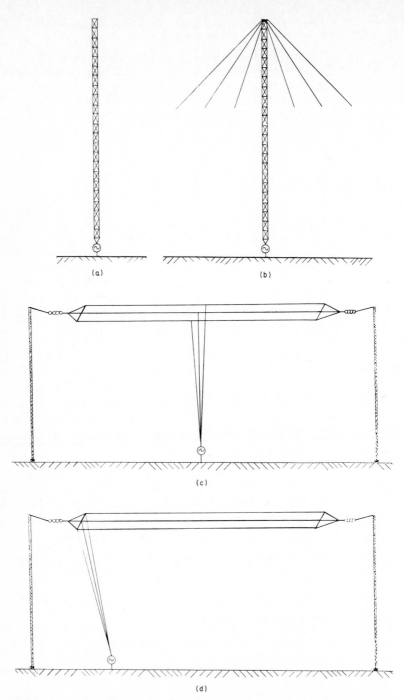

FIG. 1-2 Common types of low-frequency antennas. (*a*) Base-insulated monopole. (*b*) Top-loaded monopole. (*c*) T antenna. (*d*) Inverted-L antenna.

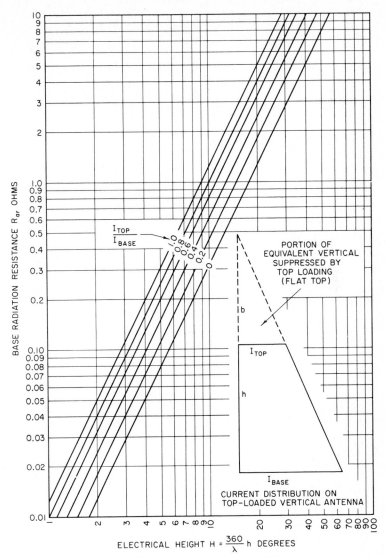

FIG. 1-3 Theoretical radiation resistance of vertical antenna for assumed linear current distribution. *(After Ref. 3.)*

should be applied to the effective height derived from the curves to account for the dead-load sag of a practical antenna.

T and Inverted-L Antennas

These antennas (Figs. 1-2c and d) require two masts from which to support the insulated top section, but they do not normally require base or guy insulators. The prac-

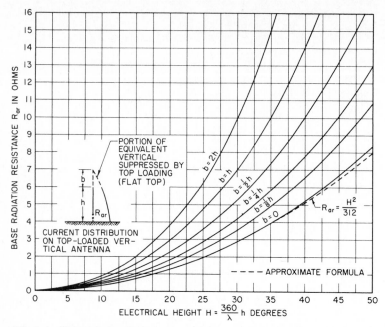

FIG. 1-4 Theoretical radiation resistance of vertical antenna for assumed sine-wave current distribution. *(After Ref. 1.)*

ticable area of the top-loading top section is somewhat limited unless additional end or side masts are supplied.

This configuration and that of the triatic antenna described below may be useful if the available site area is restricted in one dimension.

Triatic Antennas

The triatic-antenna configuration, illustrated by Fig. 1-5, consists of a relatively long antenna panel made up from parallel conductors. In addition to the masts at each end, the antenna panel is further supported at intermediate points from cross catenaries (triatics) and additional pairs of masts. The triatic antenna may be fed at any point along its length, but the design of the ground system is simplified if the download feed is near the midpoint.

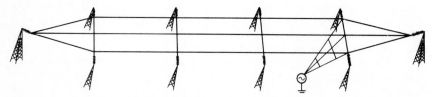

FIG. 1-5 Triatic-antenna configuration.

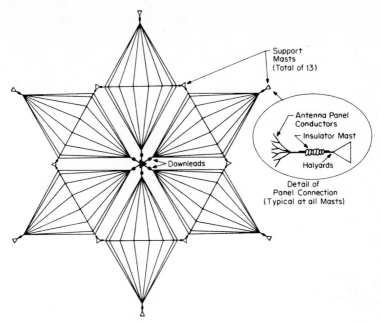

FIG. 1-6 Trideco-antenna configuration (plan view). Note: Individual down-leads are provided at each panel inner apex.

Trideco Antennas

The structural arrangement of a trideco antenna consists of one or more rhombic or triangular-shaped multiple-wire panels suspended from three or more masts. The individual downlead of each panel terminates at a common feed point (Fig. 1-6). This configuration has proved to be very effective for high-power applications, since it affords greater flexibility in the selection of both electrical-design and structural-design parameters. It is possible to design the top-hat panels so they can be individually lowered for maintenance without appreciably reducing antenna performance. This capability greatly simplifies antenna maintenance, thereby increasing system reliability and availability.

Valley-Span Antennas

These antennas have limited application. They are essentially T or inverted-L spans supported from natural geological formations such as those provided by a deep valley or mountainous ridges. The spans are interconnected and fed through a downlead extending to the valley floor. Typical examples of this type of antenna are in operation in Norway, Hawaii, and Washington State.

There are very few locations in the world where valley-spanning antennas can be effectively utilized. While the initial cost of such an antenna may be considerably less than that of a mast-supported system, the cost of ground-system maintenance and control of vegetation and erosion will generally be much greater.

The efficiency of existing valley-span antennas ranges from approximately 7 to 20 percent. Several factors contribute to their low efficiency. The topography of the typical valley site tends to reduce the effective antenna height and frequently limits the practicable area of the ground system. Soil cover may also be shallow and of poor effective conductivity.

Antenna-Design Voltage

The power-radiating capability of low-frequency antennas is proportional to the square of the antenna voltage ($P_{rad} \propto V^2$) and in most instances is limited by the allowable insulator voltage and the effective area of the antenna conductors. It appears that about 250 kV rms (rms values are commonly applied in VLF-LF antenna design) is the maximum operating voltage to which low-frequency antennas can presently be designed and put into service. This voltage level is not appreciably higher than that developed on VLF antennas 50 years earlier.

The voltage limitation is principally due to the physical limitation of available conductors and insulators. The potential surface gradient of the conductor can be controlled by providing adequate conductor surface. Insulator working voltages can be extended by the use of multiple-insulator assemblies. However, the structural design loads resulting from the increased weight and wind load of the insulators and antenna panels become increasingly difficult to accommodate. The problem is particularly severe in the case of insulated and top-loaded monopoles. The top-loading radials are generally insulated near the midpoint of their spans, at which point the insulator weight results in greatly increased sag or cable tension. The same is true of support-guy insulators, which must be capable of withstanding high tensile loads. The resulting insulator weight, distributed at several points along the guy, increases guy sag and reduces guy efficiency. The cumulative effect of insulator weight, increased cable diameters, wind load, and heavier towers is regenerative and eventually reaches a practical or economic limitation.

Triatic, trideco, and similar top-loaded antennas are somewhat less vulnerable to these problems because of their structural arrangement. The mast-support guys do not require insulators. The top-hat panel insulators can be attached relatively close to the supporting masts, where their weight is more easily supported.

Antenna Insulation

Virtually all insulators used on high-power LF-VLF antennas employ porcelain as the dielectric material. The dependability and long life of porcelain as an insulating material are well established. Reinforced-plastic insulators have been under development for many years and have recently shown promise in experimental service. They offer a considerable saving in weight.

The selection of the type of insulators to be used on the antenna depends to a large extent on structural considerations and the location of the insulator in the system. Three basic types are available: the strain, or "stick" type, in which the porcelain body is subjected to tensile loads; the fail-safe, or compression, type, in which the porcelain is supported in a frame or yoke arrangement which places the porcelain under compression; and a third type which does not subject the porcelain to any appreciable working loads. This type is a variation of the stick type in which tensile loads

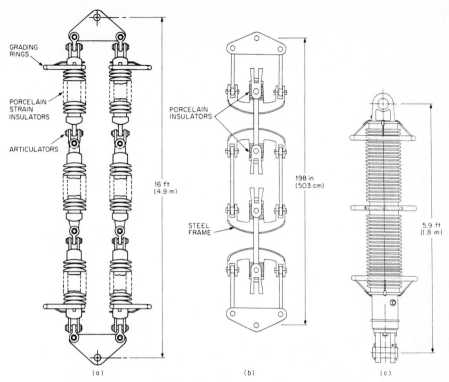

GRADING RINGS

PORCELAIN STRAIN INSULATORS

ARTICULATORS

16 ft
(4.9 m)

PORCELAIN INSULATORS

198 in
(503 cm)

STEEL FRAME

5.9 ft
(1.8 m)

(a)   (b)   (c)

FIG. 1-7   Three types of insulators suitable for use in low-frequency antennas. (*a*) Strain type. (*b*) Fail-safe type. (*c*) Fiberglass-core type.

are accommodated by a high-strength material such as fiberglass, located within the interior of the porcelain body. These insulators are illustrated in Fig. 1-7.

The voltage-handling capability of the insulators must be adequate to ensure operating reliability at design power under normal environmental conditions. The published voltage rating of the insulators must be derated to allow for rain and condensation, surface contamination, overvoltage transients from atmospheric disturbances, and deterioration from service aging. A derating factor of 2:1 has been commonly applied; that is, a wet flashover rating twice the normal operating voltage is required. This factor has not always been adequate, primarily because of the difficulty of testing the insulator at its projected operating frequency rather than at 60 Hz and in an environment similar to that to be encountered in service. If the normal antenna working voltage exceeds about 150 kV rms, insulator assemblies should be tested at radio frequency to confirm their continuous wet-withstand capability at a voltage *at least* 50 percent above the working voltage.

Individual compression-type insulators are not available for operating voltages exceeding about 50 kV. Therefore, these insulators must often be connected in series to obtain the required voltage-withstand capability. Insulation "efficiency" decreases rapidly as insulators are added in series. Voltage-grading rings or cages are useful in this respect since they improve voltage distribution across the insulator string. These devices increase the insulator weight and wind-load area and must be carefully designed to avoid vibration and early structural failure.

Structural-design codes generally require structural guys to employ compression fail-safe insulators. This requirement may also apply to panel-support insulators, in which insulator failure may jeopardize the entire antenna structure. The active non-structural radials of top-loaded monopoles may employ strain insulators provided the tower and primary antenna guy system are properly designed.

Individual strain insulators are capable of withstanding voltages up to several hundred kilovolts. However, the tensile strength of porcelain is relatively low, and individual insulators may have to be operated in parallel combinations of two or more series units in order to develop the necessary voltage and strength characteristics. The resulting insulator and hardware weight may be difficult to accommodate.

It may appear that the selection of insulators at this point in the design is premature. However, insulator selection has a considerable impact on the final antenna configuration and performance, since it affects the dead-load shape of the antenna spans, number and placement of guys, wind-load area, and connecting hardware. The usual practice is to conduct the initial design trades on typical insulator components, using the manufacturer's estimated weights. Later, during the detailed-design phase, as various factors are optimized, specific insulators and actual weights are used.

Conductors

An equally critical selection is the choice of conductors and structural cables to be used in constructing the antenna. Conductors are required for antenna panels or spans, current jumpers, and structural cables for halyards and guys. Low-frequency-antenna panels involve long spans. The conductors must have low RF resistance, adequate tensile strength, and minimum weight. The conductor diameter and total length must be sufficient to limit the surface potential to preclude corona.

The conductors are large and consist of multilayer cables stranded from aluminum or aluminum-covered steel wires. Figure 1-8 illustrates the construction of three types of conductors currently manufactured for electric power transmission and distribution systems. The ready availability of these conductors is an important consideration to the designer.

The conductor which has generally proved to be most suitable for antenna spans and panels is stranded from aluminum-clad, steel-core wires and is designated as AW (Fig. 1-8a). This conductor is listed by the manufacturer in diameters up to approximately 1⅛ in. It may be procured in diameters up to 2 in when arrangements are made for special stranding.

The size of the individual wires of the strand ranges from AWG No. 10 to AWG No. 4, depending upon the overall conductor diameter. The thickness of the outer layer of aluminum is 10 percent of the wire radius. This thickness may be less than the skin depth at the operating frequency, and RF losses are thereby increased. In applications such as downleads which are carrying a large current, it may be necessary to produce aluminum-clad wire cable with the outer strands of solid-aluminum wire (Fig. 1-8b). This type of conductor is designated AWAC, i.e., aluminum wire, aluminum-clad steel-wire core.

AWAC is not entirely satisfactory from a structural standpoint for applications in which the tensile load is high because of differences in the elasticity of the alumi-

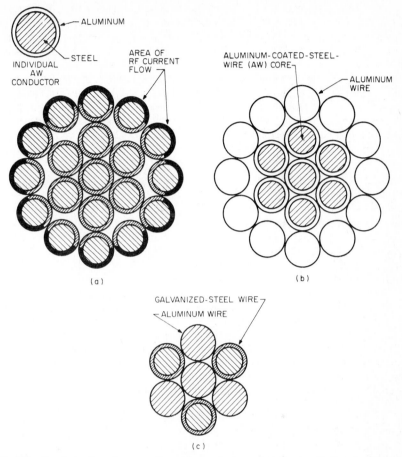

FIG. 1-8   Cross section of composite-cable conductors. (*a*) Concentric-lay stranded aluminum-clad steel conductor. (*b*) Aluminum-wire, aluminum-clad (AWAC) wire core. (*c*) Aluminum-conductor, steel-reinforced (ACSR).

num and aluminum-clad steel wires. This often results in the "basketing" of the outer wires when they are being handled or when pretension loads are relaxed.

A third type of composite conductor (Fig. 1-8*c*) is also in general use in power-transmission service. This conductor is made up in various combinations of aluminum and aluminized or galvanized steel wires and is designated ACSR, i.e., aluminum conductor, steel-reinforced. ACSR having steel wires exposed to the surface of the cable is not a suitable conductor for low-frequency conductors because of its relatively poor RF conductivity and the tendency of the aluminum wires to basket when tensile loads are released.

Conductors stranded of solid-aluminum wires are available and should be considered whenever high strength is not a factor. They are useful for current jumpers and for connections at feed-line terminal points.

When extreme flexibility and fatigue resistance are required in a jumper, stainless-steel wire rope should be considered. It should be fabricated from a nonmagnetic stainless alloy (Series 300), which will provide maximum skin depth and effective conductor area. Current jumpers are generally short, and their loss is usually insignificant.

Other specialized conductor materials have been used in high-power low-frequency antennas. Calsum bronze cables are in use at the U.S. Navy VLF station at Cutler, Maine, and hollow-core copper cable at the original VLF station at Annapolis. Future use of these specialized conductors is extremely unlikely, considering the present availability of aluminum-covered steel and the cost of procuring relatively short production runs of special cables.

Selection of Conductor Diameter   The diameter of the conductors making up the antenna spans, panels, or downleads must be adequate to avoid corona as well as to develop the necessary tensile strength. The minimum conductor diameter required at the antenna design voltage should be established during the preliminary design since conductor weight and wind load are important factors in the structural design. Generally, a conductor size adequate to handle the working loads typical of very long spans will be of sufficient diameter to limit the surface potential gradient to a safe value, provided the antenna panels are made up of an adequate number of conductors.

Experience has confirmed that limiting the surface gradient of the antenna conductors to about 0.7 kV/mm will ensure corona-free operation under normal and extreme environmental conditions.

The usual expression for computing the surface gradient of an isolated conductor above ground is not applicable to a multiconductor transmitting-antenna panel. A method developed by Wheeler[2,6] provides a solution based on the average current flowing from the wires toward the ground per unit of wire surface area. By using this method, the average voltage gradient on the conductor surface is found by

$$E_a \text{ [kV/mm]} = \frac{3\lambda^2}{2\pi H_e A_a} \sqrt{10P \text{ [watts]}} \times 10^{-6}$$

where E_a is the voltage gradient, kV/mm, and A_a is the total wire surface area, m^2.

When the antenna span or panels are made up of a number of conductors, the gradient at the outer wires is higher than that at the inner wires. The surface gradient of the wires of a multiwire panel can be partially equalized by varying their spacing with respect to each other, the more closely spaced wires being placed at the panel sides. The gradient on the outer wires can also be equalized by increasing their relative diameter.

Downleads

Base-insulated monopoles are generally fed at the base, and the antenna current is carried by the structural members of the tower. Virtually all low-frequency-antenna towers are fabricated of steel. The steel is usually galvanized except when a low-alloy steel is used. In any case, the thickness of the zinc is much less than a skin depth, and supplementary current buses must be provided up the tower to reduce losses.

It is recommended that triatics and similar wire-panel antennas be fed by one or more downleads terminating on a common feed point. A number of early low-frequency antennas used a system of multiple tuning, i.e., several downleads connected at various points to the antenna panel and separately tuned. Only one of the downleads was fed; the others terminated to the ground system. Multiple downleads substantially increase antenna cost and complicate tuning procedures. Their use has not proved to be effective in reducing antenna-system losses.

It is advantageous to use a multiwire cage (Fig. 1-9) for the antenna downlead. The total antenna current flows in the downlead, and it is more practical to control the downlead loss and potential gradient by using a multiwire cage than to use a single conductor of large diameter.

In addition, the effective diameter of a multiwire cage is much greater than that of its individual conductors. As a result, the download inductance is

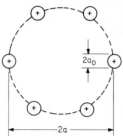

FIG. 1-9 Multiwire cage for antenna download.

reduced, thereby increasing the self-resonant frequency of the antenna. The large effective diameter of the multiwire cage also reduces the potential gradient at the surface of the conductors. This is particularly important along the lower portion of the download where it approaches the earth and grounded structures.

The effective radius of a multiwire cage is given by [7]

$$a_{eff} = a(na_0/a)^{1/n}$$

Where a_{eff} = effective radius of cage, m

n = number of cage wires

a = cage radius, m

a_0 = individual wire radius, m

and, for practical purposes, the average potential gradient E_a on the individual cage wires is found by

$$E_a \ [V/m] = V/na_0 \ln (2h/a_{eff})$$

where h = elevation of cage above ground, m

V = voltage on cage

Structural Cables

Non-current-carrying cables used for guys, halyards, and support cables are usually fabricated from bridge strand, which is available up to approximately 4 in in diameter. The strand is made up of a number of counterwound layers of galvanized high-strength steel wires. When flexibility is important, as in passing over winches and shives, wire rope fabricated from bundles of fine steel-wire strands over a hemp core is preferable to bridge strand.

Special Problems of Insulated Towers

Antennas requiring base-insulated towers can be advantageously employed in low- and medium-power applications, particularly within the upper portion of the low-frequency band and almost exclusively within the medium-frequency broadcast band.

They may not be cost-effective, however, in applications in which the antenna operating voltage is high. In addition to expensive guy and base insulators, such towers require a high-voltage obstruction-lighting isolation transformer with a voltage-withstand rating equal to that of the base insulator, a special provision for elevator power if an elevator is desired, and some provision for boarding the tower above the level of the base insulator.

In some locations, the heavy rainfall accompanying typhoons or hurricanes may flood the porcelains of base insulators and isolation transformers, causing flashover and operational interruptions. Isolated sections of tower guys may accumulate a high-potential static charge, triggering flashover of all insulators of the guy. These problems can be overcome, but the cost may not compare favorably with an alternative configuration using uninsulated masts and guys.

1-5 ANTENNA AND TUNING-NETWORK DESIGN PARAMETERS

The potential performance of a low-frequency antenna can be predicted from three basic parameters: effective height, antenna capacity, and self-resonant frequency. These values can be computed from the principal dimensions of the proposed antenna configuration or acquired by direct measurement on a scaled antenna model (Sec. 1-8). If the proposed antenna is similar to an existing system the properties of which are known, the design parameters may be appropriately scaled.

The principal performance characteristics of the antenna are related as follows. The power-radiating capability of the antenna for a chosen level of top-hat voltage V_t is equal to

$$P_{\text{rad}} = 6.95 \times 10^{-10} V_t[\text{kV rms}] \, C_0[\mu f] H_e^2 f^4[\text{kHz}] \qquad \text{kW}$$

The voltage at the antenna base or feed point V_b establishes the working-voltage requirement for the base insulator, downlead terminal, and entrance bushing:

$$V_b \, [\text{kV}] = V_t \, [\text{kV}] \, \{1 - (f \, [\text{kHz}]/f_r \, [\text{kHz}])^2\}$$

where f = operating frequency
f_r = antenna self-resonant frequency

The antenna-base current required to radiate a specified power is given by $I_b = (P/R_r)^{1/2}$, where P is expressed in watts, the antenna radiation resistance R_r in ohms, and the antenna-base current I_b in amperes. The antenna-base current largely determines the current rating of the helix conductor.

The power input to the radiation system required to radiate a given power is

$$P \, [\text{watts}] = I_{\text{base}}^2 (R_r + R_\ell)$$

where R_r = antenna radiation resistance
R_l = antenna loss resistance, Ω

The antenna inductance L_a is expressed by

$$L_a \, [\mu H] = 1/4\pi^2 f_r^2 \, [\text{kHz}] \, C_0 \, [\mu f]$$

Since the total circuit inductance required to resonate the antenna at a given operating frequency f (below self-resonance) is $1/4\pi^2 f \, [\text{Hz}]^2 \, C_0 \, [\mu f]$, the effective tuning inductance required at the antenna feed point is

$$L_{tune} \; [\mu H] \; = \; 1/4\pi^2 f^2 \; [Hz] \; C_0 \; [\mu f] \; - \; L_a \; [\mu H]$$

If the antenna is to be designed for operation over a band of frequencies, the operating conditions at the lowest frequency will be the most severe and will establish the principal design requirements of the system. As the operating frequency is increased, the bandwidth will be improved and the current and voltage levels will be appreciably reduced.

Tuning Low-Frequency Antennas

Low-frequency antennas are generally designed to be operated well below their self-resonant frequency. The antenna impedance will then be capacitive, and the antenna may be tuned with a series inductor. A schematic drawing illustrating the arrangement of a typical tuning and matching network is given in Fig. 1-10. At frequencies above self-resonance the antenna impedance is inductive, and a series capacitor must be used. Capacitors of the size and rating required for series-tuning, high-power VLF antennas are not readily available. In addition, an inductor is inherently more reliable than a capacitor and less subject to failure from lightning or overvoltage transients. Therefore, in order to avoid the necessity of matching into an inductive reactance, the self-resonant frequency should be designed to be at least 10 percent and preferably 15 percent above the highest anticipated operating frequency. This objective may be difficult or impossible to achieve with very large antennas. In such cases, the downlead inductance should be minimized or alternative feed arrangements considered.

The helix is usually designed as a single-layer inductor. In this form, it is less difficult to install and to provide with means for taps. The voltage difference across such an inductor is quite evenly distributed. As a result, the turn-to-turn voltage is minimized and can be accommodated by means of normal turn-to-turn spacing (generally about two wire diameters). The helix is most conveniently fed at the bottom. Maximum voltage exists at the top, where it may be connected to a high-voltage exit bushing extending to the downlead.

Variometers

Low-frequency antenna-tuning circuits generally require one or more variable inductors termed *variometers*. Variometers consist of two concentric coaxial (partially

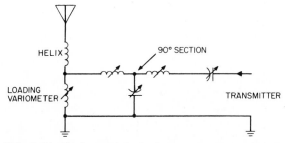

FIG. 1-10 Typical tuning circuit for use with a low-frequency antenna operating below self-resonant frequency.

spherical) coils connected in series. The smaller coil can be rotated about the common axis by means of a motor-driven shaft. The coefficient of coupling of the inner and outer coils is relatively high. Rotation of the inner coil through one-half revolution results in smooth variation of the effective inductance. A total inductance variation of about 8:1 can be achieved with large transmitting units.

Variometer performance can be estimated by computing the inductance, mutual inductance, and coefficient of coupling between the concentric inductors. Judgment is required in the selection of the effective radius and length of the coils, since the physical requirements of the variometer frame require the inductors to be incomplete spherical coils with few turns of varying pitch.

Inductance of helix (solenoid):

$$L = \pi\mu_0 a^2 n^2/b + 0.9\, a$$

Mutual inductance of two concentric coaxial helices if inner coil is slightly smaller than outer coil:

$$L_{12} = L_2(a_1^2 n_1)/(a_2^2 n_2)$$

Coefficient of coupling of two concentric spherical coils:

$$K_{12} = (a_1/a_2)^{3/2}$$

where a = coil radius, m
$\quad b = n \times$ pitch of winding, axial length, m
$\quad n$ = number of turns
$\quad \mu_0 = 1.257 \times 10^{-6}$ = magnetivity of free space, H/m
subscript 1 = inner coil
subscript 2 = outer coil
subscript 12 = mutual

1-6   RADIATION-SYSTEM LOSS BUDGET

The radiation resistance of a low-frequency radiator is relatively small; consequently, its efficiency is critically dependent upon antenna-system losses. It is useful to prepare a *loss budget* in which identifiable loss items are individually evaluated. The major losses are likely to occur in the ground system and tuning network. Among other losses are those in the antenna conductors and support guys and in the insulator dielectric. The sum total of all radiation-system losses must not exceed a *maximum allowable value* if the antenna efficiency goal is to be achieved.

Ground Losses

It is possible to some extent to allocate certain losses so as to minimize the overall cost of the radiation system. For example, if the site topography or soil geology makes installation of the ground system difficult and unusually expensive, it may be more economical to accept somewhat higher ground losses, provided the loss in the tuning network can be reduced accordingly. Calculation of ground loss is discussed in Sec. 1-7.

Tuning-Inductor (Helix) Loss

The principal loss in the tuning network will usually be in the main tuning inductor used to tune the antenna. It is possible to design a VLF inductor having a computed Q of several thousand, but the necessity of providing for taps and connectors makes it very difficult to achieve a practical value exceeding approximately 2500. Values of 1500 to 2000 are typical. The helix Q is primarily related to its overall size. Other factors include the conductor area and material, the inductor form factor, and the volume and material of the shield.

The most effective helix conductor is a composite cable stranded of Litz wire as illustrated in Fig. 1-11. The Litz-wire strands are made up of numerous insulated copper wires approximately a skin depth in diameter. The Litz strands are bundled together into larger groups and jacketed over a central jute core. The wire and wire bundles are transposed within the jacket so as to distribute the current density uniformly among the individual wires.

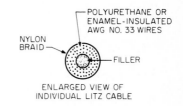

ENLARGED VIEW OF
INDIVIDUAL LITZ CABLE

The number of individually insulated wires making up the cable is determined from the maximum value of the antenna current on the basis of 1000 circular mils per ampere. If the current requirement exceeds about 1500 A, the resulting cable diameter (typically 3½ in) becomes difficult to handle. In this case, two or three parallel cables of smaller diameter may be used. An excellent summary of the design of Litz inductors is given by Watt in Chap. 2.5 of *VLF Radio Engineering.*[8]

Copper tubing of appropriate size may also be used for low-frequency inductors in applications in which the additional losses may be tolerated and the power level permits the use of tubing of a reasonable diameter (6 in or less). Large-tubing inductors must be assembled from sections bolted or brazed together during installation.

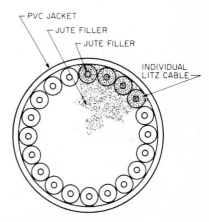

FIG. 1-11 Cross section of a large Litz conductor used in low-frequency inductor.

The helix-room shield may be fabricated from brazed or welded copper or aluminum sheets having a thickness of two to three skin depths at the lowest operating frequency. A square or rectangular helix room is usually more practical than one of cylindrical form.

The additional losses contributed by the shield will be small if the size of the helix building provides an inductor-to-shield separation of approximately one coil diameter.[9] The separation at the floor and ceiling may be somewhat less. It is advisable to limit the surface gradient of the Litz cable to a maximum of 0.5 kV/mm (rms).

Conductor Losses

Antenna current flows in downlead conductors, top-hat panels, or top-loading radials and, to a lesser extent, in guy cables and tower members. To compute the losses in these items, it is necessary to evaluate the RF resistance of each at the frequency of operation.

Concentric Lay-Stranded Aluminum-Clad Steel Conductors (AW)

This type of conductor is suitable for antenna panels and long spans. The individual wire size of the conductor depends upon the diameter of the finished cable and ranges from AWG No. 10 through AWG No. 4. The thickness of the aluminum cladding over the steel core is 10 percent of the finished wire radius, or approximately 0.25 mm for the largest-diameter wire (AWG No. 4). For frequencies below about 100 kHz, this thickness is less than a skin depth. The current penetration into the steel-wire core will be negligible owing to its relative permeability. Therefore, essentially all the current is forced to flow in the aluminum surface. This is illustrated by Fig. 1-8. Since the number and size of the individual wires are selected so as to provide a tightly grouped outer layer, the current will flow only in the outer half surface of each exposed wire.

For example, a nominal 1-in cable may consist of four layers of AWG No. 7 wire with 18 wires in the outer layer. The total effective cross-sectional area of the current-carrying aluminum is

$$A = (r_1 + r_2)(r_1 - r_2)\pi N \times 10^{-6}/2$$

where r_1 = radius of individual wire, mm
$r_2 = 0.9\, r_1$ (Thickness of aluminum on steel wire is 10 percent of wire radius.)
N = number of wires in outer wrap (lay), or 18.04×10^{-6} m² in this example
The resistance per meter length is

$$R_{rf} = 1/\sigma A$$

where A = area, m²
σ = conductivity of aluminum (3.5×10^7 S/m)

$$R_{rf} = 1.6 \text{ m}\Omega/\text{m at 15 kHz}$$

Aluminum Wire, Aluminum-Clad Wire Core (AWAC)

In some applications, downlead cages for example, the loss in aluminum-clad steel wires may exceed the allowable loss budget. It is possible to procure cable similar to AW in which the outer lay consists of solid-aluminum wires. RF loss in this cable will be equivalent to that of all-aluminum strand.

Since the wire diameter will be very much greater than a skin depth, the resistance can be expressed as

$$R_{ac} = 1/\sigma\pi\, d\delta$$

Where σ = conductivity, S/m
d = wire diameter, m
δ = skin depth, m, at frequency of interest: $\delta = 1/2\pi(f\sigma\mu_r \times 10^{-7})^{1/2}$
μ_r = relative permeability of conductor material

Bridge Strand

Guy cables are generally made up from bridge strand. The individual steel wires of bridge strand are protected from corrosion by a thin layer of zinc. The zinc is too thin to be a factor in lowering the resistance of the cable. If a value of 200 for relative permeability and 5×10^6 S/m for conductivity of the steel is assumed, the skin depth for bridge strand at 15 kHz is approximately equal to 0.13 mm. The RF resistance calculation is straightforward, taking into account skin depth and exposed-surface area.

Antenna Insulators

Low-frequency insulators employ porcelain dielectric, and the loss factor at operating frequency is relatively low. The insulator loss for the antenna system is estimated from the applied working voltage, dielectric power factor, and number of insulators in the system. The dielectric loss of the insulator can also be estimated by determining the porcelain heat rise during preproduction testing at low frequency.

1-7   DESIGN OF LOW-FREQUENCY GROUND SYSTEMS

The radiation resistance of electrically small antennas is very small, and if an antenna is to have a useful efficiency, system losses must be carefully controlled. The principal circuit losses occur in the tuning network and the ground system. Tuning losses can be minimized by using high-Q components. Ground losses within the site area can be reduced by providing a low-reactance, low-resistance ground system. This may consist of a radial network of wires buried a half meter or so below the surface and extending in all directions to some distance beyond the antenna.

Ground losses result from ground-return currents flowing through the lossy soil and from E-field displacement current in the soil or vegetation above the level of the ground wires. Generally, the E-field losses are not significant and may be neglected except in locations where the soil may freeze to a substantial depth or acquire a deep snow cover. E-field loss can be evaluated by methods outlined by Watt (Secs. 2.4.37 and 2.4.38).[8]

The radiation-system ground loss is derived by separately computing the E-field and H-field losses of small sectors within the radian circle (a circle having a radius of $\lambda/2\pi$) and summing them to get the total loss. The loss within the radian distance is attributed to ground-system loss; that beyond the radian circle is considered to be propagation loss. Since it is rarely practical to extend the ground system to the radian distance, ground losses can be further separated into *wired-area* and *unwired-area* losses.

Early VLF-LF ground systems varied widely in concept, depending upon the antenna configuration and whether single or multiple tuning was employed. In most cases, the ground wires were buried, although several designs returned earth currents from dispersed ground terminals on overhead conductor systems. Present practice is to design the ground system around a single-point feed by using an essentially radial configuration.

There are many ways in which ground-loss computations can be carried out. Since ground-conductivity values may vary considerably over the radian-circle area, a method that allows individual computation of relatively small areas is desirable. The appropriate conductivity values can then be applied. For example, the radian circle can be divided into 36 ten-degree sectors centered at the feed point, each sector being further divided into 50 or more segments. The loss is computed for each sector; when summed, these losses give the total loss.

Once the magnitude and orientation of the normalized* H field within the antenna region have been determined, the H-field losses are calculated in the following manner. The *longitudinal loss within the wired area* is found from the product of the square of the normalized longitudinal component of the H field, the resistance of a 1-m^2 area, and the area. The longitudinal loss is computed for each segment of the sector with the exception of the last few segments, which are treated as termination losses. The total loss for each sector is the sum of all its segments.

The unit resistance is the resistance component of the parallel impedance of the ground wires and the earth to a current flowing parallel to the ground wires. Losses due to the wire resistance (usually small) are covered by this method. The earth impedance, the wire impedance, and the parallel impedance are computed from Watt.[8]

The *transverse loss in the wired portion* is equal to the product of the square of the normalized transverse component of the H field, the earth resistance of 1 m^2, and the area (Watt,[8] Sec. 2.4.10). The transverse loss for a sector is the sum of the segment losses for all wired segments.

The *termination loss* occurs in the last few segments of the wired area, where the transition from the wired to the unwired area occurs. The number of segments is chosen to represent the end portion of the radial wires with a length equal to the skin depth of the earth. The resistance of a single wire of this length to ground is computed from Sunde[10] (Sec. 3.36). A computed or empirically derived mutual-resistance factor is employed to obtain the effective resistance of a single wire in proximity to adjacent wires. The resistance of the terminating ground rods, if used, is then computed from Sunde (Sec. 3.30). The termination resistance for the wire–ground-rod combination is then computed, using an estimated factor for the mutual resistance, and paralleling the two components. This figure is adjusted to a resistance per square meter and multiplied by the square of the longitudinal component of the normalized H field and by the area to provide the termination loss.

The *unwired losses* apply to the unwired area extending from the ground-system boundary to the radian circle. Losses are computed for each segment as the product of the H field squared, the earth resistance (Watt,[8] Sec. 2.4.10), and the area. The segment losses are summed to provide the total sector loss.

Calculations of ground-system resistance, if done manually, are laborious and time-consuming. As a result, it may be difficult to determine the most cost-effective configuration. The computations described above should be programmed for a computer, thereby enabling various ground-system configurations to be quickly evaluated.

---

*The H field is generally normalized for an antenna current of 1 A so as to make the loss resistance equal to the power loss in watts.

Practical ground systems are centered in the exit bushing or downlead terminal. The surrounding area is generally covered with a dense copper ground mat, lightly covered with earth and extending out to a distance of 50 to 100 m. The mesh area is enclosed by a rugged copper cable (secured by ground rods at corners and intermediate points), to which the ground radials are connected. All connections are brazed by hand or welded by using an exothermic process. The ground radials terminate to ground rods driven, if possible, to a depth of a few meters. The ground radials may be interconnected at various radii with circumferential buses to reduce transverse losses within the wired area. All support masts, foundation-reinforcing steel, and guy systems are connected to a number of ground-system radials. Guy-anchor foundations outside the ground-system area are grounded to multiple ground rods.

The ground system should be installed only after tower erection has been completed to avoid damage from heavy machinery. The ground system may be installed with a wire plow or placed manually, buried to a depth of approximately ½ m.

Soil-conductivity values generally available from published Federal Communications Commission (FCC) conductivity maps may not be suitable for use in VLF-LF ground-system design, since the skin depth in the earth at the frequency for which they are applicable is relatively shallow. The depth of penetration in average soil for frequencies within the VLF-LF region may extend to 50 m. Soils to this depth are usually composed of several layers having appreciably different values of conductivity. As a result, the effective conductivity may not be that of the surface layer but depend upon the conductivity and thickness of the individual layers at the frequency of interest.

A conductivity map of North America showing effective earth-conductivity values for a 10-kHz propagating wave has been prepared.[11] The work was extended at a later date to include the preparation of a worldwide map showing effective conductivity values of the earth at frequencies of 10 to 30 kHz.[12] The data presented in these reports are of a general nature. Specific conductivity values must be obtained at the proposed antenna location for use in the design of the ground system.

Ground conductivity can be determined within the antenna-site area from data acquired by the use of a four-terminal array.[13,14] Soil-conductivity values may vary greatly when the soil is frozen.[12] Sufficient measurement radials are required to establish the soil characteristics within the immediate antenna region and to establish an average value to apply to calculations beyond the wired area. The conductivity values can be entered onto the site map and tabulated into a data file for use with a computer.

1-8   SCALE-MODEL MEASUREMENTS

Scale-model measurements provide direct confirmation of the effective height, static capacity, and self-resonant frequency of a proposed antenna configuration from which the performance of the full-size antenna can be accurately predicted. Modeling is most useful as a supplement to the design of VLF-LF antennas having extensive top loading and numerous support masts.

The *E*-field and *H*-field distribution of a simple monopole can be accurately predicted from theory, and the corresponding ground-system design is thereby simplified.

This is not the case for a complex top-loaded antenna in which the field beneath the antenna is influenced by overhead-conductor currents and field distortion resulting from grounded masts and guys.

To evaluate ground losses within this high-current area, a knowledge of the H-field magnitude and orientation is necessary. These data can be acquired by probing the conductive ground plane upon which the model is erected with a small shielded loop.

Model Techniques

VLF-LF antennas operate over an image plane provided by the earth. To avoid excessive ground losses and to provide an effective image plane for the model, a large, flat, highly conductive surface is required. This can be simulated by using metal-wire mesh 1 cm or so per square, preferably galvanized after weaving. The area of ground plane required depends upon the model scale factor and antenna size. A scale factor of 100 is convenient for low-frequency antennas. The ground plane should extend beyond the antenna in all directions to accommodate E-field fringing and to approximate a semi-infinite surface.

It is advisable to elevate the ground plane to a height that will permit measurements to be taken from below. This will provide a degree of instrument shielding and avoid unnecessary disturbance of the antenna field.

The antenna model should be scaled as accurately as possible. Frame towers can be simulated with tubing of equivalent electrical diameter. Multiple-wire cages can be simulated with tubing or solid wire having an equivalent radius. Particular care should be taken to avoid stray shunt capacity at the antenna feed point, as this may reduce measurement accuracy.

A reference antenna with which to refer measurements of effective height should be provided. This can be a slender monopole located on its own ground plane some distance away.

A source antenna is necessary to provide an essentially plane radiation field of uniform intensity over the model ground plane for measurement of effective height. This may be provided by a nearby broadcast station if its frequency and field intensity are satisfactory or by a transmitting antenna set up several wavelengths distant. The measurement frequency should be scaled up approximately by the antenna model factor but need not correspond exactly to the design frequency. The general arrangement of an antenna range for scale-model measurements is shown in Fig. 1-12.

The effective height of the model antenna is derived from measurement of its open-circuit voltage. To reduce instrument loading, the voltage measurement should be taken directly at the antenna terminal by using a field-effect-transistor (FET) emitter follower or a similar very-high-impedance probe. The voltage at the reference antenna should be measured in a similar manner when each model reading is taken.

The effective height H_e of the antenna is determined by

$$H_e = K \frac{V_{oc} \text{ test model}}{V_{oc} \text{ reference model}} \times \text{scale factor}$$

where K is the range-calibration factor, which is determined by relating the open-circuit voltage of a reference monopole to the intensity of the incident field and its electrical height (the electrical height of a short, thin monopole is one-half of its physical height).

The antenna *static capacitance* C_0 is here defined as the apparent capacitance measured at the antenna terminal as the frequency approaches zero. In practice, the measurement can be taken at a frequency not greater than a few percent of the frequency of self-resonance without introducing a significant error. An accurate measurement can be taken by using a Q meter, work coil, and standard variable capacitor. The measurement value is multiplied by the scale factor to derive the full-size value.

The antenna self-resonant frequency is determined by noting the frequency at which the antenna terminal reactance is zero or at which the terminal current and voltage are in phase. The equivalent resonant frequency of the full-size antenna is equal to the model resonant frequency divided by the scale factor.

E- and *H*-Field Distribution

To obtain the *E*- and *H*-field distribution about the antenna, it is necessary to drive the model antenna with sufficient base current to permit measurements to be taken out to a distance well beyond the top-loaded panels.

The *H*-field probe can consist of a well-balanced shielded loop as small as possible ($<5 \times 10^{-4} \lambda$ diameter at the measurement frequency). The *H*-field magnitude, normalized to the antenna-base current, and *H*-field orientation with respect to a chosen reference point should be measured along the surface of the ground plane. Current direction is found by rotating the loop to the null position which corresponds to 90° from current maximum.

E-field distribution (normalized to base-current amperes) can be taken with a short *E*-field probe calibrated at the measurement frequency.

Potential-Gradient Evaluation

The scale model can also be used to identify areas of the antenna structure in which the potential gradient is higher than average. This condition may be due to panel-conductor configuration or conductor proximity to grounded structural members. The scale model is raised to a high alternating-current potential and observed or photographed after dark by using an appropriate time exposure. Successive photographs are taken as the voltage is increased in increments to the corona level. The test voltage may be provided by a variable high-voltage transformer or by a high-potential test set. Test voltages up to approximately 35 kV rms may be required.

These tests may reveal "hot spots" that will require additional grading. Scaling the corona on-set voltage to actual operating conditions, however, is not practical.

Electrolytic-Tank Model

Similitude modeling can also be conducted on a scale model immersed in an electrolytic tank. This technique has been found to be most useful in confirming the proper position of insulators in the support guys of insulated towers.

A scaled model (1:1000 is a convenient scale factor for large antennas) is erected

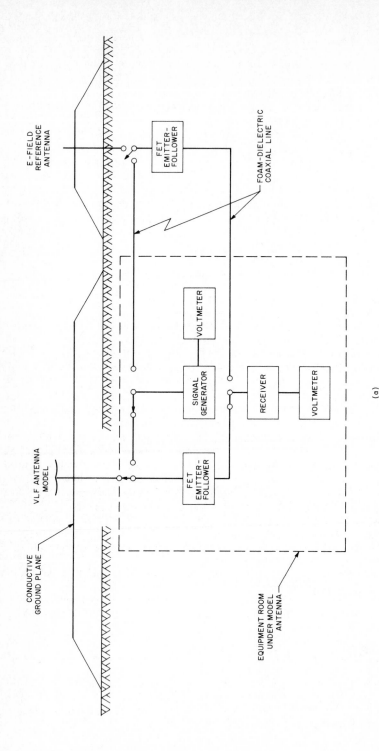

FIG. 1-12 Antenna range for scale-model measurements. (*a*) Sectional schematic. (*b*) Plain view.

(*a*)

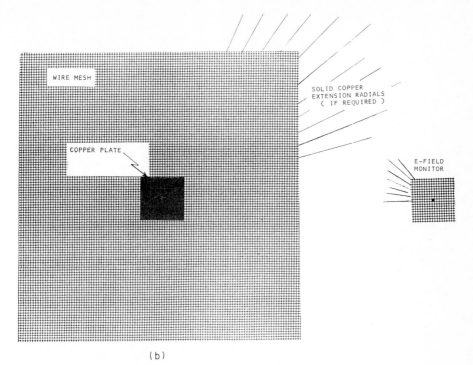

(b)

FIG. 1-12   (*Continued.*)

over a conductive sheet which forms the base for the antenna and guy system. Insulators can be simulated by small glass beads. The antenna is immersed to the bottom of the electrolytic tank and alternating-current-measurement voltages applied between the antenna-base insulator and the ground plane. The measurement voltage must be sufficient to enable accurate measurement. A convenient measurement frequency is 1 kHz. Ordinary tap water is usually a satisfactory electrolyte.

The voltage, relative to ground, on each side of each insulator is measured by using a probe that is insulated except at the tip. The number and location of the guy insulators can be varied as necessary to ensure that all units are working at their rated voltage.

Since the model simulates a static condition, this technique is applicable only to electrically small antennas.

1-9   TYPICAL VLF-LF ANTENNA SYSTEMS

The principal characteristics of a number of current VLF-LF antenna systems are summarized in Table 1-1. The table includes examples of monopoles, top-loaded monopoles, multiple-panel triatics, valley-spans, tridecos, and several hybrid variations.

TABLE 1-1 Characteristics of Typical VLF-LF Antenna Systems

Example	Location	Nominal operating frequency, kHz	Effective height, m	Static capacity, μF	Self-resonant frequency, kHz	Radiated power
Monopole, 800 ft (244 m)	Annapolis, Md.	51	104	0.0031	270.0	20 kW at 51 kHz
Top-loaded monopole, base-insulated tower, 366 m	Trelew, Argentina	10.2–13.6 (Omega)	193	0.0279	53.2	10 kW at 10.2 kHz
Hybrid top-loaded monopole, grounded tower, 427 m	St-Denis, Réunion	10.2–13.6 (Omega)	163	0.0348	27.17	10 kW at 10.2 kHz
Valley span, six 1920-m spans	Haiku, Hawaii	10.2–13.6 (Omega)	169	0.047	39.0	10 kW at 10.2 kHz
Valley span, 10 spans, average length 2200 m	Jim Creek, Wash.	15–27	114	0.078	34.0	15 kW at 24.8 kHz
Array of two top-loaded monopoles, base-insulated, 457 m	Lualualei, Hawaii	15.5–30	192	0.0399	37.5	460 kW at 23.4 kHz
Top-loaded monopole, base-insulated, 366 m	Hawes, Calif.	27–60	228	0.01448	72.5	50 kW at 27 kHz
Top-loaded monopole, base-insulated tower, 213 m	Aberdeen, Md.	179	152	0.00636	153.0	40 kW at 179 kHz
Hybrid trideco-triatic, 266-m base-insulated tower, eleven 183-m grounded towers	Annapolis, Md.	15.5–28	148	0.053	33.5	266 kW at 21.4 kHz
Array of two multiple-panel trideco	Cutler, Me.	15.5–30	152	0.225	38.2	1 MW at 17.8 kHz
Single-multiple-panel trideco	North West Cape, Australia	15.5–30	192	0.1626	34.19	1 MW at 22.3 kHz

FIG. 1-13   U.S. Navy megawatt VLF antenna located at North West Cape, Australia.

The frequency range represented extends from 10 to 179 kHz; input power, from 100 kW to 2 MW. A megawatt VLF antenna is illustrated in Fig. 1-13.

REFERENCES

1   C. E. Smith and E. M. Johnson, "Performance of Short Antennas," *IRE Proc.*, vol. 35, October 1947, pp. 1026–1038. Discusses characteristics of short vertical radiators. Presents low-frequency measurements of antenna impedance and field strength using an existing broadcast tower with various amounts of top-loading umbrella added.

2   H. A. Wheeler, "Small Antennas," in R. C. Johnson and H. Jasik (eds.), *Antenna Engineering Handbook,* 2d Ed., McGraw-Hill Book Company, New York, 1984, Chap. 6. Defines properties and limitations of small antennas.

3   E. A. Laport, *Radio Antenna Engineering,* McGraw-Hill Book Company, New York, 1952, chap. 1. Explains characteristics of small antennas. Describes low-frequency application of Beverage (wave) and Adcock antennas. Describes scale-model measurement of impedance characteristics. Presents measured impedance characteristics of several representative low-frequency configurations, some obtained from existing antennas and others from scale models.

4   E. C. Jordan, *Electromagnetic Waves and Radiating Systems,* Prentice-Hall, Inc., Englewood Cliffs, N.J., 1950, chap. 14. Discusses characteristics of short antennas.

5   T. E. Devaney, R. F. Hall, and W. E. Gustafson, "Low-Frequency Top-Loaded Antennas," R&D Rep. U.S. Navy Electronics Laboratory, San Diego, Calif., June 1966.

6   H. A. Wheeler, "Fundamental Relations in the Design of a VLF Transmitting Antenna," *IRE Trans. Antennas Propagat.,* vol. AP-6, January 1958, pp. 120–122. Defines concepts

and formulas for the fundamental relations that govern the design of a high-power VLF antenna.

7 S. A. Schelkunoff and H. T. Friis, *Antennas: Theory and Practice,* John Wiley & Sons, Inc., New York, 1952, app. I.

8 A. D. Watt, *VLF Radio Engineering,* Pergamon Press, New York, 1967. Textbook containing a detailed coverage of the fields involved in VLF radio engineering; a compendium of basic antenna, propagation, and system engineering.

9 A. G. Bogle, "Effective Resistance and Inductance of Screened Coils," *J. IEE,* vol. 87, 1940, p. 299.

10 E. D. Sunde, *Earth Conduction Effects in Transmission Systems,* D. Van Nostrand Company, Princeton, N.J., 1949.

11 R. R. Morgan and E. L. Maxwell, "Omega Navigational System Conductivity Map," Rep. 54-F-1, Office of Naval Research, December 1965.

12 R. R. Morgan, "Preparation of a Worldwide VLF Conductivity Map," Mar. 15, 1968, and "Worldwide VLF Effective Conductivity Map," Westinghouse Electric Corp., Environmental Science & Technology Department, Jan. 15, 1968. Gives VLF conductivity values for many major land areas of the world.

13 J. R. Wait and A. M. Conda, "On the Measurement of Ground Conductivity at VLF," *IRE Trans. Antennas Propagat.,* vol. AP-6, no. 3, July 1958.

14 G. V. Keller and F. C. Frischknecht, *Electrical Methods in Geophysical Prospecting,* Pergamon Press, New York, 1965.

BIBLIOGRAPHY

Abbott, F. R.: "Design of Optimum Buried-Conductor RF Ground System," *IRE Proc.,* vol. 40, July 1952, pp. 846–852. Derives a design procedure for a radial-wire ground system to obtain maximum power radiated per unit overall cost.

Alexanderson, E. F. W.: "Trans-Oceanic Radio Communications," *IRE Proc.,* vol. 8, August 1920, pp. 263–286. Explains the theory of multiple tuning and states the results obtained when applied to the New Brunswick, N.J., VLF antenna.

Ashbridge, N., H. Bishop, and B. N. MacLarty: "Droitwich Broadcasting Stations," *J. IEE (London),* vol. 77, October 1935, pp. 447–474. Contains a description in some detail and measured performance characteristics pertaining to a 1935, 150-kW, LF, BBC broadcast antenna of the T type. Explains the procedure used in the design of a broadbanding network and shows the performance of the network.

Bolljahn, J. R., and R. F. Reese: "Electrically Small Antennas and the Low-Frequency Aircraft Antenna Problem," *Trans. IRE Antennas Propagat.,* vol. AP-1, no. 2, October 1953, pp. 46–54. Describes methods of measuring patterns and effective height of small antennas on aircraft by means of models immersed in a uniform field.

Brown, G. H., and R. King: "High Frequency Models in Antenna Investigations," *IRE Proc.,* vol. 22, April 1934, pp. 457–480. Describes and justifies the use of small-scale models to investigate problems of vertical radiators.

Brown, W. W.: "Radio Frequency Tests on Antenna Insulators," *IRE Proc.,* vol. 11, October 1923, pp. 495–522. Discusses design and electrical tests of porcelain insulators for VLF antennas.

——— and J. E. Love: "Design and Efficiencies of Large Air Core Inductances," *IRE Proc.,* vol. 13, December 1925, pp. 755–766. Describes several designs and characteristics of VLF antenna-tuning coils.

Buel, A. W.: "The Development of the Standard Design for Self-Supporting Radio Towers for the United Fruit and Tropical Radio Telegraph Companies," *IRE Proc.,* vol. 12, February 1924, pp. 29–82.

Doherty, W. H.: "Operation of AM Broadcast Transmitter into Sharply Tuned Antenna Systems," *IRE Proc.*, vol. 37, July 1949, pp. 729–734. Shows the impairment of bandwidth of a broadcast transmitter caused by a high-Q antenna.

Feld, Jacob: "Radio Antennas Suspended from 1000 Foot Towers," *J. Franklin Inst.*, vol. 239, May 1945, pp. 363–390. Describes the mechanical design of a flattop antenna on guyed masts.

Grover, F. W.: "Methods, Formulas, and Tables for Calculation of Antenna Capacity," *Nat. Bur. Stds. Sci. Pap.*, no. 568. Discusses and employs Howe's average-potential method. Out of print but available in libraries.

Henney, K. (ed.): *Radio Engineering Handbook*, McGraw-Hill Book Company, New York, 1950. Section entitled "Low-Frequency Transmitting Antennas (below 300 Kc)," by E. A. Laport, pp. 609–623, explains characteristics of small antennas. Describes low-frequency application of Beverage (wave), Adcock, and whip antennas. Describes scale-model measurement of impedance characteristics.

Hobart, T. D.: "Navy VLF Transmitter Will Radiate 1,000 KW," *Electronics*, vol. 25, December 1952, pp. 98–101. Describes U.S. Navy VLF installation of J. R. Redman, "The Giant Station at Jim Creek."

Hollinghurst, F., and H. F. Mann: "Replacement of the Main Aerial System at Rugby Radio Station," *P.O. Elec. Eng. J.*, April 1940, pp. 22–27. Describes changes made to Rugby VLF antenna in 1927 and 1937.

Howe, G. W. O.: "The Capacity of Radio Telegraphic Antennae," *Electrician*, vol. 73, August and September 1914, pp. 829–832, 859–864, 906–909; "The Capacity of Aerials of the Umbrella Type," *Electrician*, vol. 75, September 1915, pp. 870–872; "The Calculation of Capacity of Radio Telegraph Antennae, Including the Effects of Masts and Buildings," *Wireless World*, vol. 4, October and November 1916, pp. 549–556, 633–638; "The Calculation of Aerial Capacitance," *Wireless Eng.*, vol. 20, April 1943, pp. 157–158. Discuss application of Howe's average-potential method for situations stated in the titles.

IRE Standards on Antennas, Modulation Systems, and Transmitters, Definitions of Terms, New York, 1948; *Standards on Antennas, Methods of Testing*, New York, 1948.

Jordan, E. C.: *Electromagnetic Waves and Radiating Systems*, Prentice-Hall, Inc., Englewood Cliffs, N.J., 1950, chap. 14. Discusses characteristics of short antennas.

Knowlton, A. E. (ed.): *Standard Handbook for Electrical Engineers*, 7th ed., McGraw-Hill Book Company, New York, 1941, sec. 13, p. 261. Gives design equation for sleet-melting current required on power transmission lines.

Lindenblad, N., and W. W. Brown: "Main Considerations in Antenna Design," *IRE Proc.*, vol. 14, June 1926, pp. 291–323. Discusses VLF antenna features and design problems and methods.

Miller, H. P., Jr.: "The Insulation of a Guyed Mast," *IRE Proc.*, vol. 15, March 1927, pp. 225–243. Discusses the value of mast and guy insulation in VLF antennas and describes procedures to determine their voltage duties and best placement.

Pender, H., and K. Knox (eds.): *Electrical Engineers' Handbook*, vol. 5: *Electric Communications and Electronics*, 3d ed., John Wiley & Sons, Inc., New York, 1936. Refers to sleet-melting facilities for antennas.

Pierce, J. A., A. A. McKenzie, and R. H. Woodward: *Loran*, MIT Rad Lab. ser., McGraw-Hill Book Company, New York, 1948, chap. 10. Contains a description and performance characteristics of temporary 1290-ft balloon-supported antenna used by the U.S. Air Force at 180 kHz.

Redman, J. R.: "The Giant Station at Jim Creek," *Signal*, January–February 1951, p. 15. Describes U.S. Navy 120-kW VLF installation.

Sanderman, E. K.: *Radio Engineering*, vol. I, John Wiley & Sons, Inc., New York, 1948, chap. 16. Treats antenna-coupling networks in general. Explains the procedure used in the design of the 1935 Droitwich broadbanding network. Describes mechanical design methods for flattops.

Shannon, J. H.: "Sleet Removal from Antennas," *IRE Proc.*, vol. 14, April 1926, pp. 181–195. Describes the sleet-melting and weak-link provisions for the Rocky Point VLF antenna.

Shaughnessy, E. H.: "The Rugby Radio Station of the British Post Office," *J. IEE (London),* vol. 64, June 1926, pp. 683–713. Contains detailed description and measured performance characteristics of a VLF antenna.

Stratton. J. A.: *Electromagnetic Theory,* McGraw-Hill Book Company, New York, 1941, sec. 9.3. Treats electrodynamic similitude and the theory of models.

Sturgis, S. D.: "World's Third Tallest Structure Erected in Greenland," *Civil Eng.,* June 1954, pp. 381–385. Describes structural details and installation procedures for a U.S. Air Force 1205-ft low-frequency vertical radiator.

Weagand, R. A.: "Design of Guy-Supported Towers for Radio Telegraphy," *IRE Proc.,* vol. 3, June 1915, pp. 135–159.

Wells, N.: "Aerial Characteristics," *J. IEE (London),* part III, vol. 89, June 1942, pp. 76–99. Contains a description and performance characteristics of a low-frequency antenna employing a variation of the multiple-tuning principle.

West, W., A. Cook, L. L. Hall, and H. E. Sturgess: "The Radio Transmitting Station at Criggion," *J. IEE (London),* part IIIA, vol. 94, 1947, pp. 269–282. Contains description and measured performance characteristics of a VLF antenna.

Chapter 2

Medium-Frequency Broadcast Antennas

Howard T. Head
John A. Lundin

A. D. Ring & Associates

2-1   INTRODUCTION

General

Medium-frequency broadcast transmitting antennas are generally vertical radiators ranging in height from one-sixth to five-eighths wavelength, depending upon the operating characteristics desired and economic considerations. The physical heights vary from about 150 ft (46 m) to 900 ft (274 m) above ground, making the use of towers as radiators practical. The towers may be guyed or self-supporting; they are usually insulated from ground at the base, although grounded shunt-excited radiators are occasionally employed.

Scope of Design Data

The design formulas and data in this chapter are applicable primarily to broadcast service (535- to 1605-kHz band). However, the basic design principles are valid for transmitting antennas for other services in the medium-frequency band (300 to 3000 kHz).

Characteristics of Radiators

Maximum radiation is produced in the horizontal plane, increasing with radiator height up to a height of about five-eighths wavelength. The radiated field from a single tower is uniform in the horizontal plane, generally decreases with angle above the horizon, and is zero toward the zenith. Radiators taller than one-half wavelength have a minor lobe of radiation at high vertical angles. For radiators with a height in excess of about 0.72 wavelength, the energy in this lobe is maximum, and there is a reduction in horizontal radiation with increasing height. For a height of one wavelength, negligible energy is radiated in the horizontal plane. Radiators taller than five-eighths wavelength may be utilized by sectionalizing the tower approximately each half wavelength (Franklin type) and supplying the current to each section with the same relative phase. Figure 2-1 shows vertical radiation patterns for several commonly employed antenna heights, both for constant power and for constant radiated field in the horizontal plane.

Ground Currents

Current return takes place through the ground plane surrounding the antenna. High earth-current densities are encountered and require metallic ground systems to minimize losses.

Choice of Plane of Polarization

Vertical polarization is almost universally employed because of superior ground-wave and sky-wave propagation characteristics. Ground-wave attenuation is much greater for horizontal than for vertical polarization, and ionospheric propagation of horizon-

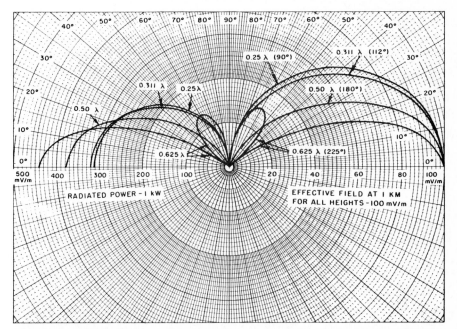

FIG. 2-1 Vertical radiation patterns for different heights of vertical antennas (FCC AM Technical Standards).

tally polarized signals is more seriously influenced by geomagnetic latitude and direction of transmission. Horizontal antennas immediately above ground produce negligible fields in the horizontal plane but radiate relatively large fields at high vertical angles for low antenna heights. This high-angle radiation is desirable only under special circumstances.

Performance Required of Medium-Frequency Broadcast Antennas

The performance required is determined by class of station and channel assignment. In North America three types of broadcast channels are established: clear, regional, and local. Class I stations, operating on clear channels, are assigned to provide ground-wave service during daytime and both ground-wave and sky-wave service at night. All other classes of station are assigned to render ground-wave service only. Class II stations are secondary stations operating on clear channels and employing directional antennas, when necessary, to protect the service areas of Class I stations. Class III stations are assigned to regional channels and employ directional antennas, when required, for mutual protection. Class IV stations are assigned to local channels; both directional and nondirectional antennas are authorized.

The station classes recognized by the Region 2 Hemispherical Agreement (North, Central, and South American countries) are Classes A, B, and C. A Class I station corresponds to Class A, Classes II and III correspond to Class B, and Class IV corresponds to Class C.

Minimum required antenna performance and power limitations for each class of broadcast station in the United States are established by the Federal Communications Commission (FCC) AM Technical Standards (Table 2-1).

Class I stations are assigned to provide both ground-wave and sky-wave service at night. The vertical radiation pattern for night operation of a Class I station should be designed to provide maximum sky-wave signal beyond the ground-wave service area and minimum sky-wave signal within the ground-wave service area (antifading antennas).[1]

TABLE 2-1   Antenna Performance and Power Limitations

Class of station	Required minimum effective field for 1-kW power (unattenuated field), mV/m		Minimum and maximum power, kW
	At 1 km*	At 1 mi	
I	362	225	10 −50
II	282	175	0.25−50
III	282	175	0.5 −5
IV	241	150	0.25− 1.0

*1 mV/m at 1 km = 1 V/m at 1 m.

2-2   CHARACTERISTICS OF VERTICAL RADIATORS

Assumptions Employed in Calculating Radiator Characteristics

The characteristics of tower antennas are ordinarily computed by assuming sinusoidal current distribution in a thin conductor over a perfectly conducting plane earth, with the wavelength along the radiator equal to the wavelength in free space. The effect of the earth plane is represented by an *image* of the antenna as shown in Fig. 2-2. These assumptions provide sufficiently accurate results for most purposes, but in the determination of base operating resistance and reactance the finite cross section of the tower must be taken into account. Also, the finite cross section modifies the vertical radiation patterns slightly. Generally, this effect is of significance only in antifading antennas for clear-channel stations and in certain directional-antenna systems (discussed in Sec. 2-4). Except as noted, all formulas and data in this chapter are based on the assumptions stated above.

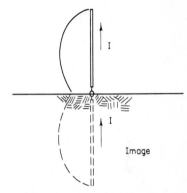

FIG. 2-2 Current distribution and image for a vertical antenna over a perfectly conducting plane.

Field Produced by Vertical Radiator

The field in the horizontal plane is a function of the current flowing and the electrical height. For uniform current in a vertical radiator over a perfectly conducting plane earth, the radiated field is 1.048 V/m (unattenuated field at 1 m) per degree-ampere. For other current distributions, the radiation is proportional to the maximum current and the *form factor K* of the antenna. For sinusoidal current distribution, the radiated field (unattenuated field, volts per meter, at 1 m) is

$$E_r = 60\, I_0\, (1 - \cos G) \qquad \textbf{(2-1)}$$

where I_0 = loop current, A
$(1 - \cos G)$ = K, form factor for sinusoidal current distribution in a vertical radiator
G = electrical height of radiator above ground

Radiation Resistance

The loop current I_0 is related to the radiated power P_r by the *loop radiation resistance* $R_r = P_r/I_0^2$. Figure 2-3 shows the radiated field for a power of 1 kW and the radiation resistance as a function of antenna height G. The radiation resistance may be calculated from

$$R_r = 15\, [4 \cos^2 G \operatorname{Cin}(2G) - \cos 2G \operatorname{Cin}(4G)$$

$$- \sin 2G\, [2 \operatorname{Si}(2G) - \operatorname{Si}(4G)]] \qquad \Omega \quad \textbf{(2-2)}$$

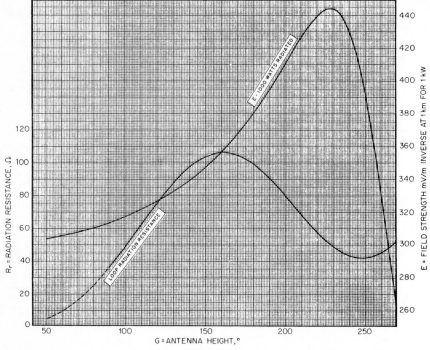

FIG. 2-3   Radiated field and radiation resistance as a function of antenna height G.

where $\text{Cin}(x) = \int_0^x \dfrac{1 - \cos x}{x}\, dx$ (cosine integral)

$\qquad\qquad\quad = \ln x + C - \text{Ci}(x)$

$\text{Ci}(x) = -\int_x^{\infty} \dfrac{\cos x}{x}\, dx$ (cosine integral)

$C = 0.5772\cdots$ (Euler's constant)

$\text{Si}(x) = \int_0^x \dfrac{\sin x}{x}\, dx$ (sine integral)

Operating Base Resistance

For sinusoidal current distribution, the *base radiation resistance* is related to the loop radiation resistance by $R_{r(\text{base})} = [R_{r(\text{loop})}]/\sin^2 G$. However, the actual base resistance of a practical tower radiator may vary widely from this value because of the finite cross section of the tower and other effects.[2,3]

Vertical-Radiation Characteristic

The relative field pattern in a vertical plane through the radiator is known as the *vertical-radiation characteristic*. It is defined as having unit value in the horizontal plane ($\theta = 0°$).* Based on the assumptions stated above, the vertical-radiation characteristic is

$$f(\theta) = \frac{\cos (G \sin \theta) - \cos G}{(1 - \cos G) \cos \theta} \qquad (2\text{-}3)$$

General Formulas for Calculating Radiating Characteristics

The form factor and vertical-radiation characteristic establish the radiated field in the horizontal and vertical planes. For any current distribution, they may be determined from

$$Kf(\theta) = 1.048 \int_0^G I(z) \cos \theta \cos (z \sin \theta) \, dz \qquad \text{V/m at 1 m} \qquad (2\text{-}4)$$

where z = height of current element dz, electrical degrees
   $I(z)$ = current at z, A
   θ = elevation angle
   G = antenna height
The radiation resistance is†

$$R_r = 60 \int_0^{\pi/2} [Kf(\theta)]^2 \cos \theta \, d\theta \qquad \Omega\ddagger \qquad (2\text{-}5)$$

Effects of Finite Cross Section

Current Distribution   The effect of finite tower cross section on the current distribution and vertical-radiation characteristic may be taken into account by assuming a current distribution of the following form suggested by Schelkunoff:[2]

$$I(z) = I_0 \sin [\gamma(G - z)] + jkI_0 (\cos \gamma z - \cos \gamma G) \qquad (2\text{-}6)$$

where I_0 = maximum in-phase current, A
   z = height of current element
   γ = λ_0/λ, ratio of wavelength in free space to wavelength along tower
   G = height of tower, electrical degrees
   $k \approx 50/(Z_0 - 45)$, for antenna heights near $\lambda/2$
   Z_0 = 60 [ln $(2G/a_{eq}) - 1$], characteristic impedance of tower $\qquad (2\text{-}7)$
   a_{eq} = $\sqrt{naa_0}$, equivalent tower radius
   n = number of tower legs
   a = cage radius (distance from tower center to tower leg)
   a_0 = tower-leg radius

---

*The symbol θ is used in this chapter to denote angle above the horizontal plane.

†An alternative method of calculation is given by Carter.[4]

‡Note that the radiated power is equal to $I^2 R_r$. For $E = Kf(\theta)$, mV/m at 1 km, P (watts) = $1/60 \int_0^{\pi/2} E^2 \cos \theta \, d\theta$.

When the cross section is vanishingly small, Z_0 is very large, $k \to 0$, and the current distribution approaches the simple sinusoidal.

For short towers, this current distribution is very nearly the simple sinusoidal; for a quarter-wave tower, the two are identical except for the reduced wavelength. For towers higher than a quarter wavelength, the departure becomes significant.

Vertical-Radiation Characteristic and Form Factor   Applying Eq. (2-4) to the Schelkunoff current distribution, the form factor K_2 and the vertical-radiation characteristic $f_2(\theta)$ become

$$
\begin{aligned}
K_2 f_2(\theta) = {} & \frac{\gamma \cos \theta}{\gamma^2 - \sin^2 \theta} \left[\cos (G \sin \theta) - \cos \gamma G \right] \\
& + jk \cos \theta \left[\frac{\gamma \sin \gamma G \cos (G \sin \theta)}{\gamma^2 - \sin^2 \theta} \right. \\
& \quad - \frac{\sin \theta \cos \gamma G \sin (G \sin \theta)}{\gamma^2 - \sin^2 \theta} \\
& \quad \left. - \frac{\cos \gamma G \sin (G \sin \theta)}{\sin \theta} \right]
\end{aligned}
\tag{2-8}
$$

The variations of the vertical-radiation characteristic, based on the sinusoidal and Schelkunoff assumptions, are relatively small and would have only minor practical effects in nondirectional radiators or in directional-antenna systems employing identical towers. However, for directional-antenna systems employing tall towers of different heights, the variations in amplitude and phase of the vertical-radiation characteristics must be taken into account [see Sec. 2-4, Eq. (2-45)].

Shunt-Fed Radiators

Energy may be supplied to a grounded tower by shunt excitation, using a slant-wire feed as shown in Fig. 2-4. The dimensions h/λ and d/λ shown in the figure determine the impedance at the end of the slant wire.[5,6]

The current in the slant wires and between the tap point and ground modify the vertical radiation pattern of the shunt-fed tower and result in a slight nonuniformity in radiation in the horizontal plane. Notwithstanding this, the radiation field from a shunt-fed tower is essentially the same as that from a series-fed tower of the same height.

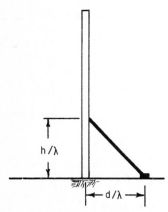

h/λ

$\leftarrow d/\lambda \rightarrow$

FIG. 2-4 Shunt-fed element.

Top-Loaded Radiators

The radiating characteristics may be modified by altering the current distribution. Top loading is generally accomplished on guyed towers by connecting a portion of the top guy wires directly to the tower instead of having it connected through an insulator. Another method of top loading that is sometimes employed is a capacity "hat" mounted on the tower top. These methods of top loading have essentially the same effect (within limits) as increasing the tower height (Fig. 2-5). For guy-wire top loading, the length of the top portion of the guy wire connected directly to the tower can be used as a rough approximation for B.

A capacity disk of radius r at the tower top is equivalent (within practical limits) to an electrical length of

$$B = \frac{rZ_0}{15\pi} \qquad \textbf{(2-9)}$$

Z_0 is defined in Eq. (2-7).

The vertical-radiation characteristic for the top-loaded antenna is given by[7,8,9]

$$f(\theta) = \frac{\cos B \cos (A \sin \theta) - \sin \theta \sin B \sin (A \sin \theta) - \cos G}{\cos \theta (\cos B - \cos G)} \qquad \textbf{(2-10)}$$

where A = electrical height of vertical portion of antenna
B = equivalent electrical height of top loading
$G = A + B$

Sectionalized Radiators

Effects of Sectionalization   Antenna heights above five-eighths wavelength may be employed to obtain increased effective field by dividing the tower with insulators into sections of approximately one-half wavelength. The currents in each section are maintained in phase.

For heights less than one wavelength, the tower may be sectionalized in either of the arrangements shown in Fig. 2-6. The vertical-radiation characteristic for the two

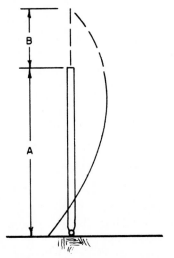

FIG. 2-5   Top-loaded element.

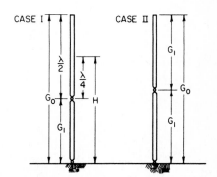

FIG. 2-6   Two methods of sectionalizing vertical radiators.

arrangements, based on simple sinusoidal current distribution, is

Case I: $f(\theta) = \dfrac{2 \cos (90 \sin \theta) \cos (H \sin \theta) + \cos (G_1 \sin \theta) - \cos G_1}{\cos \theta (3 - \cos G_1)}$ (2-11)

Case II: $f(\theta) = \dfrac{\cos (G_1 \sin \theta) [\cos (G_1 \sin \theta) - \cos G_1]}{\cos \theta (1 - \cos G_1)}$ (2-12)

If top loading is employed on a sectionalized antenna (Fig. 2-7), the vertical-radiation characteristic is expressed as[8,9]

$$f(\theta) = \dfrac{\begin{aligned}&\sin \Delta \, [\cos B \cos (A \sin \theta) - \cos G] + \sin B \, [\cos D \cos (C \sin \theta) \\ &\quad - \sin \theta \sin D \sin (C \sin \theta) - \cos \Delta \cos (A \sin \theta)]\end{aligned}}{\cos \theta \, [\sin \Delta \, (\cos B - \cos C) + \sin B \, (\cos D - \cos \Delta)]}$$ (2-13)

where A = height of lower section
$\quad B$ = equivalent top loading of lower section
$\quad C$ = height of entire antenna
$\quad D$ = equivalent top loading of top section
$\quad G = A + B$
$\quad H = C + D$
$\quad \Delta = H - A$

Effects of Finite Cross Section of Sectionalized Radiators The actual current distribution on sectionalized towers varies from simple sinusoidal in the manner discussed above. In this case, the upper section of the tower is a half-wavelength (or very nearly so), and the base termination is adjusted to provide the same current dis-

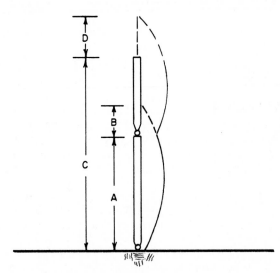

FIG. 2-7 Top-loaded sectionalized vertical radiator.

tribution on the lower section as on the upper section. For these conditions, the currents are (see Fig. 2-6, Case I)

$$I_1(z) \text{ (top section)} = I_0 \sin [\gamma(G_0 - z)]$$
$$+ jkI_0 \{1 + \cos [\gamma(G_1 - z)]\} \quad \textbf{(2-14}\textbf{\textit{a}}\textbf{)}$$

$$I_2(z) \text{ (bottom section)} = I_0 \sin [\gamma(G_1 - z)]$$
$$+ jkI_0 \{1 + \cos [\gamma(G_1 - z)]\} \quad \textbf{(2-14}\textbf{\textit{b}}\textbf{)}$$

where $\gamma = \lambda_0/\lambda$, ratio of wavelength in free space to wavelength along tower
  I_0 = maximum in-phase current
  G_0 = overall height of tower (both sections)
  z = height of current element
  $k \approx 50/(Z_0 - 45)$ for antenna heights near $\lambda/2$
See Eq. (2-7) for Z_0.

The form factor and the vertical-radiation characteristic are given by

$$K_0 f_0(\theta) = \frac{\gamma \cos \theta}{\gamma^2 - \sin^2 \theta} [\cos (G_0 \sin \theta) - \cos \gamma G_1$$
$$- 2 \cos \gamma\pi \cos (G_1 \sin \theta)]$$
$$+ jk \cos \theta \left\{ \frac{\sin (G_0 \sin \theta)}{\sin \theta} \right. \qquad \textbf{(2-15)}$$
$$+ \frac{\gamma}{\gamma^2 - \sin^2} \left[\sin \gamma G_1 - 2 \sin \gamma\pi \cos (G_1 \sin \theta) \right.$$
$$\left. \left. + \frac{\sin \theta \sin (G_0 \sin \theta)}{\gamma} \right] \right\}$$

2-3 GROUND SYSTEMS

General Requirements

The ground system for a medium-frequency antenna usually consists of 120 buried copper wires, equally spaced, extending radially outward from the tower base to a minimum distance of one-quarter wavelength. In addition, an exposed copper-mesh ground screen may be used around the base of the tower when high base voltages are encountered.

Wire size has a negligible effect on the effectiveness of the ground system and is chosen for mechanical strength; AWG No. 10 or larger is adequate. A depth of 4 to 6 in (102 to 152 mm) is generally adequate, although the wires may be buried to a depth of several feet if desired in order to permit cultivation of the soil. When such deep burial is required, the wires should descend to the required depth on a smooth, gentle incline from the tower base, reaching the ultimate depth some distance from the tower. If this precaution is observed, the deep burial will have relatively little effect on the effectiveness of the ground system for typical soil conditions.

Nature of Ground Currents

Ground currents are conduction currents returning directly to the base of the antenna. The total earth current flowing through a cylinder of radius x concentric with the antenna is known as the *zone current*. It is a function of tower height and is given by

$$I_{\text{zone}} = I_0 [\sin r_2 - \cos G \sin x + j (\cos r_2 - \cos G \cos x)] \qquad \textbf{(2-16)}$$

where I_0 = loop antenna current
       G = electrical height of antenna
       $r_2 = \sqrt{x^2 + G^2}$

Effect of Ground-System Losses on Antenna Performance

Ground-system losses dissipate a portion of the input power and reduce the field radiated from the antenna. These losses are equivalent to the power dissipated in a resistor in series with the antenna impedance.[10] Computed values of radiated field based on an assumed series loss resistance of 2 Ω give results for typical installations in good agreement with actual measured effective field intensities. FCC rules require that the assumed series loss resistance for each element of a directional-antenna system shall be no less than 1 Ω. For directional-antenna systems, a series loss of 1 Ω is assumed for each radiator.

Effect of Local Soil Conductivity on Ground-System Requirements

A less elaborate ground system may be effective in soil of high local conductivity,[11,12] although adequate local conductivity data are rarely available.

For a seawater site (conductivity, approximately 4.6 S/m*), the salt water provides a adequate ground system; a submerged copper ground screen is employed to make contact with the salt water.

Ground Systems for Multielement Arrays

Individual ground systems are required for each tower of a multielement array. If the individual systems would overlap, the adjoining systems are usually terminated in a common bus. Complete systems may be installed around each tower when high ground currents are expected. Figure 2-8 shows a typical multielement ground system.

Recent theoretical studies have shown that the ground currents of directional antennas vary in both magnitude and direction from those which would be associated with single radiators considered individually. The magnitudes of the ground currents depend upon the directional-antenna parameters, and the directions of the ground currents vary over the individual radio-frequency cycle. This leads to the conclusion that the ground system for a directional antenna should in general be modified from the conventional nondirectional ground system in two ways: (1) the ground wires should be concentrated in the regions of higher current density, and (2) the paths of the

---

*Siemens/meter or mho/meter.

120 RADIALS SPACED EVENLY
AROUND EACH TOWER

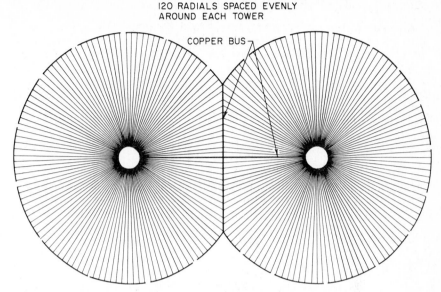

COPPER BUS

FIG. 2-8 A typical ground system for a two-element directional antenna.

ground wires should be chosen to favor the direction of maximum current flow at the peak of the radio-frequency cycle.[13]

2-4 DIRECTIONAL ANTENNAS

Purpose of Directional-Antenna Systems

A directional antenna employs two or more radiators to produce radiation patterns in the horizontal and vertical planes different from those produced by a single radiator. Directional antennas are used principally to reduce the radiated signal toward other stations on the same or adjacent channels in order to avoid interference, although the resulting concentration of radiated signal in other directions may be utilized to improve service to specific areas.

Permissible Values of Radiation

Methods of computing interference and establishing maximum permissible values of radiation to avoid interference are described in detail in the FCC *Rules and Regulations,* Volume III, Part 73.[8] For interference from ground-wave signals, radiation in the horizontal plane only is considered. For interference from sky-wave signals, radiation throughout a specified range of vertical angles must be considered. Suppression of radiation is required over sufficient arc to subtend the protected station's service area.

Computation of Pattern Shape

Fundamental Considerations: Two-Element Systems The principles under-lying the computation of shape of radiation patterns from two-element arrays are fun-damental and may be extended to the computations of shape of patterns from mul-tielement arrays.

The field strength at any point from two radiators receiving radio-frequency energy from a common source is the vector sum of the fields from each of the two radiators. At large distances, the antenna system may be considered to be a point source of radiation. Referring to Fig. 2-9 and considering tower 1 to be reference or zero phase, the theoretical radiated field from the array at the angle ϕ, θ is

$$E_{th}(\theta) = E_1 f_1(\theta) \underline{/0} + E_2 f_2(\theta) \underline{/\alpha_2} \qquad \textbf{(2-17)}$$

where α_2, the difference in phase angle between the two fields, is the sum of the *time-phase-angle* difference A_2 and the apparent *space-phase-angle* difference:

$$\alpha_2 = s \cos \phi \cos \theta + A_2 \qquad \textbf{(2-18)}$$

where E_1 = field radiated by element 1
E_2 = field radiated by element 2
$f_1(\theta)$ = vertical-radiation characteristic of element 1
$f_2(\theta)$ = vertical-radiation characteristic of element 2
s = spacing of element 2 from element 1
ϕ = angle between element line and azimuth of calculation
For towers of equal height with $F_2 = E_2/E_1$, Eq. (2-17) simplifies to

$$E_{th}(\theta) = E_1 f(\theta) \sqrt{2F_2} \left[\frac{1 + F_2^2}{2F_2} + \cos (s \cos \phi \cos \theta + A_2) \right]^{1/2} \qquad \textbf{(2-19)}$$

The magnitude of E_{th} is a minimum when

$$\cos (s \cos \phi \cos \theta + A_2) = -1 \qquad \textbf{(2-20)}$$

This occurs when $(s \cos \phi \cos \theta + A_2) = 180° \pm 360°$; if $F_2 = 1$, E_{th} will be zero.
For $F_2 = 1$, Eq. (2-19) simplifies to

$$E_{th}(\theta) = 2E_1 \cos \left(\frac{s}{2} \cos \phi \cos \theta + \frac{A_2}{2} \right) \qquad \textbf{(2-21)}$$

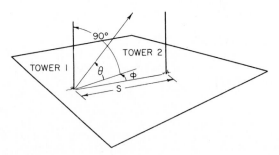

FIG. 2-9 Two-element array.

SPACING

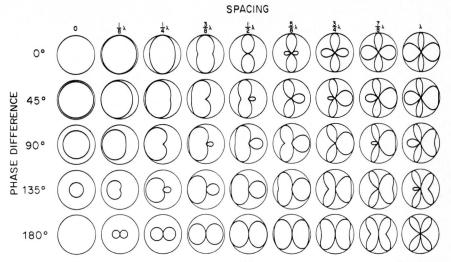

FIG. 2-10 Two-element horizontal plane patterns.

Combinations of equations of the two-element form are widely used in establishing and computing patterns for multielement arrays. Figure 2-10 shows horizontal plane patterns for two-element arrays for a variety of spacings and phasings.[14] The patterns shown are for equal fields in the two radiators ($F_2 = 1$); the effect of unequal fields is to "fill" the nulls of the pattern. For a given spacing, the angular position of the minimum is changed only by changing the phase angle. The value of $E_{th}(\theta)$ is unchanged by substituting the inverse for F_2, $F_2 = 1/F_2$.

Two or More Elements: General Equations   The pattern shape of directional arrays of two or more elements may be computed by an extension of Eq. (2-17):

$$E_{th}(\theta) = E_1 \sum_{i=1}^{n} F_i f_i(\theta) \ \underline{/\alpha_i} \qquad \textbf{(2-22)}$$

where E_1 = field radiated by element 1
  F_i = field ratio of ith element, $F_i = E_i/E_1$
  $f_i(\theta)$ = vertical-radiation characteristic of ith element
  $\alpha_i = s_i \cos \phi_i \cos \theta + A_i$
  s_i = spacing of ith element relative to a reference point (usually element 1)
  ϕ_i = angle between orientation of ith element and azimuth of calculation
  θ = vertical angle
  A_i = phase angle of the ith element

Three or More Elements: Simplified Formulas for Special Cases   The following are formulas for many commonly used array configurations which, through the combination of element pairs, provide a convenient means for establishing the angular position and depth of the minima.

For these examples:

r_i = assumed field ratio for two-element pair

a_i = phase angle for two-element pair

F_i = resulting field ratio for ith element in array

A_i = resulting phase angle for ith element in array

Elements are assumed to be of equal height.

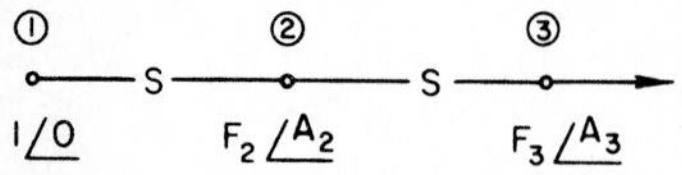

FIG. 2-11 Linear array of three elements.

$$E_{th}(\theta) = E_1 f(\theta) \sqrt{4r_2 r_3} \left\{ \left[\frac{1 + r_2^2}{2r_2} + \cos (s \cos \phi \cos \theta + a_2) \right] \right.$$

$$\left. \left[\frac{1 + r_3^2}{2r_3} + \cos (s \cos \phi \cos \theta + a_3) \right] \right\}^{1/2} \qquad \textbf{(2-23)}$$

$$F_2 \underline{/A_2} = r_2 \underline{/a_2} + r_3 \underline{/a_3}$$

$$F_3 \underline{/A_3} = r_2 r_3 \underline{/a_2 + a_3}$$

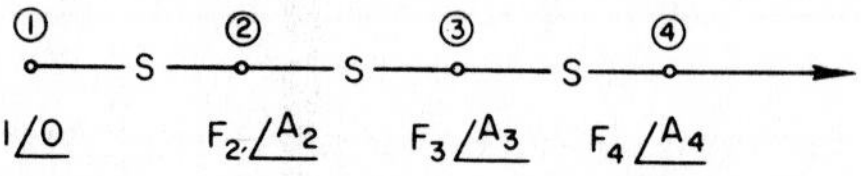

FIG. 2-12 Linear array of four elements.

$$E_{th}(\theta) = E_1 f(\theta) \sqrt{8r_2 r_3 r_4} \left\{ \left[\frac{1 + r_2^2}{2r_2} + \cos (s \cos \phi \cos \theta + a_2) \right] \right.$$

$$\left[\frac{1 + r_3^2}{2r_3} + \cos (s \cos \phi \cos \theta + a_3) \right]$$

$$\left. \left[\frac{1 + r_4^2}{2r_4} + \cos (s \cos \phi \cos \theta + a_4) \right] \right\}^{1/2} \qquad \textbf{(2-24)}$$

$$F_2 \underline{/A_2} = r_2 \underline{/a_2} + r_3 \underline{/a_3} + r_4 \underline{/a_4}$$

$$F_3 \underline{/A_3} = r_2 r_3 \underline{/a_2 + a_3} + r_2 r_4 \underline{/a_2 + a_4} + r_3 r_4 \underline{/a_3 + a_4}$$

$$F_4 \underline{/A_4} = r_2 r_3 r_4 \underline{/a_2 + a_3 + a_4}$$

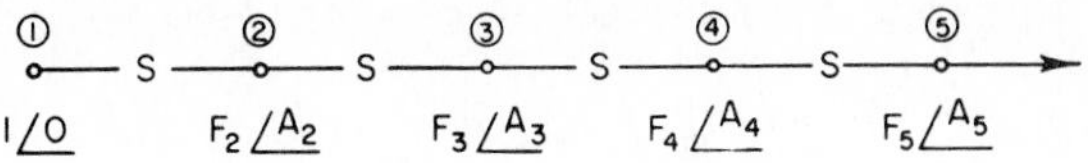

FIG. 2-13   Linear array of five elements.

$$E_{th}(\theta) = E_1 f(\theta) \sqrt{16 r_2 r_3 r_4 r_5} \left\{ \frac{1 + r_2^2}{2 r_2} + \cos (s \cos \phi \cos \theta + a_2)] \right.$$

$$\left[\frac{1 + r_3^2}{2 r_3} + \cos (s \cos \phi \cos \theta + a_3) \right]$$

$$\left[\frac{1 + r_4^2}{2 r_4} + \cos (s \cos \phi \cos \theta + a_4) \right]$$

$$\left. \left[\frac{1 + r_5^2}{2 r_5} + \cos (s \cos \phi \cos \theta + a_5) \right] \right\}^{1/2}$$

$$F_2 \underline{/A_2} = r_2 \underline{/a_2} + r_3 \underline{/a_3} + r_4 \underline{/a_4} + r_5 \underline{/a_5} \qquad \text{(2-25)}$$

$$F_3 \underline{A_3} = r_2 r_3 \underline{/a_2 + a_3} + r_2 r_4 \underline{/a_2 + a_4} + r_2 r_5 \underline{/a_2 + a_5}$$

$$+ r_3 r_4 \underline{/a_3 + a_4} + r_3 r_5 \underline{/a_3 + a_5} + r_4 r_5 \underline{/a_4 + a_5}$$

$$F_4 \underline{/A_4} = r_2 r_3 r_4 a_2 + a_3 + a_4 + r_2 r_3 r_5 \underline{/a_2 + a_3 + a_5}$$

$$+ r_2 r_4 r_5 \underline{/a_2 + a_4 + a_5} + r_3 r_4 r_5 \underline{/a_3 + a_4 + a_5}$$

$$F_5 \underline{/A_5} = r_2 r_3 r_4 r_5 \underline{/a_2 + a_3 + a_4 + a_5}$$

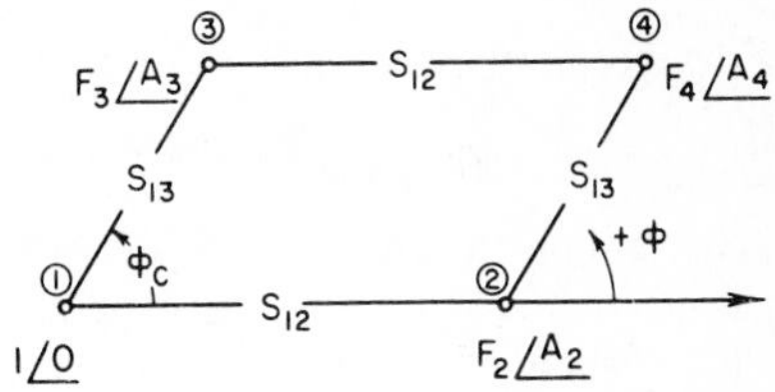

FIG. 2-14   Parallelogram array of four elements.

$$E_{th}(\theta) = E_1 f(\theta) \sqrt{4 F_2 F_3} \left\{ \left[\frac{1 + F_2^2}{2 F_2} + \cos (s_{12} \cos \phi \cos \theta \right. \right.$$

$$\left. \left. + A_2) \right] \left[\frac{1 + F_3^2}{2 F_3} + \cos (s_{13} \cos [\phi - \phi_c] \cos \theta + A_3) \right] \right\}^{1/2} \qquad \text{(2-26)}$$

provided $F_4 \underline{/A_4} = F_2 F_3 \underline{/A_2 + A_3}$.

In these formulas, the field is the product of several two-element-array expressions and will be zero whenever any of the component expressions is zero. For $r_n \neq 1$, the position of the minima may be shifted slightly; however, this shift is usually small.

Formulas (2-23) through (2-26) are valid only for computing patterns in the vertical planes for towers of equal height.

Computation of Pattern Size

General Formulas (2-17) through (2-26) establish the shape of radiation patterns. To determine the magnitude of the radiation, the reference field E_1 must be evaluated.

Total Resistance The current I_1 in the reference element is related to the total radiated power P_r by the *total resistance* R_t of the array:

$$R_{t1} = \frac{P_r}{I_1^2} \quad \Omega \qquad (2\text{-}27)$$

where the subscript 1 indicates that the total resistance is referred to the current in element 1. The radiated field is directly proportional to the current, and

$$E_1 = E_0 \left(\frac{R_{r1}}{R_{t1}}\right)^{1/2} \quad \text{mV/m at 1 km} \qquad (2\text{-}28)$$

where E_0 is the effective field of the reference element, operating independently and without loss, and R_{r1} is the radiation resistance. The total resistance is calculated from the operating resistances of each element of the array, as discussed in the following subsection.

Mutual Impedance: Definition for Antenna Case The mutual impedance between two antennas is defined in the usual manner:

$$Z_{12} = Z_{21} = V_1/I_2 = V_2/I_1$$

The mutual impedance is a complex quantity:

$$Z_{12}\underline{/\sigma_{12}} = R_{12} + jX_{12} = \sqrt{R_{12}^2 + X_{12}^2}\ \underline{/\tan^{-1} X_{12}/R_{12}}$$

The mutual impedance is a function of the spacing between the antennas and the height of the elements. The general equations for mutual resistance and reactance between two antennas of unequal height have been derived by Cox.[15]

Operating Resistance of Radiators in a Directional-Antenna System The *operating resistance* of a radiator in a directional array is the sum of the radiation resistance R_r, an assumed loss resistance R_a (see Sec. 2-3), and the *coupled resistance* R_c.

$$R_0 = R_r + R_a + R_c \qquad (2\text{-}29)$$

The coupled resistance of element 1 is

$$R_{c1} = M_{12}Z_{12}\cos(\mu_{12} + \sigma_{12}) + M_{13}Z_{13}\cos(\mu_{13}$$
$$+ \sigma_{13}) + \cdots M_{1n}Z_{1n}\cos(\mu_{1n} + \sigma_{1n}) \qquad (2\text{-}30)$$

The coupled reactance is

$$X_{c1} = M_{12}Z_{12} \sin(\mu_{12} + \sigma_{12}) + M_{13}Z_{13} \sin(\mu_{13}$$

$$+ \sigma_{13}) + \cdots M_{1n}Z_{1n} \sin(\mu_{1n} + \sigma_{1n}) \qquad \textbf{(2-31)}$$

where M_{1n} = current ratio between tower n and tower 1
  Z_{1n} = mutual impedance between tower n and tower 1
  μ_{1n} = phase angle of current between tower n and tower 1
  σ_{1n} = phase angle of Z_{1n}

Note that M_{12} is the *current* ratio of tower 2 referred to tower 1. The loop-current ratio and the *field* ratio are related by the ratio of the form factors. For loop-current reference,

$$\frac{M_{12}}{F_{12}} = (1 - \cos G_1)/(1 - \cos G_2) \qquad \textbf{(2-32)}$$

The base resistance of a radiator in a directional array will usually be different from the operating resistance computed from Eq. (2-29). The determination of base operating resistance is discussed below.

Computation of Total Resistance: Formulas   The total resistance R_{t1} is

$$R_{t1} = R_{01} + M_{12}^2 R_{02} + M_{13}^2 R_{03} + \cdots M_{1n}^2 R_{0n} \qquad \textbf{(2-33)}$$

Determination of Reference Field by Mechanical Integration (Hemispherical Integration)[*]   The value of E_1 may be established by integration of the power flow from the antenna system over the hemisphere. The resulting hemispherical root-mean-square (rms) value is then compared with an isotropic reference, in which the power flow is uniform in all directions over the hemisphere:

$$E_1 = 245\sqrt{P} \qquad \text{mV/m at 1 km} \qquad \textbf{(2-34)}$$

where P = power, kW. The hemispherical rms value of relative field strength e_h is determined by integration over the hemisphere. The integration may be approximated by the trapezoid method:[†]

$$e_h = \sqrt{\frac{\pi\Delta}{180} \frac{[e_{\text{rms}}(\theta = 0)]^2}{2} + \sum_{n=1}^{N} \{[e_{\text{rms}}(\theta)]^2 \cos\theta\}} \qquad \textbf{(2-35)}$$

where Δ = interval in degrees of vertical angle θ for equally spaced calculations
  θ = vertical angle (0 = horizontal)
  $N = 90/\Delta - 1$

$$e_{\text{rms}}(\theta) = \sqrt{\sum_{i=1}^{k} \sum_{j=1}^{k} F_i f_i(\theta) F_j f_j(\theta) \cos \mu_{ij} J_0(S_{ij} \cos\theta)} \qquad \textbf{(2-36)}$$

---

[*]In the discussion which follows, lowercase e is used to denote relative fields. Actual electric field strengths are indicated by capital E.

[†]Other numerical approximations, such as Simpson's or Gauss's, may provide some improvement in accuracy.

where k = number of elements
F = field ratio
$f(\theta)$ = vertical-radiation characteristic
μ_{ij} = phase-angle difference between the ith and jth elements
S_{ij} = spacing between ith and jth elements, rad
$J_0(s)$ = zeroth-order Bessel function

The no-loss value of E_1 is then calculated from

$$E_{01} = \frac{E_i\sqrt{P}}{e_h} \quad \text{mV/m at 1 km} \tag{2-37}$$

Power Losses   If the individual currents are known (i.e., the operating resistances are known), the total power lost in an array by dissipation in the antenna system (ground system, element-coupling system, etc.) can be calculated from

$$P_\ell = \frac{\displaystyle\sum_{i=1}^{k} R_{ai} I_{oi}^2}{1000} \quad \text{kW} \tag{2-38}$$

where k = number of elements
R_{ai} = assumed loss resistance of ith element
I_{oi} = loop current of ith element (base current if ith-element height is less than $\lambda/4$)

The reference field E_1 is directly proportional to the current in the reference tower.

If, on the other hand, the individual operating resistances are not known, as would be the case when e_h is determined by hemispherical integration, the losses may be calculated and the reference field E_1 with loss determined by reference to the root-sum-square (rss) currents of the elements in the array. The rss currents are related to the rss fields by the form factor K [see Eq. (2-1)] of the individual elements. If all the elements in the array are of the same height, the two are directly proportional. To determine the losses for this condition, first determine the horizontal-plane no-loss rss of the fields from the elements of the array:

$$E_{\text{rss}} \text{ (no loss)} = \frac{245\sqrt{P}}{e_h} \sqrt{\sum_{i=1}^{k} F_i^2} \tag{2-39}$$

Next, determine the no-loss *current* rss of the array by using the following:

$$I_{\text{rss}} \text{ (no loss)} = \frac{4.08\sqrt{P}}{e_h} \sqrt{\sum_{i=1}^{k} \left(\frac{F_i}{K_i}\right)^2} \tag{2-40}$$

where K_i = form factor of element i.

For towers of different heights, the form factor K_i must be applied for each height of tower in determining I_{rss}. The reference field E_l with loss may be determined from

$$E_l \text{ (loss)} = E_l \text{ (no loss)} \left(\frac{P}{P + I_{\text{rss}}^2 \text{ (no loss)} \dfrac{R_a}{1000}}\right) \tag{2-41}$$

Base Operating Impedance of Radiators

General The base operating resistance of a tower in a directional-antenna array is usually different from the computed operating resistance. The base operating resistance and reactance may be estimated as follows:

Estimate the base resistance (R_{base}) and reactance (X_{base}) of the tower. Then assume the base mutual impedance to be $Z_{m(base)} = Z_m(R_{base}/R_r)$,[16] and substitute these values in Eqs. (2-30) and (2-31).

Parasitic (Zero-Resistance) Elements A radiating element operating at zero resistance and not supplied with power by connections from the distribution circuits is said to be parasitic. This condition obtains when the negative coupled resistance is equal to the sum of the radiation resistance and loss resistance. A parasitic element properly tuned will operate in phase-and-field-ratio relationships approximating those computed. In critical multielement arrays, independent control of phase and amplitude is required and parasitic radiators should be avoided. However, they may be employed in antennas designed primarily for power gain.

Negative-Resistance Elements A negative-resistance element receives more power by coupling to the other elements than is required to obtain the desired field from the element. The excess power is sometimes dissipated in a resistor but is usually returned to the positive-resistance elements through the power-distributing circuits.

Effective Field

Definition The rms value of the radiation pattern in the horizontal plane ($\theta = 0$, in unattenuated field at 1 km) is referred to as the *effective field*. The effective field of a directional antenna is modified from that for a single radiator by directivity at vertical angles and higher ground and circuit losses.

Calculation of Directional-Antenna Effective Field The rms field of a directional-antenna pattern at any vertical angle may be calculated from

$$E_{eff} = E_1 e_{rms}(\theta) \qquad \text{mV/m at 1 km} \qquad \textbf{(2-42)}$$

where E_1 is the reference field and $e_{rms}(\theta)$ is as defined in Eq. (2-36).

FCC Standard Radiation In the United States the FCC *Rules and Regulations* require that the radiation from a medium-wave directional-antenna system be depicted by a standard pattern. The standard radiation pattern is an envelope around the theoretical radiation pattern. It is intended to provide a tolerance within which the actual operating pattern can be maintained. The standard radiation values are calculated from

$$E_{std}(\theta) = 1.05 \sqrt{[E_{th}(\theta)]^2 + [Qg(\theta)]^2} \qquad \text{mV/m at 1 km} \qquad \textbf{(2-43)}$$

where $E_{th}(\theta)$ = theoretical radiation [Eqs. (2-17) through (2-26)]
$\quad Q$ = (10.0) or ($10\sqrt{P}$) or ($0.025 E_{rss}$), whichever is largest
$\quad P$ = power, kW
$\quad f_s(\theta)$ = vertical-radiation characteristic of shortest element in array

$g(\theta) = f_s(\theta)$ if shortest element is shorter than $\lambda/2$

$g(\theta) = \dfrac{\sqrt{[f_s(\theta)]^2 + 0.0625}}{1.030776}$ if shortest element is taller than $\lambda/2$

$E_{\text{rss}} = E_1 \sqrt{\displaystyle\sum_{i=1}^{n} F_i^2}$, horizontal plane pattern root sum square

   n = total number of elements
   F_i = field ratio of ith element
   E_l = reference field

FCC rules include a provision to modify or augment the standard pattern to accommodate actual operating patterns when radiation is in excess of the standard radiation pattern. Radiation is augmented over a specified azimuthal span and is calculated from

$$E_{\text{aug}}(\theta) = \sqrt{[E_{\text{std}}(\theta)]^2 + A\left[g(\theta) \cos\left(\frac{180 D_A}{S}\right)\right]^2} \qquad \textbf{(2-44)}$$

where $E_{\text{std}}(\theta)$ = standard radiation [See Eq. (2-43).]
   A = $\{[E_{\text{aug}}(\theta)]^2 - [E_{\text{std}}(\theta)]^2\}$ at central azimuth of augmentation
   S = azimuthal span for augmentation. The span is centered on central azimuth of augmentation.
   D_A = absolute difference between azimuth of calculation and central azimuth of augmentation. D_A cannot exceed $S/2$ for augmentation within a particular span.

$g(\theta)$ is as defined above.

   If there are overlapping spans of augmentation, the augmentations are applied in ascending order of central azimuth beginning with true north (0° true). In this case, there is in essence an augmentation of an augmentation. If the central azimuth of an earlier augmentation overlaps the central azimuth of a later augmentation, the value of A is adjusted in the latter case to provide for the specified resulting radiation at the later central azimuth.

Choice of Orientation and Spacing in Directional-Antenna Design

The required placement of towers in a directional array is determined by the general shape of pattern desired. In-line arrays produce patterns having line symmetry; other configurations may or may not exhibit symmetry, depending upon the operating parameters. Closely spaced towers have mutual impedances which are relatively large compared with the self-resistances. This may result in low operating resistances and high circulating currents, which make for instability and high losses and should be avoided. In general, these effects may occur when spacings less than about one-quarter wavelength are introduced.

Choice of Tower Height

The choice of tower height is governed by the effective field required, the need for adequately high base resistances, the desired vertical patterns, aeronautical restrictions, and economic limitations. These requirements are usually met by heights on the order of one-quarter wavelength. Shorter towers are sometimes employed on the lower frequencies for practical reasons, as long as the required FCC minimum effective field is obtained and adequately high base resistances are provided. Taller towers produce higher effective field strengths and may reduce radiation at high vertical angles.

Effect of Finite Radiator Cross Section on Directional Antennas

The modified current distribution due to the finite cross section of practical tower radiators results in a complex vertical-radiation characteristic different from that computed by using the simplified assumptions. When towers of unequal height are employed, the difference in amplitude (and sometimes in phase) of vertical-radiation characteristics may be taken into account. The radiation from an array employing towers of unequal height may be computed by using Eqs. (2-17) and (2-22). The difference in phase angle requires the substitution for α_n with the angle α'_n, which includes an additional term δ_n:

$$\alpha'_n = s_n \cos \phi_n \cos \theta + A_n + \delta_n \qquad \textbf{(2-45)}$$

where δ_n = phase-angle difference between $f_1(\theta)$ and $f_n(\theta)$ at angle θ.

2-5 CIRCUITS FOR SUPPLYING POWER TO DIRECTIONAL AND NONDIRECTIONAL ANTENNAS

General

Radio-frequency power must be supplied to the individual radiators of the directional-antenna system in the proper proportions and phase-angle relationships to produce the desired radiation patterns. Means for controlling the current ratios and phase angles are required to permit adjustment and maintenance of the patterns. The circuits must provide a load into which the transmitter will operate properly.

The required functions are shown in block form in Fig. 2-15. The antenna-tuning units transform the operating base impedance of the radiators to the characteristic impedance of the transmission lines and provide a portion of the required phase shift. Additional phase shift is introduced by the transmission lines. The phase-control networks contain variable components for phase control. The power-dividing network supplies variable voltages to each line for power control. The networks should be

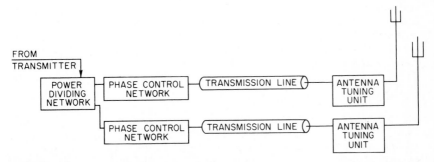

FIG. 2-15 Basic functions of array feed system.

designed for minimum power loss and for a broadband-frequency response. (See Reference 17 for more detailed information.)

Transmission-Line Requirements

Transmission lines may be either of the concentric or the open-wire, unbalanced type. Concentric lines have a lower characteristic impedance, usually requiring less transformation between tower and line. Their complete shielding eliminates any radiation from the line.

2-6   ADJUSTING DIRECTIONAL-ANTENNA ARRAYS

General Requirements

A directional antenna must be adjusted to produce a radiation pattern substantially in accordance with its design. In the United States, the construction permit issued by the FCC specifies maximum permissible values of radiation in pertinent directions.

Sampling System

An indication of the field-ratio and phase-angle relationships among the radiators is needed to adjust a directional-antenna system. Depending on the tower heights, this is provided by permanently installed *sampling transformers* in the antenna-tuning unit or *sampling loops* mounted on the towers. These devices are connected by *sampling lines* to an antenna monitor.

　　The individual sampling lines may be cut to the same length, making the phase delay on all lines equal. Excess line should be stored so as to be exposed to the same weather conditions as the longer portions of the other lines. Section 73.68 of the FCC *Rules and Regulations* specifies the requirements for an approved sampling system for medium-wave directional-antenna systems in the United States.

Initial Adjustment

The initial adjustment of a directional-antenna system is generally accomplished by setting the reactances of the antenna-tuning and power-dividing components to their computed values and supplying power to the antenna system. Usually, the phase and field-ratio indications will be different from those desired. The adjustable tuning components are then varied until the phase and field-ratio indications correspond closely to the computed values.

　　Preliminary field-intensity measurements are then made to determine the approximate shape of the pattern. These data are analyzed to determine if changes are required to bring the operating pattern shape into agreement with the computed pattern shape.

Proof of Performance

A *proof of performance* is required in the United States for all directional-antenna systems before regular operation is authorized. Part 73 of the FCC *Rules and Regu-*

lations describes the procedures and required measurements for an acceptable proof of performance.

After the pattern has been properly adjusted and the adjustment confirmed by the proof of performance, *monitoring points* are established in directions specified by the construction permit. The field intensities for directional operation are measured at each monitor point at regular intervals to provide an indication of pattern performance and stability.

2-7 MISCELLANEOUS PROBLEMS

Guy-Wire Insulation

Guy wires supporting tower radiators must be insulated from the tower and from ground and must be broken up into sections sufficiently short so that the induced currents do not distort the radiation pattern. Strain insulators are installed at the guy anchor, the point of attachment to the tower, and at intervals along the guy wire. The maximum length of any individual guy-wire section should not exceed $\lambda/8$ to $\lambda/10$. Occasionally, portions of the topmost guy wires are not insulated from the tower in order to provide top loading to the element.

Circuits across Base Insulators

It is usually necessary to cross the base insulator of a tower antenna with alternating-current power circuits. Power for aeronautical-obstruction lighting of the tower or other purposes may be supplied by means of chokes or transformers. Lighting chokes are wound of ordinary insulated copper wire on a suitable form, with a sufficient number of turns to provide a reactance at the operating frequency, which is high compared with the tower-base impedance. Radio-frequency bypass condensers are installed between individual windings.

Alternating current can be supplied by an Austin transformer, consisting of linked toroidal cores mounting the primary and secondary windings. There is an air gap of several inches between the two cores, and the only effect at the medium frequency is that due to the shunt capacity.

FM and television transmission lines (and other metallic circuits) may be isolated from the tower at the medium frequency by a quarter-wave isolation section of line. The outer conductor of the FM-TV coaxial cable is grounded immediately before the rise up the tower and is supported on insulated hangers to a point approximately one-quarter wavelength from ground, where it is connected to the tower. The outer conductor of the FM-TV line and the tower constitute a shorted quarter-wavelength transmission line at the medium frequency, resulting in a high impedance at the tower base. A method for towers shorter than $\lambda/4$ is to install the quarter-wavelength section along the ground rather than up the tower.

Simultaneous Use of Single Tower at Two or More Frequencies

It is occasionally desired to use a single tower radiator simultaneously at two or more different frequencies. This may be done by employing suitable filters to isolate the transmitters from each other.

Selection of Transmitter Sites

Transmitter sites should be selected in an area providing sufficient ground which is reasonably flat and level, of high local conductivity, free of obstructions which might interfere with the proper functioning of the radiating system, and so located as to provide maximum signal to the principal city and the service area. This last requirement, applied to operations with a directional-antenna system, usually dictates a choice of site which will place the main radiation lobe in the direction of the city.

Effect of Signal Scattering and Reradiation by Nearby Objects

General   Structures and terrain features near the transmitter site may reflect the signal from the antenna or may reradiate sufficient signal to affect the performance of the antenna. Large buildings near the transmitter site, mountains, or rugged terrain may distort the radiation pattern of directional and nondirectional antennas. The effects may be serious in the case of a directional-antenna system requiring a high degree of signal suppression, particularly if the buildings or hills are in the main radiation lobe of the antenna. Such objects are usually too irregular to permit application of analytic methods to a determination of reradiation. Their effect may often be estimated on the basis of experience with similar objects.

Tall Towers: General   It is frequently desired to erect tall towers to support FM and television transmitting antennas or for other purposes in the immediate vicinity of a medium-frequency antenna. These structures are usually of sufficient electrical height to be capable of substantial reradiation in high incident fields.

Tall Towers: Control of Reradiation   The tower location should be chosen to have minimum effect on the medium-frequency antenna. If the tower is to be installed in the immediate vicinity of a medium-frequency directional antenna, the field intensities should be computed at a number of locations, and a position chosen for the tower where the incident field is a minimum.

The reradiation from a tower of this height may be controlled by insulating the tower from ground and installing sectionalizing insulators at one or more levels. Guy wires must be insulated at suitable intervals, and transmission and alternating-current lines must be isolated.

Reradiation from tall towers may be controlled to some extent without sectionalizing insulators by insulating the tower from ground and installing a suitable reactor between the tower and ground.

Protection against Static Discharges and Lightning

In the absence of suitable precautions, static charges accumulate on towers and guy wires and may discharge to ground. This discharge ionizes the path, and a sustained radio-frequency arc may follow. Protection can be provided to minimize the accumulation or quench the arc, but it is difficult to provide protection against direct lightning hits. A lightning rod or rods extending above the beacon on the tower top will provide some protection to the beacon, and horn or ball gaps at the tower base will provide protection to the base insulator. However, depending on the magnitude of lightning current, some damage may result to the meters and tuning components.

A direct-current path from the tower to ground will minimize static accumulation. A separate radio-frequency choke, the tuning inductor, or the sampling-line inductor may be connected to maintain the tower at direct-current ground potential. Difficulty may occasionally be experienced with charges accumulating on the individual guy wires and arcing across the guy insulators. This may be eliminated by installing static drain resistors across each guy insulator. These resistors may have a value of 50,000 to 100,000 Ω and should have an insulation path somewhat longer than provided by the guy insulator.

REFERENCES

1 C. L. Jeffers, "An Antenna for Controlling the Nonfading Range of Broadcasting Stations," *IRE Proc.,* vol. 36, November 1948, pp. 1426–1431.
2 S. A. Schelkunoff, "Theory of Antennas of Arbitrary Size and Shape," *IRE Proc.,* vol. 29, September 1941, pp. 493–521.
3 G. H. Brown and O. M. Woodward, Jr., "Experimentally Determined Impedance Characteristics of Cylindrical Antennas," *IRE Proc.,* vol. 33, April 1945, pp. 257–262.
4 P. S. Carter, "Circuit Relations in Radiating Systems and Applications to Antenna Problems," *IRE Proc.,* vol. 20, June 1932, pp. 1004–1041.
5 J. F. Morrison and P. H. Smith, "The Shunt-Excited Antenna," *IRE Proc.,* vol. 25, June 1937, pp. 673–696.
6 P. Baudoux, "Current Distribution and Radiation Properties of a Shunt-Excited Antenna," *IRE Proc.,* vol. 28, June 1940, pp. 271–275.
7 G. H. Brown, "A Critical Study of the Characteristics of Broadcast Antennas as Affected by Antenna Current Distribution," *IRE Proc.,* vol. 24, January 1936, pp. 48–81.
8 *Rules and Regulations of the Radio Broadcast Series,* part 3, Federal Communications Commission, 1959.
9 *Final Acts of the Regional Administrative MF Broadcasting Conference (Region 2),* Rio de Janeiro, 1981.
10 C. E. Smith and E. M. Johnson, "Performance of Short Antennas," *IRE Proc.,* vol. 35, October 1947, pp. 1026–1038.
11 J. R. Wait and W. A. Pope, "The Characteristics of a Vertical Antenna with a Radial Conductor Ground System," *IRE Conv. Rec.,* part 1, *Antennas and Propagation,* 1954, p. 79.
12 F. R. Abbott, "Design of Optimum Buried-Conductor RF Ground System," *IRE Proc.,* vol. 40, July 1952, pp. 846–852.
13 O. Prestholdt, "The Design of Non-Radial Ground Systems for Medium-Wave Directional Antennas," NAB Eng. Conf., 1978, unpublished.
14 C. E. Smith, *Theory and Design of Directional Antennas,* Cleveland Institute of Radio Electronics, Cleveland, 1949.
15 C. R. Cox, "Mutual Impedance between Vertical Antennas of Unequal Heights," *IRE Proc.,* vol. 35, November 1947, pp. 1367–1370.
16 G. H. Brown: "Directional Antennas," *IRE Proc.,* vol. 25, January 1937, pp. 78–145.
17 D. F. Bowman, "Impedance Matching and Broadbanding," in R. C. Johnson and H. Jasik (eds.), *Antenna Engineering Handbook,* McGraw-Hill Book Company, New York, 1984, Chapter 43.

BIBLIOGRAPHY

Brown, G. H.: "The Phase and Magnitude of Earth Currents near Radio Transmitting Antennas," *IRE Proc.*, vol. 23, February 1935, pp. 168–182.

————: "A Consideration of the Radio-Frequency Voltages Encountered by the Insulating Material of Broadcast Tower Antennas," *IRE Proc.*, vol. 27, September 1939, pp. 566–578.

————, R. F. Lewis, and J. Epstein: "Ground Systems as a Factor in Antenna Efficiency," *IRE Proc.*, vol. 25, June 1937, pp. 753–787.

Chamberlain, A. B., and W. B. Lodge: "The Broadcast Antenna," *IRE Proc.*, vol. 24, January 1936, pp. 11–35.

Effective Radio Ground-Conductivity Measurements in the United States, Nat. Bur. Stds. Circ. 546, February 1954.

Terman, F. E.: *Radio Engineers' Handbook,* McGraw-Hill Book Company, New York, 1943.

Fine, H.: "An Effective Ground Conductivity Map for Continental United States," *IRE Proc.*, vol. 42, September 1954, pp. 1405–1408.

Hansen, W. W., and J. G. Beckerley: "Concerning New Methods of Calculating Radiation Resistance, Either with or without Ground," *IRE Proc.*, vol. 24, December 1936, pp. 1594–1621.

King, R.: "Self- and Mutual Impedances of Parallel Identical Antennas," *IRE Proc.*, vol. 40, August 1952, pp. 981–988.

Laport, E. A.: *Radio Antenna Engineering,* McGraw-Hill Book Company, New York, 1952.

Moley, S. W., and R. F. King: "Impedance of a Monopole Antenna with a Radial-Wire Ground System," *Radio Sci. (MBS),* vol. 68D, no. 2, February 1964.

Moulton, C. H.: "Signal Distortion by Directional Broadcast Antennas," *IRE Proc.*, vol. 40, May 1952, pp. 595–600.

Nickle, C. A., R. B. Dome, and W. W. Brown: "Control of Radiating Properties of Antennas," *IRE Proc.*, vol. 22, December 1934, pp. 1362–1373.

Norton, K. A.: "The Calculation of Ground-Wave Field Intensity over a Finitely Conducting Spherical Earth," *IRE Proc.*, vol. 29, December 1941, pp. 623–639.

Schelkunoff, S. A., and H. T. Friis: *Antennas: Theory and Practice,* John Wiley & Sons, Inc., New York, 1952.

Smeby, L. C.: "Short Antenna Characteristics: Theoretical," *IRE Proc.*, vol. 37, October 1949, pp. 1185–1194.

Spangenberg, K.: "Charts for the Determination of the Root-Mean-Square Value of the Horizontal Radiation Pattern of Two-Element Broadcast Antenna Arrays," *IRE Proc.*, vol. 30, May 1942, pp. 237–240.

Williams, H. P.: *Antenna Theory and Design,* vol. 2, Sir Isaac Pitman & Sons, Ltd., London, 1950.

Chapter 3

High-Frequency Antennas

Ronald Wilensky
William Wharton

Technology for Communications International

3-1   GENERAL DESIGN REQUIREMENTS

High-frequency (HF) antennas are used in the range of frequencies from 2 to 30 MHz for communications and broadcasting by means of ionospheric propagation. Ionospheric transmission over large distances usually involves high overall transmission losses, especially under unfavorable ionospheric propagation conditions. High initial and operating costs of transmitting equipment therefore make high-gain transmitting and receiving antennas desirable to provide reliable communications. High-gain antennas are also in common use at international broadcasting stations, where transmissions are beamed to specific geographical areas. In additition to providing an increased signal in the target area, the use of high-gain directional antennas decreases radiation in undesired directions, thereby reducing potential interference to other cochannel services.

Important Design Parameters

The important design parameters of an HF transmitting or receiving antenna include its frequency range, vertical and horizontal angles at which maximum radiation is desired, gain or directivity, voltage standing-wave ratio (VSWR), input power, and environmental and mechanical requirements.

The frequency range should be established by a propagation study to determine the optimum working frequency (OWF) for the path involved, which varies according to distance and location, time of day, season of the year, and sunspot activity. Details of such a study are beyond the scope of this book but may be found in the literature.[1,2] There are available several computer programs which determine path losses and optimum frequencies on the basis of stored ionospheric data.[3,4] Actual ionospheric characteristics may also be measured in real time by using commercially available ionospheric sounders and frequency-management systems.

The vertical angle of maximum radiation of the antenna is known as the takeoff angle (TOA). The range of vertical angles pertinent to ionospheric propagation depends on distance, effective layer height, and mode of propagation (i.e., one-hop, two-hop, etc.). Figure 3-1 is a graph showing estimated TOA for one-hop transmission as a function of distance and virtual layer heights. The height of the E layer may be considered constant at 100 km, but the height of the $F2$ layer varies widely according to time of day, season of the year, and location. A detailed propagation study is required to determine the required range of TOA. As a rough estimate, however, a range of $F1$ and $F2$ virtual layer height of approximately 250 to 400 km may be used in conjunction with Fig. 3-1 to determine approximate TOA range. Multiple-hop propagation can also be assessed approximately by using Fig. 3-1 for each hop. For transmission or reception over distances of more than 4000 km, maximum signal results from low-angle transmission in the range from 2 to 15°, with the lower angles generally providing better results. Propagation over very long paths (greater than 10,000 km) is more complicated than the simple multiple-hop theory, and sophisticated methods of propagation analysis must be used.[4]

The horizontal range of angles or beamwidth (defined as the azimuth angle between the points at which radiation is 3 dB below beam maximum) is determined by the geographical area to be covered. Wide areas require large beamwidths, and narrow areas or point-to-point circuits require small beamwidths. The minimum beamwidth for point-to-point coverage depends on irregularities in the ionosphere and effects of magnetic storms, which cause deviations from great-circle paths. Directions of arrival of HF signals may vary by as much as $\pm 5°$ in the horizontal plane because of these effects.[6] This makes the use of antennas with extremely narrow horizontal beamwidths undesirable, especially on circuits which skirt or traverse the auroral zone. International broadcasting requires that the beamwidth subtend the target area with due allowance for path-deviation effects.

Antenna directivity (the gain, neglecting losses) depends primarily on the vertical and azimuthal beamwidths of the radiation pattern. Antenna gain is the directivity multiplied by the antenna radiation efficiency. High gain is necessary when large effective radiated power is needed to overcome large ionospheric transmission losses.

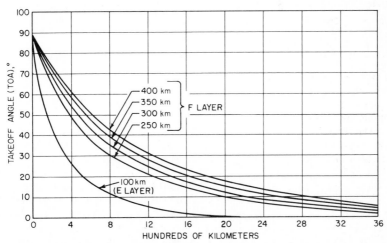

FIG. 3-1 Takeoff angles for one-hop ionospheric transmission as a function of virtual layer height. (*After Ref. 5.*)

High directivity is desirable for reception in the presence of interference. Low radiation efficiencies can generally be accepted for reception when there is a high level of ambient noise and the high antenna noise figure does not degrade performance. At very quiet receiving sites, high radiation efficiency (i.e., low noise figure) is necessary to prevent the antenna from degrading circuit performance.

For transmitting antennas, the acceptable level of reflected power from the antenna is usually determined by the characteristics of the transmitter. Reflected power is usually specified in terms of the VSWR. Modern solid-state transmitters operating in the 0.1- to 10-kW range tolerate maximum VSWR of 2.5:1. Many such transmitters will tolerate a higher VSWR of 3:1 or 4:1 by automatically reducing their output power. High-power broadcasting transmitters in the 100- to 500-kW range of carrier-power levels tolerate maximum VSWRs of 2:1. Older transmitters have much lower maximum VSWR limits of 1.5:1 or 1.4:1.

The power-handling capacity of a transmitting antenna is specified by both an average power and a peak envelope power (PEP). The average power determines the required current-carrying capacity of the antenna conductors. The peak power determines the voltages and electric fields on the antenna, which set requirements on insulator and conductor size to avoid arcing or corona discharge.

Mechanical requirements depend on environmental effects (wind, ice loading, temperature variation, corrosion-causing conditions), limitations on antenna size or tower height, special restrictions imposed by the site, or other considerations such as transportability.

Types of High-Frequency Antennas

Most HF antennas are very broadband and usually require little or no tuning. Several techniques are used to make HF antennas broadband: (1) lowering antenna Q, (2) using log-periodic arrays of monopole or dipole elements, and (3) automatic tuning. Antenna Q may be lowered by adding loss resistance or using "fat" radiators to decrease reactance. Examples of antennas with resistors are rhombics, terminated V's, and resistively loaded dipoles. Typical fattened antennas are fan dipoles, conical monopoles, and wide-bandwidth dipole arrays used in HF broadcasting. The bandwidth of a log-periodic antenna (LPA) is limited only by the number of radiators used, and it is easy to design such an antenna to operate over a 2- to 30-MHz range. Automatic tuning units are used with whips or loop antennas, which by themselves have unacceptably high VSWR over wide frequency ranges.

For special applications, narrow-bandwidth antennas, such as simple resonant dipoles or monopoles or Yagi-Uda arrays, are used because they are often smaller and less expensive than broadband antennas.

Analysis of Antennas

Analysis of most practical HF antennas is not possible by using simple mathematical techniques. Antennas are designed by using sophisticated computer programs which solve Maxwell's equations numerically for arbitrary antenna geometries.[7–10] Such programs calculate the current distribution on the antenna, the most important and hardest-to-calculate property of an antenna. The radiation patterns calculated by the most comprehensive computer programs are very accurate, and the results are acceptable substitutes for measured patterns. Driving-point impedances are also calculated well, but often to somewhat less accuracy than radiation patterns because impedance depends critically on the mathematical representation of the fields in the region of the driving point.

3-2   SIMPLE ANTENNAS MOUNTED ABOVE GROUND

Effect of the Ground Plane

HF antennas are usually operated within one or two wavelengths of the ground or, in the case of base-fed monopoles, directly on the ground. The proximity of the ground modifies the input impedance and the radiation pattern of an antenna. In most practical cases, an accurate evaluation of the effect of the ground is difficult. However, many fundamental properties of HF antennas can be understood in terms of the simple theory of idealized monopoles or dipoles situated over an infinite, flat ground plane. When the ground plane is perfectly conducting, it can be represented exactly by a single image (with the ground plane removed). For ground of finite conductivity, simple image theory does not give an exact solution but will often provide useful approximate results.

Antenna Impedance

The currents in the conductors of an antenna induce currents in the ground plane, and these (or, alternatively, the currents in the image antenna) modify the antenna currents from the values that would occur in free space. The interaction between the antenna and the ground is given by the mutual impedance, which varies with the height of the antenna above ground. The input impedance of the antenna is given by

$$Z_{ant} = Z_{fs} + Z_m \qquad \textbf{(3-1)}$$

where Z_{fs} = self-impedance of antenna (impedance when isolated in free space)
 Z_m = mutual impedance between antenna and its image
The mutual impedance can be either positive or negative. Reference 28 contains curves of mutual impedance for several simple but useful geometries. More complex arrangements are best analyzed by using antenna-analysis computer programs.

The mutual impedance of a thin center-fed half-wave vertical dipole whose lower point is at height $S/2$ above ground is given in Fig. 4-29* of Ref. 28. In this case, mutual impedance adds to self-impedance so that total antenna impedance increases as the height of the dipole decreases. The mutual impedance of a thin center-fed horizontal dipole at height $d/2$ above ground is given in Fig. 4-28* of Ref. 28. In this case, the currents in the image and the antenna are 180° out of phase so that the mutual impedance must be deducted from the self-impedance. For vanishingly small heights self-impedance and mutual impedance cancel, and radiation resistance approaches zero. This means that a horizontal dipole placed close to the ground has a low radiation resistance. Steps must be taken to modify its impedance so that the transmitter is presented with an acceptably low VSWR. The input resistance can be increased by adding series resistance, which improves the VSWR but reduces radiation efficiency (and hence gain). The radiation efficiency with added resistance is given by $R_{rad}/(R_{rad} + R_{loss})$, where R_{rad} is the radiation resistance and R_{loss} is the series loss

---

*This figure is reproduced at the end of this chapter, p. 3-42.

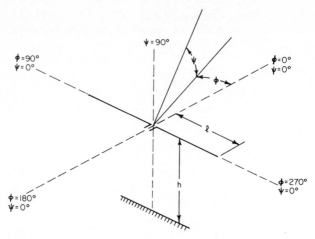

FIG. 3-2 Geometry of horizontal dipole above ground.

resistance. Over ground of finite conductivity, an extra resistance must be added in the denominator to account for losses in the ground.

Radiation Patterns of Horizontally Polarized Antennas

In terms of the geometry of Fig. 3-2, the free-space radiation pattern of a horizontal dipole of length 2ℓ is given by

$$d(\varphi') = \frac{\cos(2\pi\ell \sin \varphi'/\lambda) - \cos(2\pi\ell/\lambda)}{(1 - \cos 2\pi\ell/\lambda) \cos \varphi'} \qquad \textbf{(3-2)}$$

where φ' = angle between line of propagation and equatorial plane
$\sin \varphi' = \sin \varphi \cos \psi$
$\quad \varphi$ = azimuth angle
$\quad \psi$ = elevation angle
$\quad \lambda$ = wavelength

When the dipole is placed above ground as shown in Fig. 3-2, the radiation reflected from the ground adds to that from the dipole (i.e., the dipole and image fields add). For the practical situation $h/\lambda > 0.2$, the total radiation pattern,[11] with the small effect of mutual impedance ignored, is

$$D(\varphi,\psi) = d(\varphi')[1 + r_H^2 + 2r_H \cos(\rho_H + 4\pi h \sin \psi/\lambda)]^{1/2} \qquad \textbf{(3-3)}$$

where h is the height above ground and r_H and ρ_H are respectively the magnitude and the phase of the complex reflection coefficient[12]

$$r_H \,\underline{/\rho_H} = \frac{\sin \psi - (\epsilon' - \cos^2 \psi - j60\sigma\lambda)^{1/2}}{\sin \psi + (\epsilon' + \cos^2 \psi - j60\sigma\lambda)^{1/2}} \qquad \textbf{(3-4)}$$

where $\epsilon' = \epsilon/\epsilon_0$, the relative dielectric constant
$\quad \sigma$ = ground conductivity, S/m

At grazing incidence ($\psi = 0°$), $r_H = 1$ and $\rho_H = 180°$ for all ground constants, so that $D(\varphi,0) = 0$. In the HF band, for most practical values of ϵ' and σ, ρ_H is nearly equal to $180°$ for all elevation angles and r_H decreases from 1 at $\psi = 0°$ to about 0.5 to 0.6 directly overhead at $\psi = 90°$. In most HF applications the TOA is in the range $0 < \psi \le 30°$, and the radiation pattern is given accurately by setting $r_H = 1$ and $\rho_H = 180°$ in Eq. (3-4), which is equivalent to assuming that the ground is perfectly conducting. Equation (3-3) therefore becomes

$$D(\varphi,\psi) = 2d(\varphi') \sin (2\pi h \sin \psi/\lambda) \qquad \textbf{(3-5)}$$

The maxima (TOA) and minima (nulls) of this expression occur in the equatorial plane ($\phi = 0$ and $180°$) at elevation angles given by

$$\psi_{max} = \sin^{-1} (n\lambda/4h) \qquad n = 1, 3, 5, \ldots \qquad \textbf{(3-6)}$$

$$\psi_{min} = \sin^{-1} (n\lambda/2h) \qquad n = 1, 2, 3, \ldots \qquad \textbf{(3-7)}$$

At the maxima, $D^2 = 4d^2$, so that the gain is 6 dB above that obtained in free space. This value is exact for perfectly conducting ground and is nearly exact for low elevation angles above ground of finite conductivity. Thus, low-angle radiation from horizontal antennas suffers negligible loss owing to the finite conductivity of the ground, and a radiation-pattern ground screen in front of the antenna is not required. For high values of TOA, the ground losses are not always negligible. As an example, consider an antenna mounted 0.25λ above ground, for which from Eq. (3-6) $\psi_{max} = 90°$. For a typical set of ground constants, $r_H = 0.5$ and $\rho_H = 170°$, so Eq. (3-3) becomes $D^2 = 2.24d^2$, which represents a reduction of gain of 2.24/4 or 2.5 dB compared with that for a perfect ground. However, because a high TOA is used for transmission over short paths, it is not generally necessary to use a ground screen to minimize the loss. An exception is a high-TOA, high-power HF broadcasting antenna which must deliver the maximum signal in the target service area, and in this situation, a ground screen is occasionally used.

Radiation Patterns of Vertically Polarized Antennas

The elevation pattern of a vertical dipole whose center point is at height h above ground has the same form as Eq. (3-3). For the vertical dipole the angles φ and ψ are interchanged in Eq. (3-2), and r_H and ρ_H in Eq. (3-3) are replaced respectively by r_V and ρ_V, the reflection coefficients for vertical polarization:[12]

$$r_V \underline{/\rho_V} = \frac{(\epsilon' - j60\sigma\lambda) \sin \psi - (\epsilon' - \cos^2 \psi - j60\sigma\lambda)^{1/2}}{(\epsilon' - j60\sigma\lambda) \sin \psi + (\epsilon' - \cos^2 \psi - j60\sigma\lambda)^{1/2}} \qquad \textbf{(3-8)}$$

This reflection coefficient behaves very differently from that for horizontal polarization [Eq. (3-4)]. The phase ρ_V varies from $-180°$ at grazing incidence to a value approaching $0°$ at high elevation angles. The magnitude r_V is 1 at grazing incidence, drops to a value of a few tenths, and then increases to the value equal to r_H at $\psi = 90°$. The elevation angle at which r_V reaches a minimum occurs when $\rho_V = 90°$ and is known as the pseudo-Brewster angle, which at HF is typically in the range $5° \le \psi \le 15°$.

The most important consequence of this behavior is that radiation at low elevation angles is drastically reduced over imperfectly conducting ground. Finite ground conductivity can cause a gain reduction of as much as 8 to 10 dB at low elevation angles, as shown later in Fig. 3-18 and in Ref. 12. For receiving antennas, losses of this magnitude are usually tolerable, but for transmitting antennas they are not. For transmission, the loss must be reduced by having the antenna radiate over seawater (which has a high conductivity) or by installing in front of the antenna a large metallic ground screen with a mesh small in relation to the skin depth of the ground underneath the mesh.

3-3   HORIZONTALLY POLARIZED LOG-PERIODIC ANTENNAS

Single Log-Periodic Curtain with Half-Wave Radiators

The basic curtain, illustrated in Fig. 3-3, comprises a series of half-wave dipoles spaced along a transposed transmission line. Successive values of the dipole length and spacing have a constant ratio τ. In a practical LPA, τ has a value between 0.8 and

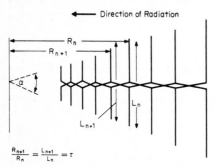

$$\frac{R_{n+1}}{R_n} = \frac{L_{n+1}}{L_n} = \tau$$

FIG.   3-3   Basic   half-wave   transposed dipole LPA.

0.95. The bandwidth of the LPA is limited only by the lengths of the longest and shortest dipoles. The number of radiators depends on the frequency range and the value of τ. The parameters τ and α, as discussed in more detail in Refs. 13, 14, and 29, determine the gain, impedance level, and maximum VSWR of the antenna. It is important to choose τ and α carefully because of the possibility of unwanted resonant effects. Sometimes, energy that normally radiates from the active region travels back to a second active region where the radiators are $3\lambda/2$ long. Excitation of this second active region causes undesirable impedance and pattern perturbations.

In a well-designed horizontally polarized LPA (HLPA), at any given frequency only dipoles whose lengths are approximately $\lambda/2$ long are excited, and these form a single active region. The active region moves along the curtain from the longest to the shortest dipole as the frequency is increased. Because of this movement, it is possible to mount an HLPA so that the electrical height of the active region is constant or variable in a controlled way, as shown in Fig. 3-4a and b. In other types of horizontal

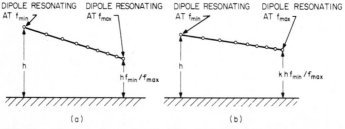

FIG. 3-4   Control of electrical height of LPA. (a) Constant electrical height. (b) Uniform variation of electrical height by a factor of K from f_{min} to f_{max}.

antennas, the physical height is constant, so the designer cannot control the variation of electrical height or TOA. A major advantage of the HLPA is, therefore, the ability to maximize the gain at the elevation angles required for reliable communications.

In the basic LPA shown in Fig. 3-3, the dipoles are conventional rods. It is also possible to form the dipoles into a sawtooth shape as shown in Fig. 3-5. The larger effective diameter of the sawtooth (compared with the rod) increases the bandwidth of each dipole so that a greater number are excited to form the active region at each frequency. This increases the gain and power-handling capacity and reduces the VSWR.

The basic HLPA using half-wave radiators has a gain of 10 to 12 dBi and an azimuthal beamwidth of 60 to 80°.

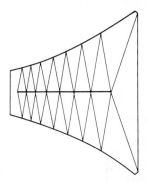

FIG. 3-5 LPA with sawtooth wire elements.

Clamped-Mode Log-Periodic Curtain with Full-Wave Radiators

The gain of a half-wave LPA can be increased by 2 to 3 dB by doubling its horizontal aperture to one wavelength, thereby decreasing its azimuthal beam-width to 30 to 40°.

The obvious way to increase the horizontal aperture is to use two half-wave curtains in broadside array as shown in Fig. 3-6a, but the resulting structure is relatively complex. The same horizontal aperture can be achieved with a much simpler structure by use of the *clamped-mode technique*[15] illustrated in Fig. 3-6b.

The clamped-mode arrangement is equivalent to taking one of the LPA curtains in Fig. 3-6a and pulling it apart sideways, leaving its radiators connected to the same half of the transmission line. When properly designed, a clamped-mode antenna has a gain and radiation pattern nearly identical to those of the broadside array, but it requires only half the number of radiator and transmission-line wires.

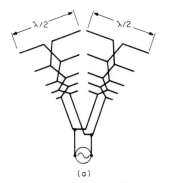 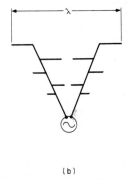

(a) (b)

FIG. 3-6 Clamped-mode technique for increasing horizontal aperture of LPA. (a) Two LPA curtains in broadside array. (b) Clamped-mode LPA curtain with equivalent performance.

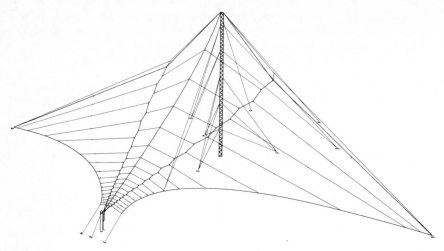

FIG. 3-7 Practical clamped-mode LPA using radiators made from single wires. (*Courtesy of TCI.*)

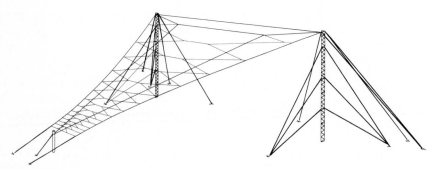

FIG. 3-8 Practical clamped-mode LPA using sawtooth radiators made from wires. (*Courtesy of TCI.*)

Two practical clamped-mode antennas are shown in Figs. 3-7 and 3-8. The antenna in Fig. 3-7 uses simple rod radiators made from single wires. It has only one tower in contrast to the two towers required for a conventional half-wave dipole HLPA, and consequently the radiators slope down toward the ground. This slope partially offsets the larger horizontal aperture and yields a gain of 12 dBi and an azimuthal beamwidth of 55 to 80°, the values for an equivalent conventional two-tower LPA. Figure 3-8 shows a clamped-mode antenna using sawtooth radiators and two towers. The horizontal curtain is flat, and the aperture is about 1.25λ, so the azimuthal beamwidth is 38° and the gain 14 to 15 dBi.

Multiple-Curtain Log-Periodic Arrays

LPAs can be stacked horizontally and/or vertically to decrease their beamwidths and increase their gain. Horizontal stacking can be done simply by using the clamped-mode technique described previously. Vertical stacking can be applied to simple LPAs with half-wave dipole elements, as shown in Fig. 3-9, or in conjunction with clamped-mode antennas, as in Fig. 3-10.

Vertical stacking raises the average radiating height of the antennas to a value given by the average electrical height of the curtains, thereby reducing the TOA and vertical beamwidth. Vertical stacking with spacings of about $\lambda/2$ also suppresses high-elevation secondary lobes, which are fully formed when the radiation height is large and can substantially reduce antenna gain. The number of elements in the vertical stack depends on the TOA and gain requirements and the maximum allowed tower height.

To achieve maximum gain in a vertically stacked LPA, it is necessary that the active regions of each curtain line up vertically at each frequency. This will not occur automatically because the lengths of the curtains are different, the upper curtains being longer than the lower curtains. This effect is most noticeable in antennas with more than two curtains in the stack. One technique for aligning the active regions is to control the velocity of propagation along the curtains, making the wave go more slowly on the shorter, lower curtains than on the longer, upper curtains. Wave velocity on a curtain can be controlled by adjusting the length, effective diameter, and spacing of the dipole elements.

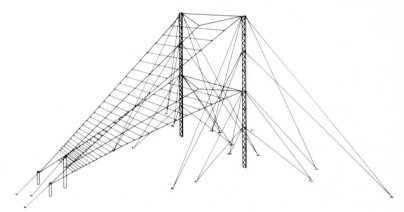

FIG. 3-9   Vertically stacked transposed dipole LPA curtains. *(Courtesy of TCI.)*

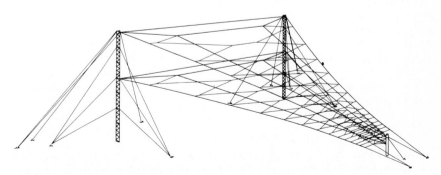

FIG. 3-10   Vertically stacked clamped-mode LPA curtains. *(Courtesy of TCI.)*

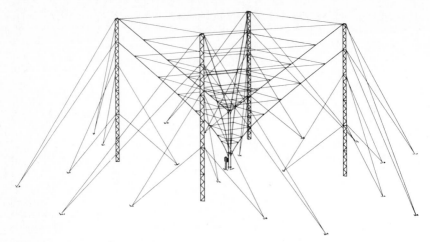

FIG. 3-11   Omnidirectional LPA with optimum TOA. *(Courtesy of TCI.)*

Omnidirectional Log-Periodic Array with Optimum Vertical Radiation Pattern

It has been pointed out that an HLPA can be arranged to give the optimum TOA at each frequency. A practical example of this technique is shown in Fig. 3-11, which illustrates an antenna designed for shore-to-ship and ground-to-air communication, for which an onmidirectional azimuth radiation pattern is normally required. Such an antenna must have a vertical radiation pattern in which the main lobe has a high TOA at low frequencies, dropping to a low TOA at high frequencies, in the HF band.[16] This is due to the fact that the low frequencies are used for short-distance communication, for which a high TOA is required, and the high frequencies are used for long-distance communication, for which a low TOA is required.

To meet these requirements, a number of horizontal loops which radiate omni-directionally in azimuth are stacked vertically and connected in log-periodic form in the inverted-cone configuration shown in Fig. 3-11. The active region of the array, which comprises loops which are about two wavelengths in perimeter, is at the top of the array at the lowest frequency and at the bottom of the array at the highest frequency. A typical array operates over a 4- to 30-MHz-frequency range. The highest loop is placed 37 m above the ground, so that at 4 MHz the antenna has a broad vertical lobe with a TOA of about 30° and a gain of 7 dBi. The lowest loop is placed 10 m above the ground, so that the vertical lobe is narrow with a TOA of 10° and a gain of 10 dBi at 30 MHz.

A single inverted cone of loops suffers from the disadvantage that the vertical radiation pattern at high frequencies contains grating lobes. (See above, "Multiple-Curtain Log-Periodic Arrays.") To reduce the level of the unwanted lobes and max-imize gain, a second inverted cone of loops is installed inside the first. The upper loops of the two inverted cones are nearly coincident, and the vertical radiation pattern at the lower frequencies is therefore the same as that of a single cone. However, because the lowest loops of the inner cone are at a greater height than those of the outer cone, the antenna behaves as two stacked loops at the higher frequencies. The stacking factor of the two loops modifies the vertical radiation pattern by suppressing the high-elevation grating lobes, thereby increasing the gain of the antenna.

Rotatable Log-Periodic Antenna

When a high-gain, steerable azimuthal beam is necessary, a rotatable LPA (RLPA) can be used. An example is illustrated in Fig. 3-12. This antenna is approximately 33 m high and has a boom length of about 33 m. Its operating frequency range is 4 to 30 MHz, its gain about 12 dBi, and its azimuthal beamwidth 60°. The antenna does not require a very large area for installation and provides complete azimuthal coverage. However, such antennas have major mechanical and electrical disadvantages. The major mechanical disadvantage is their complexity, resulting from the need to rotate a very large structure. These antennas require regular maintenance to ensure mechanical reliability and are prone to failure in harsh environments. The electrical disadvantage is that the antennas have constant physical height and, therefore, their electrical height increases with increasing frequency. As a result, grating lobes form at the higher frequencies, reducing gain in the principal lobe and introducing elevation nulls at TOAs required for communications.[16]

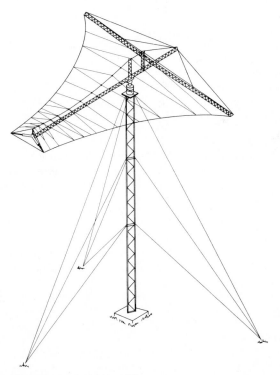

FIG. 3-12   Rotatable LPA.

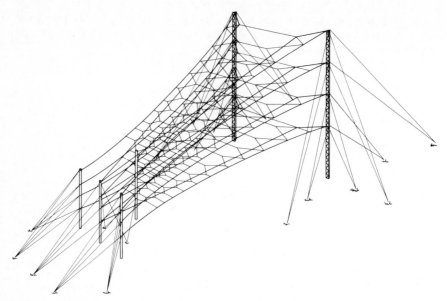

FIG. 3-13   Eight-curtain high-power LPA for HF broadcasting. *(Courtesy of TCI.)*

High-Power Log-Periodic Antenna

For most HF communication links, the transmitter power can be relatively low (usually 0.1 to 40 kW PEP) because high-gain receiving antennas and sophisticated receiving equipment are used at the other end of the circuit. However, for broadcasting
much higher transmitter powers are necessary to provide an adequate signal for lis-
teners using simple transistor receivers and low-gain antennas. As a result, transmitters with carrier powers between 100 and 500 kW are used, and the transmitting
antenna must be designed to operate with the resulting high voltages and currents.

A high-power HLPA is useful for broadcast purposes because it has an extremely
wide bandwidth and its power gain is not diminished by ground losses. In principle, a
high-power HLPA can be designed to cover the whole HF band from 2 to 30 MHz.
However, cost and land-area considerations require that the antenna size be minimized by selecting the highest possible low-frequency cutoff. A broadside array of
horizontally stacked HLPA curtains can be slewed in azimuth electrically by setting
up a phase difference between the stacks. If the electrical spacing between the radiation centers of the stacks is kept constant with frequency, the phase difference must
also remain constant. Networks providing constant phase delay can be realized by
using coupled transmission lines.

At the power levels used in HF broadcasting, the voltages and electric fields in
an HLPA are very high. To minimize the radiator-tip voltages, which can be as high
as 30 to 35 kV rms, the radiation Q of the dipoles (defined in Sec. 3-7) is reduced by
constructing the dipoles in the form of a fat cylindrical cage of wires. A simpler technique, illustrated in Figs. 3-10 and 3-13, forms the wires into a sawtooth arrangement
to increase their effective diameter. Large radial electric fields on the surface of the
wires can initiate arcing and corona discharge, which can burn through the wire and

cause structural failure of the antenna curtain. These fields can be reduced to safe levels by using wires with the largest diameter acceptable on structural grounds. All insulators used in the antenna, and particularly those at the radiator tips, must be able to withstand high voltages and fields with a large factor of safety.

Well-designed two- and four-curtain HLPAs can handle between 100- and 250-kW carrier power with 100 percent amplitude modulation. Eight- and sixteen-curtain HLPAs can handle up to 500-kW carrier power. The VSWR of these antennas is less than 2:1 over their entire frequency range and is generally less than 1.5:1 at most frequencies. As can be seen from Fig. 3-13, HLPAs with more than four curtains are very complicated and therefore are difficult to install and are expensive.

3-4 VERTICALLY POLARIZED ANTENNAS

Vertically polarized antennas are useful when a broad target area must be covered with a wide azimuthal beamwidth. An omnidirectional radiation pattern is obtainable with a single monopole. A directional pattern with wide beamwidth is obtainable by using vertical log-periodics. However, these antennas have a vertical null which militates against short-distance sky-wave communications. A very important disadvantage of vertical antennas is that, unlike horizontally polarized antennas, their power gain is diminished by losses in the surrounding ground. To overcome these losses partially, high ground conductivity or large ground screens are necessary, particularly if low TOAs are required for long-range communications.

Conical Monopole and Inverted Cone

Examples of the inverted cone and conical monopole are shown in Fig. 3-14a and b; their principal use is to provide omnidirectional coverage over a wide frequency band. The azimuthal pattern of these antennas is nearly circular at all frequencies, but the elevation pattern changes with frequency as shown in Fig. 4-14* of Ref. 28. The impedance bandwidth of the antennas is very large. The 22-m-high inverted cone illustrated in Fig. 3-14a has a VSWR of less than 2:1 over the entire 2- to 30-MHz range. The major disadvantage of this antenna is that the maximum radius of the cone is approximately equal to the height. The necessary support structure must consist of up to six nonconducting poles, which are usually expensive, and occupies a relatively large ground area.

The inverted-cone antenna can also be realized by using a single metallic support tower instead of the six nonconducting poles. The tower is placed along the cone axis on an insulated base. When designing the single-support antenna, great care must be taken to avoid internal resonances between the tower and the cone.

A more compact antenna, which also uses a single metallic mast and can be accommodated in a smaller ground area, is the conical monopole illustrated in Fig. 3-14b. The radiating structure comprises two cones, one inverted and one upright, connected at their bases. Here, the tower is grounded and the lower cone wires are insulated and fed against the ground. Midway up the tower the cone wires are connected to the tower to form an inductive loop. This acts as a built-in impedance-matching circuit, allowing the antenna to present a low VSWR when it is only 0.17λ tall. A

---

*This figure is reproduced at the end of this chapter, p. 3-41.

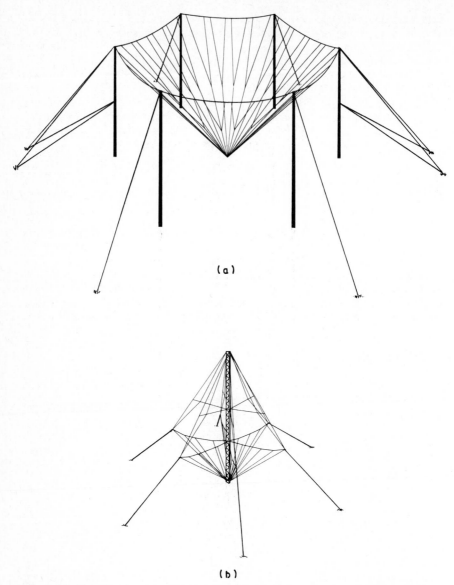

(a)

(b)

FIG. 3-14 Broadband monopole antennas. (a) Inverted cone. (b) Conical monopole. (*Courtesy of TCI.*)

conical monopole covering 2 to 14 MHz is 25 m high and has a maximum radius of 20 m.

All base-fed antennas usually require an impedance ground screen, which typically consists of 60 radials, each about $\lambda/4$ long at the lowest operating frequency.

Monopole Log-Periodic Antenna

At low frequencies in the HF band the height of a half-wave vertical dipole becomes undesirably large. Thus, for operation at the lower end of the HF band and in cases when directive gain is required, an LPA comprising quarter-wave monopoles operated over an impedance ground screen is a convenient antenna. A typical LPA covering 2 to 30 MHz is about 43 m tall and has a directive gain of 9 to 10 dBi and an azimuthal beamwidth of 140 to 180°.

Dipole Log-Periodic Antenna

When greater tower height can be accommodated (81 m for a 2- to 30-MHz antenna), an LPA comprising half-wave dipoles can be used, as shown in Fig. 3-15. The larger vertical aperture yields a gain of 12 to 13 dBi, which is about 3 dB greater than that of a monopole LPA. In contrast to the requirements of the monopole LPA, an impedance ground screen is not essential to achieve an acceptable VSWR. To achieve higher power gain and lower TOAs, a long radiation-pattern ground screen is necessary.

Monopole-Dipole Hybrid Log-Periodic Antenna

In an LPA covering the 2- to 30-MHz HF band, mast height can be minimized by using monopoles for the 2- to 4-MHz radiators. Above 4 MHz, dipole radiators can

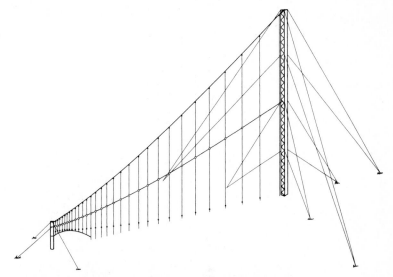

FIG. 3-15 Vertically polarized dipole LPA. *(Courtesy of TCI.)*

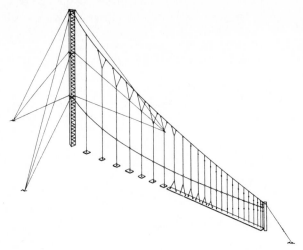

FIG. 3-16 Vertically polarized di-monopole hybrid LPA. *(Courtesy of TCI.)*

be used because they are fairly short and have more gain than the monopoles. A *di-monopole*[17] hybrid LPA can be constructed (see Fig. 3-16) in which the low-frequency elements are monopoles, the high-frequency elements are dipoles, and the elements operating at intermediate frequencies are unbalanced dipoles in which the upper conductor is longer than the lower conductor. Lumped capacitors and inductors are used at the base of the lower conductor to equalize the resonant frequencies of the upper and lower radiators. The design of a di-monopole hybrid LPA must be done very carefully because unequal impedances between upper and lower radiators can cause unbalanced currents to flow along the balanced transposed feedline in the antenna. Such in-phase common-mode currents radiate overhead and can cause unwanted antenna resonances.

Extended-Aperture Log-Periodic Antenna

The principal advantage of a vertically polarized LPA is that it has very wide azimuthal beamwidth but high directivity. Its gain can be increased without decreasing its azimuthal beamwidth if the vertical aperture is extended by lengthening the radiators. However, if the radiator lengths are made longer than a half wavelength, there will be nulls in the current distribution along the active radiators which will produce undesirable radiation-pattern nulls and unacceptable variations in antenna impedance. Increased vertical aperture is effective only if steps are taken to maintain the current distribution characteristic of a half-wave dipole (or quarter-wave monopole) as the length of the radiator is increased. This can be achieved by use of the *extended-aperture technique*,[18] which is illustrated in Fig. 3-17 for the case of a single dipole. Capacitors are inserted in series with the dipole limbs so that the current distribution never goes to zero except at the radiator tips. An extended-aperture LPA using full-wave dipoles has about 2 to 3 dB more gain than a half-wave dipole LPA.

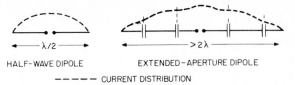

HALF-WAVE DIPOLE   EXTENDED-APERTURE DIPOLE

– – – – – CURRENT DISTRIBUTION

FIG. 3-17 Extended-aperture technique for eliminating current-distribution nulls on dipoles longer than $\lambda/2$. *(After Ref. 18.)*

Radiation-Pattern Ground Screens

The gain of a vertically polarized HF antenna is reduced and its TOA increased by the finite conductivity of the ground. To achieve maximum gain and the lowest possible TOA, a radiation-pattern ground screen extending several wavelengths in front of the antenna is necessary. The performance of a vertically polarized antenna depends on the linear dimensions of the ground screen and the conductivity of the ground. Practical ground screens are constructed by laying closely spaced wires on the earth. The wires of a ground screen must be dense enough so that the screen presents a very low resistive and reactive surface impedance. If a reactive impedance is presented, the radiation pattern is distorted because a quasi-trapped wave may be excited along the ground screen.[19]

A ground screen should not be buried more than a few centimeters beneath the earth, nor should it be allowed to be covered by snow. Burial or snow cover can com-

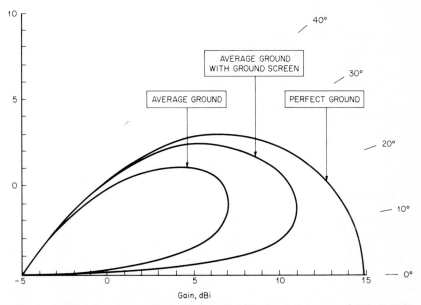

Gain, dBi

FIG. 3-18 Elevation pattern of a vertically polarized extended-aperture LPA at 10 MHz.

pletely mitigate the effectiveness of a ground screen and in some instances exacerbate the deleterious effects of the losses in the earth. There are no simple formulas for determining the radiation pattern of a vertical antenna with a ground screen. However, most of the available wire-antenna-analysis computer programs can calculate radiation patterns for antennas over lossy earth both with and without a polygonal ground screen of arbitrary size. Figure 3-18 shows the computed elevation pattern of an extended-aperture vertically polarized LPA (VLPA) over perfectly conducting ground and over ground of average conductivity, with and without a large ideal ground screen. Even with a large ideal ground screen there is a significant loss of gain owing to ground losses. Similar results are obtained for all vertically polarized antennas.

Impedance Ground Screens

For a monopole antenna, a ground screen is essential to obtain an input impedance giving an acceptable VSWR and to minimize conductive losses in the earth underneath the antenna. The input impedance of a vertical dipole is not affected by finite earth conductivity, so an impedance ground screen is not required. Impedance ground screens generally enclose an area extending no further than about $\lambda/4$ (at the lowest operating frequency) from the antenna.[20]

3-5   OTHER TYPES OF COMMUNICATIONS ANTENNAS

Vertical Whip

The vertically polarized whip antenna is a monopole 2 to 3 m long and of small diameter. It is widely used because it is simple to construct and easy to erect on a vehicle. Because the whip has narrow impedance bandwidth, an adjustable matching unit is required to tune the whip at each frequency. The less expensive matching units are preset to several frequencies. More expensive units tune automatically. The power-handling capacity of a whip and matching unit is usually limited to 1 kW average and PEP.

The whip antenna can suffer from severely low radiation efficiencies at lower HF frequencies because its radiation resistance is very low compared with the loss resistances of the tuner, whip conductor, and ground. The whip ground is usually a vehicle, and the resulting radiation pattern is not precisely predictable but is usually well behaved enough for adequate communication.

Rhombic Antenna

In the past, the horizontally polarized rhombic antenna was widely used for point-to-point communication links. It has now been almost completely superseded by the HLPA, which is smaller in size with equal or superior performance. The rhombic is a large traveling-wave antenna which is terminated at its far end by a resistive load. Radiation efficiency of a typical rhombic is 60 to 80 percent, compared with more than 95 percent for an HLPA. The rhombic behaves approximately like a matched transmission line, and this results in a nearly constant input impedance over a very wide bandwidth. However, the gain and radiation pattern vary significantly with frequency, and this reduces the useful operational bandwidth to about 2:1 (one octave). The HF band covers a frequency range of up to 15:1, so three or four rhombics are required to give complete frequency coverage. In contrast, a single HLPA can be designed to cover the whole HF band.

Fan Dipole

A simple form of wideband transmitting antenna is a horizontal dipole in the form of a fan as illustrated in Fig. 3-19. This antenna is a broadband radiator based on the same principle as the inverted-cone monopole. With careful design it is possible to achieve a VSWR of less than 2.5:1 over the whole HF band from 2 to 30 MHz. Since the height above ground (25.9 m) is fixed, the vertical radiation pattern varies with frequency. At low frequencies there is a broad vertical lobe with a high TOA suitable for short-distance transmission. As the frequency increases, the TOA decreases and the antenna becomes suitable for transmission over longer distances. The azimuth radiation pattern at low frequencies and high elevation angles is nearly omnidirectional. This is advantageous when communication with a number of short-range targets spaced randomly in azimuthal bearing is required.

For transportable applications, it is possible to use a single-mast version (height, 12.2 m) of the fan dipole as illustrated in Fig. 3-20, which is closer to the ground than the conventional fan dipole. The lower height increases the negative mutual coupling to the ground, which results in decreased radiation resistance, particularly at low frequencies. To compensate for this, it is necessary to connect resistive loads between the ends of the fan and the ground. The resistive loading yields a low VSWR over the whole HF band but results in a loss of gain of 5 to 13 dB, the greatest loss occurring at the low frequencies.

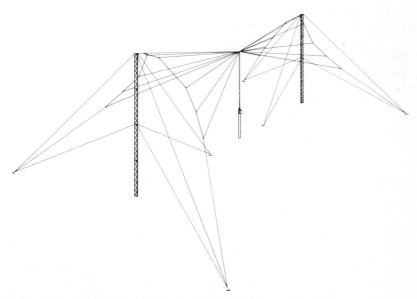

FIG. 3-19   Broadband fan dipole. *(Courtesy of TCI.)*

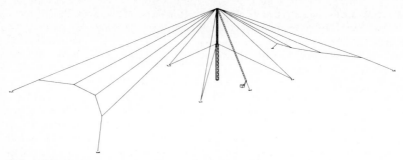

FIG. 3-20   Transportable fan dipole. *(Courtesy of TCI.)*

Transmitting Loop

When space is very restricted and an unobtrusive antenna is essential, it is possible to use a small-loop transmitting antenna. A small vertical loop in free space has radiation characteristics independent of frequency, and as is shown in Fig. 3-21 *a*, the azimuth radiation pattern becomes more circular with increasing elevation angle. Thus, unlike the whip and other monopole antennas, this antenna has no vertical null to militate against short-range communication. In practice, the loop must be mounted on the ground or possibly on the roof of a building. The performance of the free-space loop can be closely approximated by a half loop mounted on a conducting plane as shown in Fig. 3-21 *b*. The radiation resistance of the loop is proportional to $A^2 f^4$, where A is

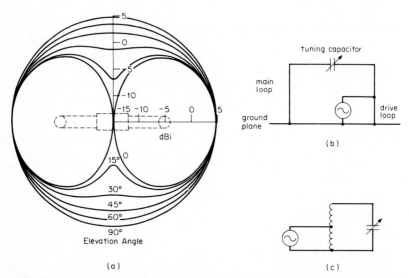

FIG. 3-21   Vertically polarized transmitting loop antenna. (*a*) Azimuth radiation patterns at different elevation angles. (*b*) Half loop mounted on ground plane. (*c*) Equivalent circuit of half loop on ground plane.

the area of the loop and f is the frequency. For small loops, therefore, the radiation resistance is very low at low frequencies. The minimum acceptable size of the loop is determined by the need to achieve a bandwidth of around 2 kHz at the lowest operating frequency, which can be 3 MHz for many applications. This requirement can be met by a loop 1 m high and 2 m wide—a small and compact device. The loop has a narrow impedance bandwidth and must be tuned at each operating frequency, as illustrated in Fig. 3-21c. A variable vacuum capacitor tunes the main radiating loop to resonance. An autotransformer, comprising a small drive loop coupled into the main radiating loop, transforms the resulting resistance to a value that maintains a VSWR of less than 2:1 over a frequency band of 3 to 24 MHz. It is possible to do this tuning automatically, using, for example, a microprocessor-based system which (1) measures the transmitter frequency, (2) monitors the current in the drive and main loops, (3) sets the capacitor to an appropriate value, using a stepping motor, and (4) determines the capacitor's position by means of a variable resistor mounted on its shaft.

The small radiation resistance of the loop at low frequencies makes it inevitable that resistive losses cause a reduction in radiation efficiency. Losses are minimized by using a high-Q vacuum capacitor and by ensuring that the radio-frequency (RF) paths' in the loop and ground screen have very high conductivity. At 3 MHz the radiation efficiency of a well-designed and carefully constructed loop will be about 5 percent. Because the radiation resistance of the loop increases with the fourth power of frequency but the resistive losses increase with the square root of frequency, efficiency rises rapidly with increasing frequency, achieving a value of about 50 percent at 10 MHz and 90 percent above 18 MHz.

3-6 HIGH-FREQUENCY RECEIVING ANTENNAS

For a passive linear antenna (one without nonlinear elements or unidirectional amplifiers) the gain, radiation pattern, and impedance are the same for transmission and reception.[21] This is a consequence of the principle of reciprocity, which applies to all passive linear four-pole networks. However, the relative importance of these antenna parameters is different for transmission and reception. For transmission it is important to maximize field strength in the target area, so the antenna should be as efficient as possible. For reception the most important parameter is the signal-to-noise ratio (SNR) at the receiver output terminals, which obviously should be as large as possible. The SNR at the receiver output depends on the gain and mismatch losses of the antenna and feeder, the internal noise generated by the receiver, and the external atmospheric, human-made, and galactic noise levels. Because the external noise level at HF is often much larger than the internal noise levels in the receiver and antenna, it is possible in these instances to obtain acceptable SNR by using inefficient receiving antennas.

Noise Figure

For a simple analysis the receiver can be assumed to be linear, and the output SNR therefore equal to the input SNR. In calculating the output SNR, the signal and noise powers can therefore be referred to the receiver input and the available noise power at the output of an HF receiver taken to be

$$n_{out} = onf \cdot kT_0 b \qquad (3\text{-}9)$$

where n_{out} = available noise power, W
   onf = operating noise figure
    k = Boltzmann's constant, 1.38×10^{-23} J/K
   T_0 = reference temperature, 288 K
    b = effective noise bandwidth of receiver, Hz
The operating noise figure[22] (or factor) is a power ratio which accounts for the total internal and external noise of the receiving system and is given by

$$onf = f_e - 1 + f_a f_t f_r \qquad (3\text{-}10)$$

where f_e = total external noise power divided by $kT_0 b$
    f_a = antenna noise figure
    f_t = transmission-line noise figure
    f_r = receiver noise figure
The product $f_t f_r$ may be replaced by a single noise figure f_{rt} of a generalized receiving system containing additional devices such as filters, preamplifiers, and power dividers and combiners. In the rest of this section, as is common practice, the decibel equivalent of these noise figures will be denoted by capital letters.

The external noise figure F_e is the level of atmospheric, human-made, and galactic noise in a 1-Hz bandwidth in decibels above kT_0, where $10 \log_{10} kT_0 = -204$ dBW. Atmospheric noise is generated by lightning discharges which produce enormous amounts of HF energy that propagate great distances. Human-made noise is produced by various kinds of electrical equipment such as motors, vehicular ignition systems, electric welders, and fluorescent lamps. Galactic noise is generated by extraterrestrial radio sources. F_e can be determined from published noise maps[22] or by direct measurement by using a calibrated receiver and antenna. Figure 3-22 shows a typical example of the frequency variation of atmospheric noise, which usually drops from a high value at low frequencies to a low value at the upper end of the HF band. Atmospheric noise level also varies with geographical location, season, and time of day. Human-made noise level depends on whether the receiving site is in an urban, suburban, or rural location; in urban areas it can equal or exceed the atmospheric noise level.

The receiver noise figure F_r is usually obtainable from the receiver's specifications table. The transmission-line noise figure F_t is the reciprocal of the attenuation, expressed in decibels. For systems with filters, preamplifiers, etc., the noise figure F_{rt} can be determined by computing the noise figure for the equivalent cascaded network.

The antenna noise figure F_a is given by

$$F_a = D - G + L_M \qquad (3\text{-}11)$$

where D = directivity, dBi
    G = gain, dBi
   L_M = mismatch loss = $10 \log_{10} (1 - |\rho|^2)$, dB
The mismatch loss is given in terms of the reflection coefficient ρ at the input terminals of the antenna, where $|\rho| = (VSWR - 1)/(VSWR + 1)$. The quantity $D - G$ can be expressed in terms of the efficiency η by using the relation $D - G = -10 \log_{10} \eta$. The efficiency must include the resistive losses of both the antenna and the sur-

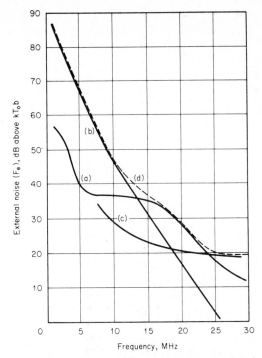

FIG. 3-22 Example of an external atmospheric and galactic noise figure. (*a*) Atmospheric daytime noise. (*b*) Atmospheric nighttime noise. (*c*) Galactic noise. (*d*) Worst noise condition.

rounding ground. For horizontally polarized antennas the ground losses are negligible. For vertical antennas ground losses may be significant, so they should be considered when there is no ground screen and earth conductivity is low.

Resistively loaded antennas, such as the small fan dipoles discussed in Sec. 3-5, usually have a VSWR of less than 2.5:1, so the mismatch loss is less than 1 dB. For these antennas the noise figure is determined primarily by the efficiency. For small nonresonant loop antennas, which will be discussed later, the mismatch loss is extremely large because the antenna has an extremely small input resistance and much larger reactance. For small loops both mismatch loss and low efficiency contribute to the noise figure. For small loops, monopoles, and dipoles the mismatch loss decreases by approximately 12-dB-per-octave increase in frequency.

Equation (3-10) has two important limiting cases. When $f_e \gg f_a f_t f_r$, the receiving system is said to be externally noise-limited because the external noise exceeds the internal noise. In this case the system performance cannot be improved by reducing antenna, feeder, or receiver noise figures. When $f_e \ll f_a f_t f_r$, the system is internally noise-limited, internal noise being greater than external noise. In this case system performance is improved by reducing the antenna or equipment noise figures.

In determining whether a receiving system is internally or externally noise-limited, it is important that actual external noise data be used. The values shown in Fig.

3-22 are for illustrative purposes only, actual variations being too extensive to display in one curve. In many geographical locations atmospheric and human-made noise levels are extremely low, and therefore receiving systems must have very low noise figures to avoid being internally noise-limited.

Signal-to-Noise Ratio

The SNR at the receiver output is the ratio s_{out}/n_{out} of the available signal and noise powers. The available signal power is given by

$$s_{out} = \frac{2.08e^2\lambda^2 d10^{-15}}{\pi^2} \quad \text{W} \tag{3-12}$$

where e = field strength of arriving signal, μV/m
       d = antenna directivity as a power ratio
Dividing this by Eq. (3-9) and expressing the quantities in decibels gives the following equation for the SNR:

$$\text{SNR} = E + D - F - \text{ONF} - B + 97 \text{ dB} \tag{3-13}$$

where $E = 20 \log_{10} e$, dBμ (dB relative to 1μV/m)
       D = antenna directivity, dBi
       $F = 20 \log_{10} f$ (in MHz)
   ONF $= 10 \log_{10}$ (onf), dB
       $B = 10 \log_{10} b$ (in Hz)

Practical Receiving Antennas

HF receiving antennas are divided into two classes. One class comprises antennas that are designed for transmission but are also used for reception. These antennas are very efficient, and their noise figures are often less than 3 dB, making them suitable for use at very quiet receiving sites. Full-sized vertically polarized antennas, when used for reception, do not require ground screens unless they are used in an emitter-locating system in which terrain variations must be neutralized by placing a metallic ground mesh under and in front of the antennas. The absence of a ground screen narrows the vertical beamwidth of the radiation pattern, thereby increasing directivity. However, ground losses reduce the gain by 5 to 10 dB and thus increase the antenna noise figure by the same amount. The resulting overall system noise level is nevertheless lower than most external noise levels, so a vertical antenna without a ground screen makes a highly suitable receiving antenna for many locations.

The second class of antennas consists of those specifically designed for reception. Again, because a high HF receiving-antenna noise figure is usually acceptable, a receive-only antenna can be inefficient and have high VSWR. Consequently, it can be small and, if necessary, be loaded resistively to improve its performance.

Typical acceptable values of antenna noise figure are given in the following illustrative analysis of a radioteletype receiving system. Assume that signal strength E is 34 dBμ (50 μV/m) and antenna directivity is 5 dBi. The required SNR for moderate character error rate is 56 dB in a 1-Hz bandwidth. The receiver noise figure F_r is 13 dB, and feeder noise figure F_t is 3 dB, both reasonable values for HF devices. The maximum allowable ONF is obtained by substituting E, D, and SNR in Eq. (3-13). The corresponding allowable antenna noise figure F_a is obtained by substituting the power ratios on f, f_e, f_r, and f_t into Eq. (26-10). In this example the external noise level F_e is assumed to be given by the dotted curve in Fig. 3-22. The accompanying table summarizes the results for several frequencies:

f, MHz	ONF, dB	F_e, dB	F_a, dB
2	74	80	· · ·
5	66	66	· · ·
10	60	46	44
15	56	37	40
20	54	29	38
30	50	19	34

Below 5 MHz the required SNR will not be obtained because the external noise level is too high. This can be overcome only by moving to a quieter site; improving equipment noise figures has no effect. Above 5 MHz the external noise level is smaller, and the required SNR can be obtained by using an antenna with a maximum noise figure at each frequency as given in the table. If the antenna VSWR is low so that the mismatch loss can be neglected, allowable antenna efficiency ranges from 0.004 percent at 10 MHz to 0.04 percent at 30 MHz.

Receiving Loops

A practical antenna having a noise figure near the upper limit allowable for most situations is the small vertical balanced loop. A typical loop is about 1.5 m in diameter and is mounted about 2 m above the ground on either fixed posts or transportable tripods. The antenna is usually made from a large-diameter metallic tube, which also serves to shield the feeder that runs through the tube to the feed point. The vertical radiation pattern of the loop has good response at both low and high TOAs, making the antenna suitable for both long- and short-distance reception of sky-wave signals. The loop also responds well to ground waves.

The noise figure of the loop depends strongly on its area and the operating frequency. A loop of 1.5-m diameter has a noise figure of about 50 dB at 2 MHz and 20 dB at 30 MHz. The noise figure can be reduced by making the area of the loop larger. However, if the loop perimeter is about a wavelength long, the loop will resonate and the radiation pattern will exhibit irregularities. These can be avoided in large loops by feeding the loop at several points.

A number of loops can be arrayed to provide a variety of directional radiation patterns. In most arrangements of small loops, mutual coupling between the loops is negligible and simple array theory can be used to predict radiation-pattern shape. Beams are formed by bringing the coaxial cable from each loop into a beam-forming unit located near the antenna. The most common arrangement places eight loops in an end-fire array with spacing between loops of 4 m, as shown in Fig. 3-23. The array can be made bidirectional by splitting the signal from each loop two ways and feeding each half into a separate beam former. Four such bidirectional arrays can be arranged in a circular rosette which provides eight independent beams covering 360° in azimuth.

FIG. 3-23 End-fire array of eight receiving loops. *(Courtesy of TCI.)*

Active Antennas

Many receiving antennas, including loop arrays, are made into *active antennas* by interposing a broadband low-noise amplifier between the output of the antenna or its beam former and the feeder line to the receiver. For some antennas the amplifier can improve the impedance match and thus reduce mismatch loss and antenna noise figure. However, for electrically small antennas little improvement in noise figure is possible with a practical amplifier.[23] The amplifier also sets the antenna-system noise figure at the point where the amplifier is placed. The amplifier has a beneficial effect in this case only if its noise figure (usually 1 to 5 dB) is less than the feeder noise figure and its gain (usually 10 to 30 dB) is much higher. The amplifier can have a deleterious effect on receiving-system performance if high ambient signal levels cause it to produce spurious intermodulation products. These are likely to occur if there are strong medium-frequency (MF), HF, or very-high-frequency (VHF) signals from nearby transmitting antennas. It is a simple matter to filter out the MF and VHF signals by using, respectively, high- and low-pass filters ahead of the amplifier. Unwanted HF signals must be rejected by using notch filters so as not to impair the usefulness of the antenna in the HF receiving system.

The availability of cheap microprocessors has made it possible to use active antenna elements as the basis of adaptive-antenna arrays. In an adaptive system the amplitudes and phases of the signals from a multielement array are adjusted continually to give a radiation pattern maximizing the wanted signal and minimizing interfering cochannel and adjacent-channel signals. Optimum weightings of component signals must be determined from a careful study of the actual signals with which the antenna must contend. These studies are not straightforward, and designing a suitable algorithm is difficult in many cases.

Circular Arrays

Arrays of circularly disposed antenna elements are used in monitoring or emitter-locating systems (see Chap. 16). In these applications there is quite often a need to receive signals from all azimuth angles and to be able to connect a number of receivers to any of a number of beams. Antennas of this type must operate over a large part of the HF band and must have highly predictable radiation patterns with low sidelobe levels. It is often advantageous to make each element of the array an LPA. Vertical LPAs should be used if the array must respond to ground waves or to low-TOA sky waves. Horizontal LPAs, placed close to ground, should be used if high-TOA sky-wave signals from short distances are to be received.

A typical array comprises 18 to 36 elements. The output of each element is fed into a beam former, containing suitable delay lines and power splitters, which forms a narrow azimuthal beam. It is possible to generate N beams simultaneously in an N-element array by splitting the power from each element N ways. The N beams are uniformly spaced around the circle. By using multicouplers, each beam can be accessed by several receivers.

In arrays of LPAs it is best to point the elements inward so that their direction of maximum sensitivity is toward the array center. This configuration keeps the electrical diameter of the array nearly constant, thereby making the azimuthal beamwidths nearly independent of frequency and eliminating unwanted azimuthal grating lobes. The disadvantage of the inward-looking array is that each element "fires through" those opposite, and this complicates the calculation of the radiation pattern.

3-7 BROADBAND DIPOLE CURTAIN ARRAYS

Broadband dipole curtain arrays are used for high-power (100- to 500-kW) HF ionospheric broadcasting. These antennas consist of square or rectangular arrays of dipoles, usually a half wavelength long, mounted in front of a reflecting screen, as shown in Fig. 3-24. Dipole arrays have many excellent performance characteristics, such as independent selectability of vertical and horizontal patterns, high power gain, slewability of beam in azimuth or elevation, wide impedance bandwidth, low VSWR, and high power-handling capacity, which have made them virtually the standard antenna at short-wave broadcasting stations.

Standard Nomenclature

Dipole curtain arrays are described by the internationally agreed nomenclature HRRS $m/n/h$,

where H denotes horizontal polarization
    R denotes a reflector curtain
    R (if not omitted) denotes that the direction of radiation is reversible
    S (if not omitted) denotes that the beam is slewable
    m is the width of the horizontal aperture in half wavelengths at the design frequency

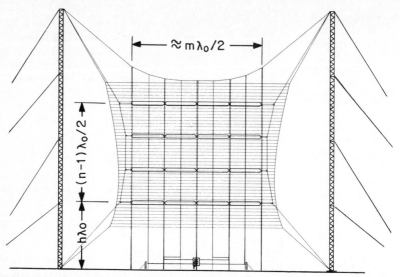

FIG. 3-24   Basic HF dipole curtain array. (Antenna shown is designated HRS 4/4/ h.)

n is the number of the dipoles in a vertical stack

h is the height of the lowest dipole above ground, in wavelengths at the design frequency

The design frequency is $f_0 = \frac{1}{2}(f_1 + f_2)$, where f_1 and f_2 are the lower and upper frequency limits; λ_0 is the corresponding design wavelength. In dipole arrays which use half-wave dipoles, the width parameter m is the number of dipoles in a horizontal row.

Radiation Pattern and Gain

The TOA and first null at frequency f depend on the average height of the dipoles in the vertical stack and are given by

$$\text{TOA} = \sin^{-1}(f_0\lambda_0/4fH_{\text{avg}}) \tag{3-14}$$

$$\text{Null} = \sin^{-1}(f_0\lambda_0/2fH_{\text{avg}}) \tag{3-15}$$

where $H_{\text{avg}} = (H_1 + H_2 + \cdots + H_n)/n$ and $H_1, H_2, \ldots, H_n$ are the physical heights above ground of the n dipoles in the stack. The TOAs at f_0 are given for several configurations in Table 3-1. The level of the unwanted minor elevation lobes is determined by the number of dipoles in the stack and their spacing.

The azimuthal half-power beamwidth (HPBW) depends primarily on the width of the array but also depends weakly on the TOA. The beamwidth at f_0 is given in Table 3-2. Beamwidth at frequency f is obtained approximately by multiplying the value in this table by f_0/f.

The gain at f_0 of an array of half-wave dipoles is given in Table 3-3. For other configurations the gain is given approximately by $G = \log_{10}(27,000/A_1A_2)$, where A_1

TABLE 3-1   Takeoff Angle of Dipole Array with Reflecting Screen

Number of elements in vertical stack *n*, half-wavelength spacing	Height above ground of lowest element in wavelengths, *h*			
	0.25	**0.5**	**0.75**	**1.0**
1	45°*	29°	19°	15 and 48°†
2	22°	17°	14°	11°
3	15°	12°	10°	9°
4	11°	10°	8°	7°
5	9°	8°	7°	6°
6	7°	7°	6°	5°

*90° without reflector.

†Two lobes present.

and A_2 are the vertical and horizontal beamwidths in degrees. The approximate gain at frequencies different from f_0 can be obtained by adding 20 log (f/f_0) to the values in Table 3-3.[24]

Slewing

Early forms of dipole arrays used full-wave dipoles as shown in Fig. 3-25. These antennas contained a very-narrow-bandwidth feed system and thin dipoles, so they were capable of operating only in one or two broadcast bands. The azimuthal beam of the antenna was slewed by using a tapped feeding arrangement. This sets up a phase difference between the two halves of the array, which slews the beam up to $\pm 10°$ in azimuth. The slew angle is limited to a small value because the dipole centers are separated by one wavelength. If greater slew angles are attempted, the horizontal pattern contains large secondary lobes which reduce gain by up to 3 dB and may interfere with cochannel transmissions.

TABLE 3-2   Horizontal Beamwidth of Dipole Array*

Number of elements in vertical stack *n*, half-wavelength spacing	Number of half-wave elements wide *m*		
	1	**2**	**4**
1	76°	54°	26°
2	74°	50°	24°
3	74°	49°	24°
4	73°	49°	24°

*Between half-power points.

TABLE 3-3 Gain in dBi of Dipole Array with Perfect Reflecting Screen Spaced 0.25λ behind Dipoles

Number of elements in vertical stack *n*, half-wavelength spacing	Number of half-wave elements wide *m*															
	1				2				3				4			
	Height above ground of lowest element in wavelengths, *h*															
	0.25	0.50	0.75	1.0	0.25	0.50	0.75	1.0	0.25	0.50	0.75	1.0	0.25	0.50	0.75	1.0
1	12.5	13.2	14.0	13.6	13.5	14.4	15.3	14.8	15.4	16.4	17.1	16.8	16.1	17.2	17.9	17.5
2	14.0	15.0	15.6	15.8	15.4	16.5	17.0	17.2	17.5	18.4	19.0	19.2	18.2	19.3	19.8	20.0
3	15.6	16.4	16.9	17.2	17.1	17.9	18.4	18.7	19.0	19.8	20.3	19.6	19.8	20.7	21.2	21.5
4	16.7	17.3	17.9	18.2	18.2	18.9	19.4	19.7	20.1	20.8	21.3	21.6	21.0	21.7	22.2	22.5

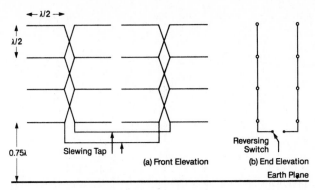

FIG. 3-25 Early form of narrowband HRS 4/4/.75 dipole array using full-wave dipoles.

To achieve larger slew angles it is necessary to reduce the horizontal spacing between the dipole stacks. In these arrays the dipole length is slightly less than $\lambda_0/2$, and the spacing between the stacks is $\lambda_0/2$. For these antennas the slewing system shown in Fig. 3-26 enables slew angles to be as large as $\pm 30°$. It is possible, in principle, to slew any dipole array with two or more stacks. In practice, two-wide arrays do not slew effectively over wide frequency ranges because of unfavorable mutual impedances between dipoles. However, four-wide arrays slew very well, and the slew has no deleterious effect on either gain or impedance.

Dipole arrays can also be slewed vertically by introducing phase shifts between vertical elements in each stack. This is usually done by introducing a 180° phase reversal between pairs of elements. Vertical slewing systems are more complicated

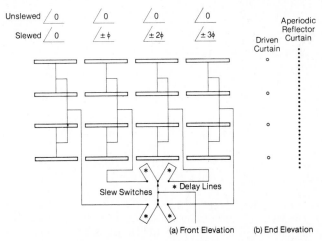

FIG. 3-26 Slewing system and corporate feed of a broadband HRS 4/4/h dipole array. Dipole length = 0.46 λ_0, dipole spacing = 0.50 λ_0 center to center, and screen-to-dipole spacing = 0.25 λ_0.

than horizontal slewing systems. The impedance bandwidth of a vertically slewed antenna is generally smaller than that of a horizontally slewed array.

Impedance Bandwidth

The VSWR of a high-power broadband dipole array must be low for two important reasons. First, the voltages on the antenna and feeder lines become excessive if the VSWR is too high. Second, although most modern high-power HF transmitters will accept a VSWR of 2:1, the feeder line and switching systems introduce discontinuities so that the antenna itself must usually have a VSWR of less than 1.6:1 or 1.5:1. Low VSWR requires that the input impedance of the antenna must remain substantially constant over its entire frequency range. The bandwidth of a dipole array depends on the inherent bandwidth of the dipoles, the bandwidth of the interdipole feeder system, and the type of reflector placed behind the dipole array.

The inherent dipole bandwidth, by analogy with simple resistance-inductance-capacitance (RLC) resonant circuits, can be characterized by a radiation $Q = f_R/\Delta f$, where f_R is the resonant frequency of the dipole and Δf is the bandwidth between the frequencies at which dipole resistance and reactance are equal. In the broadband dipole array, a very low Q of about 2 is achieved by making the dipole fat. The thin dipoles used in early arrays have Q's of about 10.

The feed system which interconnects the dipoles is made broadband by constructing it in the form of a *corporate feed,* as shown in Fig. 3-26. Wide-bandwidth transmission-line transformers, often of the optimal Chebyshev type, are used whenever impedance transformations are necessary.[25]

Early dipole arrays used parasitic dipoles in the reflector curtain. Such antennas have very narrow bandwidth because the parasites act as reflectors over only a small range of frequencies, a phenomenon familiar to designers and users of Yagi-Uda arrays. Broadband dipole arrays use an aperiodic reflecting screen consisting of closely spaced horizontal wires placed about $0.25\lambda_0$ behind the dipoles.

A well-designed broadband dipole array has a VSWR of 1.5:1 or less over a one-octave bandwidth (i.e., $f_2/f_1 = 2$). This bandwidth permits a single antenna to operate over four or five adjacent HF broadcast bands, in contrast to the one or two bands of early dipole arrays.

Practical Considerations

The design of modern dipole arrays is greatly facilitated by the use of comprehensive computer programs which enable the current distribution on the entire array to be analyzed.

To achieve low dipole Q, the half-wave dipoles are constructed as fat multiple-wire cages which are either flat, rectangular boxes or cylinders. The dipoles are usually "folded" to step up their impedance to a more useful level and provide additional impedance compensation. A folded dipole operates simultaneously in two modes. The radiating antenna mode, or unbalanced mode, depends only on the length and equivalent diameter of the dipole cage. The nonradiating transmission-line, or balanced, mode describes the currents which flow in a loop around the dipole and through the familiar short circuits at the end of the folded dipole. The impedance variations of the two modes tend to cancel, so the transmission-line mode can be used to compensate and tune the antenna mode.

This compensating effect is maximized by moving the position of the folded-dipole short circuit away from the end of the dipole and toward the feed point. The optimum position can be determined in a straightforward manner by using an antenna-analysis computer program.

The driving-point impedance of each folded dipole is about 600 Ω when mutual impedances from other dipoles are included. This allows the impedance levels in the branch feeder to be in the range 300 to 600 Ω, which is easy to realize in practice.

HRS 4/4 arrays can be designed to have a power-handling capacity of 750 kW average and 4000 kW PEP, which is large enough to accommodate two fully modulated 250-kW transmitters fed into the antenna simultaneously by using a diplexer. Smaller dipole arrays, such as HR 2/2, can handle up to 750 kW average and 2000 kW PEP, corresponding to the power output of a fully modulated 500-kW transmitter. The wire diameters in the dipole array must be large enough to carry the high currents which flow and to prevent corona discharge. Insulators must be carefully chosen to withstand voltages of up to 45 kV rms, which can occur in the high-voltage points in the antenna.

The one-octave bandwidth of a dipole array allows three such antennas, 6/7/9/11, 9/11/13/15/17, and 13/15/17/21/26 MHz, to cover the whole international broadcast spectrum, with multiple-antenna coverage of many of the bands.

3-8 SITING CRITERIA FOR HIGH-FREQUENCY ANTENNAS

In the preceding sections it has been assumed that the antenna is mounted over a smooth, level ground plane of infinite extent. However, an actual antenna site is seldom smooth and level, and sites which approximate ideal conditions usually do so only over a limited area. In addition, the effects of co-sited antennas and of natural or artificial obstructions can result in degradation of antenna performance. This section assesses the effects of imperfect sites following closely Refs. 26 and 27, and presents practical criteria for obtaining acceptable performance.

Fresnel Zone and Formation of Antenna Beam

The main lobe of a transmitting or receiving antenna (the following discussion applies to both) is formed by the interaction of the direct radiation of the antenna and the reflected radiation from the ground plane. Reflections occurring near the antenna are of greater importance than those occurring far away. With a directional antenna, radiation at or near boresight angles is more important than that occurring at azimuth angles away from boresight. From these facts, it follows that there is an elliptically shaped area (with the major axis in the direction of the main lobe, as shown in Fig. 3-27) in which the ground must be level and smooth if the first lobe is to be formed without appreciable distortion.

From simple geometrical ray theory it is clear that as the TOA is decreased, the lengths of the major and minor axes of the elliptical area increase. The curvature of the earth also affects the size of the axes, but it can be ignored if the TOA is more than 3°

The ellipse dimensions are different for vertical and horizontal polarization and

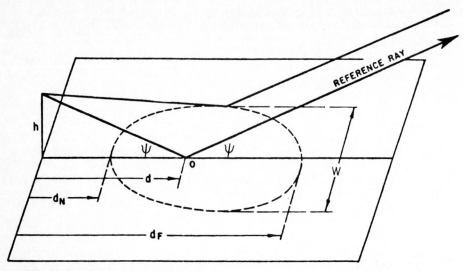

FIG. 3-27   Geometry of first Fresnel zone.

in the former case are difficult to evaluate. Fortunately, the dimensions for horizontal polarization, which are easy to evaluate, give a reasonable approximation for the vertically polarized case. The ellipse dimensions correspond to the first Fresnel zone, which is the region in front of the antenna in which direct and reflected radiation differ in phase by 180° or less. For horizontal polarization, if the antenna has a TOA of ψ_0, the radiation will appear to come from a height above ground of

$$h = \lambda/4 \sin \psi_0 \qquad (3\text{-}16)$$

The distance from this radiation point to the near and far edges of the first Fresnel zone, denoted by d_N and d_F respectively, are given by

$$d_N = \frac{h}{\tan \psi_0}\left(3 - \frac{2\sqrt{2}}{\cos \psi_0}\right) \qquad (3\text{-}17)$$

$$d_F = \frac{h}{\tan \psi_0}\left(3 + \frac{2\sqrt{2}}{\cos \psi_0}\right) \qquad (3\text{-}18)$$

The maximum width of the Fresnel ellipse is

$$w = 4\sqrt{2}h = 5.66h \qquad (3\text{-}19)$$

Equations (3-16) to (3-19) may be used to calculate, to a very good approximation, the area of flat, unobstructed land required in front of a directive antenna to ensure that the main lobe is fully formed. For very low TOA the Fresnel zone may extend for several kilometers, and it will not usually be possible to contain the zone within the boundaries of the antenna site. If the size of the controlled area is to be reduced, it is possible to limit it to the region in which the phase difference between direct and reflected radiation is 90° or less. In this case, there may be a loss of gain of up to 3 dB, but the dimensions of the ellipse will be reduced as follows:

d_F may be reduced to 0.6 times the full value.

d_N may be increased to 1.6 times the full value.

w may be reduced to 0.7 times the full value.

Roughness in First Fresnel Zone

In the zone with dimensions given in Eqs. (3-16) to (3-19), the main lobe is approximately fully formed provided the ground is flat and smooth. However, if the ground is rough or has natural or artificial obstructions or depressions, reflection will not be specular. The reflected wave will be scattered, and there will be a loss of gain and distortion of the beam. Rayleigh's criterion indicates that the transition between specular and scattered reflection occurs when the maximum height H of deviations above and below the average terrain profile does exceed

$$H = \lambda_0/16 \sin \psi_0 = h/4 \qquad (3\text{-}20)$$

This criterion may be relaxed if additional degradation in performance is acceptable. The relaxed criteria are based on the reasonable assumption that the permissible size of the obstruction may increase with distance from the antenna. It is usual to divide the first Fresnel zone into three regions, within which the heights of obstructions and depths of depressions should not exceed the values in the accompanying table.

Limit of departure from average smooth terrain	Region of first Fresnel zone
$H/4$	d_N to 0.2 d_F
$H/2$	0.2 d_F to 0.6 d_F
H	0.6 d_F to 1.0 d_F

It is difficult to predict with precision the degradation that occurs when the above criteria are applied, partly because the shapes of the obstructions have been ignored. However, it is reasonable to expect gain losses of several decibels. These criteria apply to cases in which the entire first Fresnel zone is rough. When a major portion of it is smooth, somewhat larger obstacles can be tolerated if they do not cover more than 5 to 10 percent of the zone. Caution must be exercised, however, when compromises in controlled-area size and roughness are made simultaneously because gain reductions may then become large.

Horizon Obstructions

Obstructions beyond the first Fresnel zone at distance $D_0 \gg d_F$ can reduce radiation at low angles even if the first Fresnel zone is perfectly smooth. If the height of the obstruction is H_0, antenna performance will not be noticeably degraded if the angle subtended by the obstruction as viewed from the antenna, $\psi_{obs} = \tan^{-1}(H_0/D_0)$, is less than 0.5 times the lower half-power point of the antenna elevation pattern. For TOAs of less than 30°, this criterion is given by $\psi_{obs} \leq \psi_0/4$.

Terrain Slope

If the ground at an antenna site slopes down in the direction of radiation by an angle b, the TOA is decreased from ψ_0 (the value for a flat site) to $\psi_0 - b$. Conversely, if the ground slopes up in the direction of radiation, the TOA becomes $\psi_0 + b$. Fresnel-zone theory can be applied to sloping sites by using the following modified antenna height factor:

$$h_{\text{slope}} = \lambda/4 \sin (\psi_0 \pm b) \qquad \textbf{(3-21)}$$

in Eqs. (3-16) to (3-19). The $+$ sign is used for upward slopes; the $-$ sign, for downward slopes.

Ground Conductivity

For vertically polarized antennas, high ground conductivity is important for the effective operation of transmitting antennas, as explained in Sec. 3-2. Vertical transmitting antennas mounted very near seawater, which has a high conductivity of $4\ S/m$, do not require ground screens for radiation-pattern enhancement. For most types of soil, conductivity is significantly smaller than this value, so metallic ground screens are necessary. For horizontally polarized antennas, good ground conductivity is not important unless the TOA is very high, as explained in Sec. 3-2.

Coupling between Co-Sited Antennas

When transmitting, an antenna transfers some of its radiated energy to any other antenna in its vicinity, and this transferred energy may affect the performance of the other antenna. *Antenna coupling* is defined as the ratio of the power delivered into a matched load terminating the other antenna (called for convenience the *receiving* antenna) and the input power to the *transmitting* antenna:

$$C = 10 \log_{10} (P_r/P_t) \qquad \text{dB} \qquad \textbf{(3-22)}$$

where P_r = power dissipated in matched load terminating the *receiving* antenna
P_t = input power to *transmitting* antenna
Coupling can be calculated accurately by solving the fundamental electromagnetic equations for the several antennas, using a comprehensive antenna-analysis computer program. However, good approximate coupling values can be obtained from two simple formulas, which assume that the antennas are separated by at least one wavelength λ. The magnitude of the coupling depends on the polarization of the antennas. For vertically polarized antennas the coupling factor is

$$C_v = 20 \log_{10} (\lambda/8\pi d) + G_t + G_r \qquad \text{dB} \qquad \textbf{(3-23)}$$

where d = spacing between antennas
G_t = gain of transmitting antenna, dBi
G_r = gain of receiving antenna, dBi
G_t and G_r are the gains of the two antennas in the directions toward each other. The formula neglects ground loss, which is generally negligible when d is less than a few kilometers. For horizontal polarization the coupling factor is

$$C_h = 20 \log_{10} \left[\frac{\lambda \sin (\pi L/\lambda)}{8 \pi d} \right] + G_t + G_r \quad \text{dB} \qquad \textbf{(3-24)}$$

where $L = [(H_t + H_r)^2 + d^2]^{1/2} - [(H_t - H_r)^2 + d^2]^{1/2}$
      $H_t = \lambda/4 \sin \psi_t$
      $H_r = \lambda/4 \sin \psi_r$
ψ_t and ψ_r are the TOAs of the transmitting and receiving antennas. λ, d, G_t, and G_r are defined in Eq. (3-23), except that G_t and G_r are the values at the TOAs ψ_t and ψ_r. For large values of d, C_h decreases by 12 dB each time that d is doubled.

Coupling between antennas can cause a number of effects, and one or more of the following may be relevant in a particular case.

Intermodulation   If both antennas are transmitting, the energy coupled by one into the other may give rise to intermodulation products in the transmitter connected to the *receiving* antenna. The amount of coupling that can be tolerated depends on the characteristics of the transmitter, but typical values are -20 to -25 dB.

VSWR   When energy from another antenna is coupled into a transmitting antenna, the VSWR presented to the transmitter will be changed because the apparent reflected power will be altered by the incoming energy. The reflection coefficient of the receiving antenna, ρ', has a maximum magnitude given by

$$\rho' = \left[\frac{\rho^2 P_r + C_r P_t}{P_r + C_r P_t} \right]^{1/2} \qquad \textbf{(3-25)}$$

where ρ = reflection coefficient of *receiving* antenna in absence of the other transmitter, as defined after Eq. (3-11)
      C_r = coupling factor between antennas expressed as a power ratio
P_t and P_r are the powers of the transmitters connected to the *transmitting* and *receiving* antennas. The apparent VSWR, in the worst case, is given by $V = (1 + \rho')/(1 - \rho')$.

Radiation-Pattern Distortion   Some of the power transferred from the *transmitting* antenna to the *receiving* antenna will be reradiated and may cause distortion of the *transmitting* antenna's radiation pattern. The amount of power reradiated depends on the terminating impedance of the *receiving* transmitter, which is the value presented by its output stage. In the worst case this power can be up to 4 times the coupled power, which is the power radiated by the *transmitting* antenna multiplied by the coupling factor expressed as a power ratio. However, in practice the reradiated power will seldom be this large, and it is reasonable to assume that it will be equal to the coupled power. The reradiated energy changes the gain of the *transmitting* antenna by the following approximate amount

$$\Delta G = \pm 20 \log_{10} [1 + (C_r g_r/g_t)^{1/2}] \quad \text{dB} \qquad \textbf{(3-26)}$$

where g_r, g_t = maximum gains of *transmitting* and *receiving* antennas expressed as power ratios.

This formula gives only the minima and maxima of the pattern perturbation. The detailed shape of the perturbed pattern and the locations of the azimuth angles at which the minima and maxima occur depend on the phase of the reflected energy and the relative positions of the antennas; they are difficult to calculate approximately.

Side-by-Side Antennas Horizontally polarized antennas with directive radiation patterns can be placed side by side, often using common towers, with a coupling factor of less than -20 dB being achieved. Equations (3-23) and (3-24) do not apply in this situation because the antennas are too close. Coupling must be calculated by using a comprehensive antenna-analysis computer program.

REFERENCES

1 K. Davies, *Ionospheric Radio Propagation,* Dover Publications, Inc., New York, 1966.
2 *C.C.I.R. Interim Method for Estimating Sky-Wave Field Strength and Transmission Loss at Frequencies between the Approximate Limits of 2 and 30 MHz,* CCIR Rep. 252-2, International Telecommunications Union, Geneva, 1970.
3 G. W. Haydon, M. Leftin, and R. K. Rosich, *Predicting the Performance of High Frequency Skywave Telecommunications Systems,* Rep. OT 76-102, Office of Telecommunications, U.S. Department of Commerce, September 1976.
4 J. L. Lloyd, G. W. Haydon, D. L. Lucas, and L. R. Teters, *Estimating the Performance of Telecommunication Systems Using the Ionospheric Transmission Channel,* draft report, Institute for Telecommunications Sciences, U.S. Department of Commerce, March 1978.
5 E. A. Laport, *Radio Antenna Engineering,* McGraw-Hill Book Company, New York, 1952.
6 C. B. Feldman, "Deviations of Short Radio Waves from the London–New York Great Circle Path," *IRE Proc.,* vol. 27, October 1939, p. 635.
7 R. L. Tanner and M. G. Andreasen, "Numerical Solution of Electromagnetic Problems," *IEEE Spectrum,* September 1967.
8 R. F. Harrington, *Field Computation by Moment Methods,* The Macmillan Company, New York, 1968.
9 G. A. Thiele, "Wire Antennas," in *Computer Techniques for Electromagnetics,* Pergamon Press, New York, 1973, chap. 2.
10 W. L. Stutzman and G. A. Thiele, *Antenna Theory and Design,* John A. Wiley & Sons, Inc., New York, 1982, chap. 7.
11 E. K. Miller et al., "Analysis of Wire Antennas in the Presence of a Conducting Half-Space. Part II: The Horizontal Antenna in Free Space," *Can. J. Phys.,* vol. 50, 1972, pp. 2614–2627.
12 E. C. Jordan and K. G. Balmain, *Electromagnetic Waves and Radiating Systems,* 2d ed., Prentice-Hall, Inc., Englewood Cliffs, N.J., 1968, pp. 630–644.
13 Ibid., Chap. 15.
14 Stutzman and Thiele, op. cit., pp. 287–303.
15 U.S. Patent 3,257,661.
16 H. D. Kennedy, "A New Approach to Omnidirectional High-Gain Wide-Band Antenna Design: The TCI Model 540," Tech. Note 7, Technology for Communications International, April 1979.
17 U.S. Patent 3,594,807.
18 U.S. Patent 3,618,110.
19 J. R. Wait, "On the Radiation from a Vertical Dipole with an Inductive Wire-Grid Ground System," *IEEE Trans. Antennas Propagat.,* vol. AP-18, July 1980, pp. 558–560.
20 W. L. Weeks, *Antenna Engineering,* McGraw-Hill Book Company, New York, 1968, pp. 44–52.
21 Stutzman and Thiele, op. cit., pp. 40–44.

22 *World Distribution and Characteristics of Atmospheric Radio Noise,* CCIR Rep. 322, International Telecommunications Union, Geneva, 1964.

23 R.A. Sinaiti, "Active Antenna Performance Limitation," *IEEE Trans. Antennas Propagat.,* vol. AP-30, November 1982, pp. 1265–1267.

24 Comité Consultatif International des Radiocommunications, *Antenna Diagrams,* International Telecommunications Union, Geneva, 1978.

25 G. L. Matthei, L. Young, and E. M. T. Jones, *Microwave Filters, Impedance-Matching Networks, and Coupling Structures,* McGraw-Hill Book Company, New York, 1964, chap. 6.

26 Comité Consultatif des Radiocommunications, *Handbook on High-Frequency Directional Antennae,* International Telecommunications Union, Geneva, 1966.

27 W. F. Utlaut, "Siting Criteria for HF Communications Centers," Tech. Note 139, National Bureau of Standards, April 1962.

28 C. T. Tai, "Dipoles and Monopoles," in R. C. Johnson and H. Jasik (eds.), *Antenna Engineering Handbook,* McGraw-Hill Book Company, New York, 1984, Chapter 4.

29 R. H. DuHamel and G. G. Chadwick, "Frequency-Independent Antennas," in R. C. Johnson and H. Jasik (eds.), *Antenna Engineering Handbook,* McGraw-Hill Book Company, New York, 1984, Chapter 14.

The following figures, cited on pp. 3-5 and 3-15, are from Ref. 28.

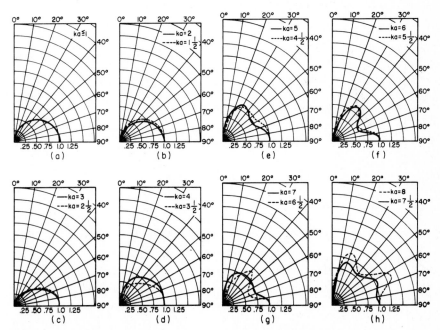

FIG. 4-14 Plots of the absolute values of the far-zone electric field as a function of the zenithal angle θ for various values of ka and with a flare angle equal to 60° ($\theta_0 = 30°$).

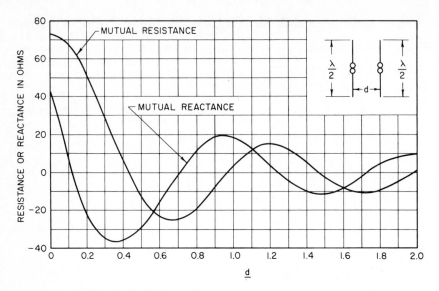

FIG. 4-28 Mutual impedance between two parallel half-wave antennas side by side.

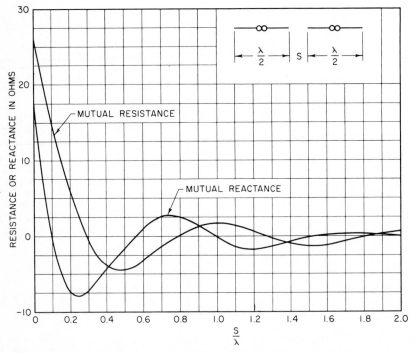

FIG. 4-29 Mutual impedance between two collinear half-wave antennas.

Chapter 4

VHF and UHF Communication Antennas

Brian S. Collins

C & S Antennas, Ltd.

4-1   INTRODUCTION

The very-high-frequency (VHF) and ultrahigh-frequency (UHF) bands are used for private and public-access services carrying speech, data, and facsimile information. The ends of a link may be installed at fixed locations or in vehicles (including ships and aircraft) or may be carried in an operator's hand. The length of a link may vary from a few tens of meters up to the maximum over which reliable communication can be obtained. This wide variety of applications generates a need for many different types of antennas. In this chapter, we examine the selection of antennas to perform various tasks, together with aspects of reliability, siting, and economics.

4-2   SYSTEM-PLANNING OBJECTIVES

The design of antennas for a radio link must provide an adequate signal-to-noise ratio at the receiver. The necessary signal-to-noise ratio will depend on the nature of the information to be transmitted and the grade of service which is required. A power budget must be drawn up to determine the total antenna gain and input power needed in the system. The antenna engineer must decide how the necessary gain can be obtained and how it should be divided between the two ends of the link. In many point-to-point applications the most economical design is obtained by using transmitting and receiving antennas of equal gain. Mobile or portable stations do not generally allow the use of high-gain antennas, so as much gain as possible must be obtained from the base-station antenna.

The electromagnetic spectrum is a limited resource, and its use is controlled by restricting the field strength which may be laid down outside the area where communication is required. This often implies a limitation on the permitted effective radiated power (ERP) both inside and outside the main-beam direction of the transmitting antenna. The sensitivity of receiving antennas must be restricted in directions outside the main beam to prevent interference being caused by the reception of signals from stations other than that intended which use the same or an immediately adjacent frequency. The system designer and antenna engineer must acquaint themselves with the requirements of the regulatory authority (Federal Communications Commission or other government agency) to make sure that a new system will work without suffering or causing interference.

To allow the largest possible number of links to be established in a given geographical area on a particular frequency band, it is desirable that each station use a very low transmitter power together with a highly directive high-gain antenna. By this means the area over which any station lays down a field which may cause interference to others is limited. A specification template for a radiation pattern defines the minimum acceptable radiation-pattern performance for an antenna and leaves scope for the designer to decide how to achieve it.

Reliability

A communication link or system must provide an adequate level of reliability. A link may become unusable if the signal-to-noise ratio falls below the design level; it is important that the design objectives for a system specify the fraction of time for which this may occur. A downtime of 0.01 percent or even less may be necessary for a link to a lifesaving emergency service, but 1 percent downtime may be as little as can be economically justified for a radiotelephone in a boat used for leisure-time fishing.

Fading due to statistical fluctuations in the propagation path is usually guarded against by a fade margin in the power budget. In a severe case a diversity system may be used to reduce the impact of fading on system reliability. This takes advantage of low correlation between fading events over two physically separate paths, at two frequencies, or for two polarizations. Other important causes of system failure are given below.

Wind-Induced Mechanical Failures The oscillating loads imposed by wind on antennas and their supporting structures cause countless failures. Aluminum and its alloys are very prone to fatigue failure, and the antenna engineer must be aware of this problem. To achieve real reliability:

1 Examine available wind-speed data for the location where the antenna is to be used.

2 Use derated permissible stress levels to allow for fatigue.

3 Check antenna designs for mechanical resonance.

4 Damp out, stiffen up, or guy parts of the antenna system which are prone to vibration or oscillation.

Reference 1 provides information on a wide range of the mechanical aspects of antenna design.

Corrosion The effects of corrosion and wind-induced stresses are synergistic, each making the other worse. They are almost always responsible for the eventual failure of any antenna system. Every antenna engineer should also be a corrosion engineer; it is always rewarding to examine old antennas to see which causes of corrosion could have been avoided by better design. The essence of good corrosion engineering is:

1 Selection of suitable alloys for outdoor exposure and choice of compatible materials when different metals or alloys are in contact. A contact potential of 0.25 V is the maximum permissible for long life in exposed conditions.

2 Specification of suitable protective processes—electroplating, painting, galvanizing, etc.

There is an enormous variety in the severity of the corrosion environment at different locations, ranging from dry, unpolluted rural areas to hot, humid coastal industrial complexes.

Plastics do not corrode, but they degrade by oxidation and the action of ultraviolet light. These effects are reduced by additives to the bulk materials. References 2 and 3 provide detailed information on corrosion mechanisms and control.

Ice and Snow The accumulation of ice and snow on an antenna causes an increase in the input voltage standing-wave ratio (VSWR) and a reduction in gain. The severity of these effects, caused by the capacitive loading of antenna elements and absorption of radio-frequency (RF) energy, increases as the frequency rises.

The fundamental design precaution is to ensure that the antenna and its mounting are strong enough to support the weight of snow and ice which will accumulate on them. This is very important, as even when the risk of a short loss of service due to the electrical effects of ice can be accepted, the collapse of even a part of the antenna is certainly unacceptable. Ice falling from the upper parts of a structure onto antennas below is a major cause of failure; safeguard against it by fitting lightweight antennas above more solidly constructed ones or by providing vulnerable antennas with shields to deflect falling ice.

In moderate conditions, antennas may be provided with radomes to cover the terminal regions of driven elements or even whole antennas. As conditions become more severe, heaters may be used to heat antenna elements or to prevent the buildup of ice and snow on the radome. A wide range of surface treatments has been tried to prevent the adhesion of ice; some of these show initial promise but become degraded and ineffective after a period of exposure to sunlight and surface pollution. Flexible radome membranes and nonrigid antenna elements have been used with some success.

Breakdown under Power   An inadequately designed antenna will fail by the overheating of conductors, dielectric heating, or tracking across insulators. The power rating of coaxial components may be determined from published data, but any newly designed antenna should be tested by a physical power test. An antenna under test should be expected to survive continuous operation at 1.5 times rated power and at 2 times rated peak voltage; for critical applications even larger factors of safety should be specified.

Lightning Damage   Antennas mounted on the highest point of a tower are particularly prone to lightning damage. The provision of a solid, low-inductance path for lightning currents in an antenna system reduces the probability of severe damage to the antenna. Electronic equipment is best protected by good antenna design and system grounding, supplemented by gas tubes connected across the feeder cables. Figure 4-1 shows a typical system with good grounding to prevent side-flash damage and danger to personnel. For detailed guidance see Refs. 4, 5, and 6.

Precipitation and Discharge Noise   This is caused when charged raindrops fall onto an antenna or when an antenna is exposed to an intense electric field in thunderstorm conditions. Precipitation noise can be troublesome at the lower end of the VHF band and may be experienced frequently in some locations. When problems arise, antenna elements may be fitted with insulating covers. These prevent the transfer of the charges from individual raindrops into the antenna circuit and reduce the energy coupled to the antenna when a charge passes between drops.

Choice of Polarization

Base stations for mobile services use vertical polarization because it is then simple to provide an omnidirectional antenna at both the mobile terminals and the base station. There is sometimes an advantage in using horizontal polarization for obstructed point-to-point links in hilly terrain, but the choice of polarization is often determined by the need to control cochannel interference. Orthogonal polarizations are chosen for antennas mounted close together in order to increase the isolation between them.

It has been found that the use of circular polarization (CP) reduces the effects of destructive interference by reflected multipath signals, so CP should be considered for any path where this problem is expected. CP has been used with success on a

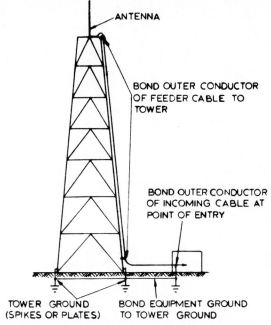

FIG. 4-1 Typical example of good grounding practice.
(© *C & S Antennas, Ltd.*)

number of long grazing-incidence oversea paths where problems with variable sea-surface reflections had been expected to be troublesome. Each end of a CP link must use antennas with the same sense of polarization.

Meeting Cost Objectives

The designer of a communications system must strive to provide the necessary overall performance for the lowest cost. A 100 percent reliability is often very difficult and costly to achieve and is only necessary for a small number of services. By comparison, 99 percent availability will entirely satisfy many users and can be provided much more readily; the user cannot justify the high cost of that extra 1 percent.

Cost-effective design is only obtained by:

1 Identifying the availability needed
2 Determining the environment at both ends of the link
3 Estimating the propagation characteristics of the path and judging the reliability of the estimates
4 Selecting the right equipment and antennas for the link to meet the communications and reliability objectives

Trade-Offs The interdependence of various parameters deserves careful consideration. For any major scheme the following checklist should always be worked through.

1 Examine the interactions of structure height, transmitter power, feeder attenuation, and antenna gain.

2 Consider using split antennas and duplicate feeders to increase reliability.

3 Consider the use of diversity techniques to achieve target availability instead of a single system with higher powers and gains.

4 Review the propagation data, especially the probability of multipath or cochannel interference. Don't engineer a system with 99.9 percent hardware availability and find 3 percent outage due to cochannel problems. Check the cost of antennas designed to reduce cochannel problems by nulling out the troublesome signals.

5 Visit the chosen sites. General wind data are useless if the tower is near a cliff edge, and a careful estimate of actual conditions must be made. Similarly, a nearby industrial area may mean a corrosive environment, and nearness to main roads indicates a high electrical noise level. Look for local physical obstructions in the propagation path.

5 Don't overdesign to cover ignorance. Find out!

4-3 ANTENNAS FOR POINT-TO-POINT LINKS

Yagi-Uda Antennas Yagi-Uda antennas are very widely used as general-purpose antennas at frequencies up to at least 900 MHz. They are cheap and simple to construct, have reasonable bandwidth, and will provide gains of up to about 17 dBi, or more if a multiple array is used. At low frequencies the gain which can be obtained is limited by the physical size of the antenna, but in the UHF band the limiting factor tends to be the accuracy with which the fed element and feeder system can be constructed.*

Yagi-Uda antennas provide unidirectional beams with moderately low side and rear lobes. The characteristics of the basic antenna can be modified in a variety of useful ways, some of which are shown in Fig. 4-2. The basic antenna (*a*) can be arrayed in linear or planar arrays (*b*). When the individual antennas are correctly spaced, an array of *N* antennas will have a power gain *N* times as large as that of a single antenna, less an allowance for distribution feeder losses. Table 4-1 indicates typical gains and arraying distances for Yagi-Uda antennas of various sizes. Different array spacings may be used when it is required to provide a deep null at a specified bearing, but the forward gain will be slightly reduced.

The bandwidth over which the front-to-back ratio is maintained may be increased by replacing a simple single reflector rod by two or three parallel rods (*c*). The back-to-front ratio of a simple Yagi-Uda antenna may be increased either by the addition of a screen (*d*) or by arraying two antennas with a quarter-wavelength axial displacement, providing a corresponding additional quarter wavelength of feeder cable to the forward antenna (*e*). A well-designed screen will provide a back-to-front ratio of as much as 40 dB, while that available from the quadrature-fed system is about 26 dB.

Circular polarization can be obtained by using crossed Yagi-Uda antennas: a

---

*For information on the design of Yagi-Uda antennas see Chap. 6. and Refs. 7, 8, 21, and 22.

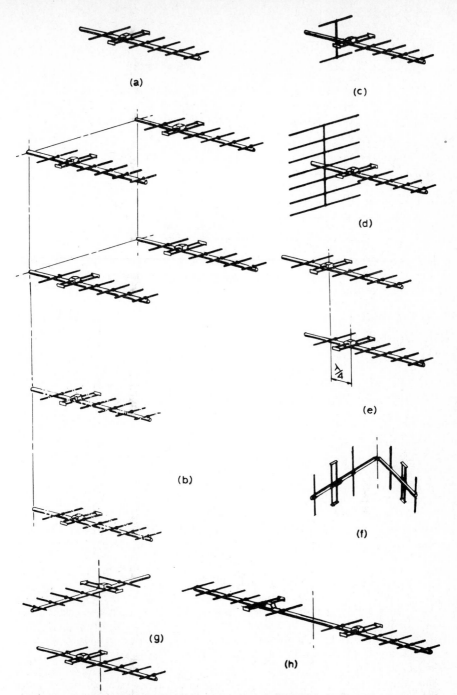

FIG. 4-2 Configurations of Yagi-Uda antennas. (*a*) Standard six-element antenna. (*b*) Stacked and bayed arrays. (*c*) Double reflector rods. (*d*) Reflector screen. (*e*) Increased *F/b* ratio by λ/4 offset. (*f*), (*g*), (*h*) Arrangements to produce aximuth radiation patterns for special applications. (© *C & S Antennas, Ltd.*)

TABLE 4-1 Typical Data for Yagi-Uda Antennas

Number of elements	Typical gain, dBi	Spacing for arraying, λ
3	7	0.7
4	9	1.0
6	10.5	1.25
8	12.5	1.63
12	14.5	1.8
15	15.5	1.9
18	16.5	2.0

pair of antennas mounted on a common boom with their elements set at right angles. The two antennas must radiate in phase quadrature, so they must be fed in quadrature or be fed in phase and mutually displaced by a quarter wavelength along the boom.

There has been an increase in interest in arrays derived from the Yagi-Uda antenna which use long, closed forms for their elements, for example, rings and squares.[9]

Log-Periodic Antennas These are widely used for applications in which a large frequency bandwidth is needed. The gain of a typical VHF or UHF log-periodic antenna is about 10 dBi, but larger gains can be obtained by arraying two or more antennas. The disadvantage of all log-periodic designs is the large physical size of an antenna with only modest gain. This is due to the fact that only a small part of the whole structure is active at any given frequency.

The most common type of design used on the VHF-UHF bands is the log-periodic dipole array (LPDA) described in Refs. 10 and 23. After selecting suitable values for the design ratio τ and apex angle α, the designer must decide on the compromises necessary to produce a practical antenna at reasonable cost. The theoretical ideal is for the cross-sectional dimensions of the elements and support booms to be scaled continuously along the array; in practice, the elements are made in groups by using standard tube sizes, and the support boom is often of uniform cross section. The stray capacitances and inductances associated with the feed region are troublesome, especially in the UHF band, and can be compensated only by experiment.

The coaxial feed cable is usually passed through one of the two support booms, thus avoiding the need for a wideband balun.

Printed-circuit techniques can be applied to LPDA design in the UHF band, as the antenna is easy to divide into two separate structures which may be etched onto two substrate surfaces. At the lower end of the VHF band the dipole elements may be constructed from flexible wires supported from an insulating catenary cord.

A typical well-designed octave-bandwidth LPDA has a VSWR less than 1.3:1 and a gain of 10 dBi.

Helices A long helical antenna has an easily predicted performance and is easy to construct and match. A VSWR as low as 1.2:1 can be obtained fairly easily over a frequency bandwidth of 20 percent. Conductive spacers may be used to support the helical element from the central support boom, so the antenna can be made very simple and robust.

Helices may be arrayed for increased gain; to obtain correct phasing the start position of each helix in the array must be the same. For further information on the design of helices see Refs. 11 and 23.

Panel Antennas   An antenna which comprises a reflecting screen with simple radiating elements mounted over it, in a broadside configuration, is generally termed a *panel antenna*. An array may comprise one or more panels connected together.

Typical panels use full-wavelength dipoles, half-wave dipoles, or slots as radiating elements (Fig. 4-3). They have several advantages over Yagi-Uda antennas:

1 More constant gain, radiation patterns, and VSWR over a wide bandwidth—up to an octave

2 More compact physical construction

3 Very low coupling to the mounting structure

4 Low side lobes and rear lobes

Panel antennas for frequencies in the UHF band lend themselves to printed-circuit design methods, as the radiating structures, feed lines, and matching system may all be produced by using stripline techniques. At lower frequencies the radiating elements are often mounted at voltage minimum points by using conducting supports, so a strong, rigid construction can be produced. A really solidly built but lightweight panel for a military application is shown in Fig. 4-4a. Here an all-welded aluminum frame and a skeleton-slot radiator are used so that the antenna will resist rough use in the field.

Multiple Arrays   Reference has been made to the use of a number of antennas arrayed together. Figure 4-5 shows a variety of simple cable harnesses which may be made from standard 50-Ω and 75-Ω coaxial cable. More complex harnesses may be constructed by combining several of the simple designs shown. For high-power applications rigid fabricated coaxial transformers are used. These may be designed to combine up to eight antenna feeds, and their impedance bandwidth may be increased by using two or three series quarter-wave matching sections designed to give Chebyshev or other chosen characteristics.[12]

Corner Reflectors   Well-designed corner-reflector antennas are capable of providing high gain and low sidelobe levels, but below 100 MHz they are mechanically cum-

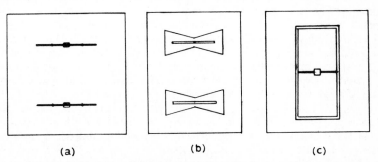

(a)   (b)   (c)

FIG. 4-3   Panel antennas. (*a*) Two full-wave dipole elements. (*b*) Two bat-wing slot elements. (*c*) Skeleton-slot elements. (© *C & S Antennas, Ltd.*)

FIG. 4-4 Robustly constructed antennas for military use. (*a*) Skeleton-slot-fed panel (225–400 MHz). (*b*) Grid paraboloid (610–1850 MHz).

bersome. Before using a corner reflector, make sure that the same amount of material could not be more effectively used to build a Yagi-Uda antenna, or perhaps a pair of them, to do the job even better.

In the UHF band, corner reflectors may be very simply constructed from solid or perforated sheet. The apex of the corner is sometimes modified to form a trough (Fig. 4-6). The provision of multiple dipoles extends the antenna aperture and increases the available gain.

Paraboloids   The design of a high-gain antenna may be reduced to a problem of illuminating the aperture necessary to develop the specified radiation patterns and gain. The size of the aperture is determined only by the gain required, whatever type of elements is used to fill it. As the cost of the feed system and the radiating elements doubles for each extra 3-dB gain, a stage is reached at which it becomes attractive to use a single radiating element to illuminate a reflector which occupies the whole of the necessary antenna aperture. The design task is reduced to choosing the size and shape of the reflector and specifying the radiation pattern of the illuminating antenna. If the antenna aperture is incompletely filled or its illumination is nonuniform, the gain which is realized decreases. The ratio of the achieved gain to the gain obtainable from the same aperture when it is uniformly illuminated by lossless elements is termed the *aperture efficiency* of the antenna. In a receiving context, this quantity represents the proportion of the power incident on the aperture which is delivered to a matched load at the terminals of the antenna.

In the VHF and UHF bands, a reflector may be made of solid sheet, perforated sheet, wire netting, or a series of parallel curved rods. As the wavelength is large, the mechanical tolerance of the reflector surface is not very demanding, and various methods of approximating the true surface required are possible. Table 4-2 indicates some of the combinations of techniques currently in use and illustrates the diversity of the methods which are successful for various purposes.

Grid paraboloids are attractive to produce because the curvature of all the rods

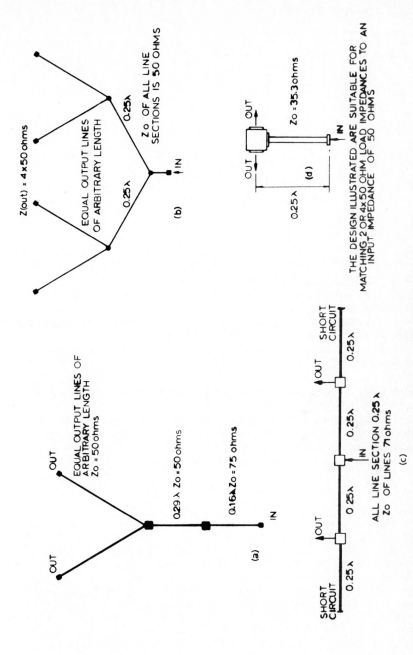

FIG. 4-5 Simple branching feeder systems. (*a*) Two-way. (*b*) Four-way. (*c*) Two-way, high-power. (*d*) Two-way, compensated. (© *C & S Antennas, Ltd.*)

Within figure (a):
OUT
OUT
EQUAL OUTPUT LINES OF ARBITRARY LENGTH
Zo = 50 ohms
0.29 λ Zo = 50 ohms
0.16 λ Zo = 75 ohms
IN

Within figure (b):
Z(out) = 4 x 50 ohms
EQUAL OUTPUT LINES OF ARBITRARY LENGTH
0.25λ
0.25λ
Zo OF ALL LINE SECTIONS IS 50 OHMS
IN

Within figure (c):
SHORT CIRCUIT
OUT
0.25 λ
0.25 λ
IN
0.25 λ
OUT
OUT
0.25 λ
SHORT CIRCUIT
ALL LINE SECTION 0.25 λ
Zo OF LINES 71 ohms

Within figure (d):
OUT
OUT
Zo = 35.3 ohms
IN
0.25 λ
THE DESIGN ILLUSTRATED ARE SUITABLE FOR MATCHING 2 OR 4x50 OHM LOAD IMPEDANCES TO AN INPUT IMPEDANCE OF 50 OHMS

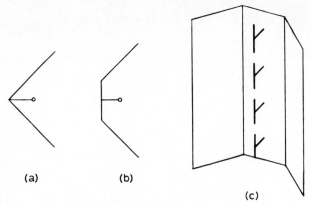

FIG. 4-6   Corner and trough reflectors. (© *C & S Antennas, Ltd.*)

is exactly the same; only their length varies across the antenna. A typical example is shown in Fig. 4-4*b*. The main deficiency of grid paraboloids is the leakage of energy through the surface, restricting the front-to-back ratio which can be achieved. For example, at 1500 MHz a front-to-back ratio of −30 dB is a typical limit. If a greater front-to-back ratio is needed, a continuous skin of solid or perforated sheet must be used and the consequent increase in wind-loaded area accepted as a necessary penalty. A mathematical treatment of grid reflectors appears in Ref. 13.

Radomes are frequently fitted to feeds or complete antennas in order to reduce the effects of wind and snow. They may be made from fiberglass or in the form of a tensioned membrane across the front of the antenna. In severe climates it is possible to heat a radome with a set of embedded wires, but this method can be applied only to a plane-polarized antenna.

Point-to-point links using tropospheric-scatter propagation require extremely high antenna gains and generally use a reflector which is an offset part of a full paraboloidal surface constructed from mesh or perforated sheet. Illumination is provided by a horn supported at the focal point by a separate tower.

TABLE 4-2   Typical Paraboloid-Antenna Configurations

Frequency, MHz	Diameter, m	f/d ratio	Construction	Feed type
200	10.0	0.5	Mesh paraboloid	4-element Yagi-Uda
700	3.0	0.25	Solid skin	Dipole and reflector
900	7.0	0.4	Perforated steel sheet	Horn
610–960	1.2	0.25	Grid of rods	Slot and reflector
1500	2.4	0.25	Solid skin	Dipole and disk
1350–1850	1.2	0.25	Grid of rods	Slot and reflector

4-4   BASE-STATION ANTENNAS

Simple Low-Gain Antennas   The simplest types of base-station antennas will provide truly omnidirectional azimuth coverage only when mounted in a clear position on top of a tower. Figure 4-7 shows standard configurations for ground-plane and coaxial dipole antennas and demonstrates that these forms are closely related. They are cheap and simple to construct and may be made to handle high power. Exact dimensions must be determined by experiment, as the stray inductance and capacitance associated with the feed-point insulator cannot be neglected. The use of a folded feed system can provide useful mechanical support and gives better control over the antenna impedance. The satisfactory operational bandwidth of the coaxial dipole d depends critically on the characteristic impedance Z_0, of the lower coaxial section formed by the feed line (radius r) inside the skirt (radius R). If this section has too small a Z_0, radiating currents will flow on the outside of the feeder line unless the skirt length is exactly $\lambda/4$. The impedance, gain, and radiation pattern of the antenna then becomes critically dependent on the positioning of the feed line on the tower, severely limiting the useful bandwidth of the antenna.

Discone Antennas   The discone and its variants are the most commonly used low-gain wideband base-station antennas. The useful lower frequency limit occurs when the cone is a little less than $\lambda/4$ high, but the upper frequency limit is determined almost entirely by the accuracy with which the conical geometry is maintained near the feed point at the apex of the cone.

Discones may be made with either the disk or the cone uppermost. The support for the upper part of the antenna usually takes the form of low-loss dielectric pillars or a thin-walled dielectric cylinder, fitted well outside the critical feed region.

Variants of the basic discone use biconical forms in place of the conventional cone and replace the disk by a cone with a large apex angle. At the lower end of the VHF band the antennas may be mounted at ground level, so a minimal skeleton disk which couples to the ground may be used if some loss of efficiency can be accepted.

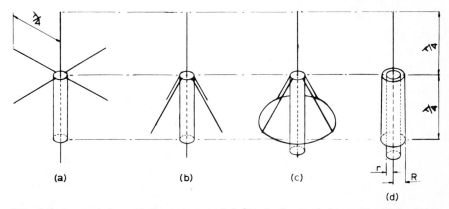

FIG. 4-7   Low-gain base-station antennas. (*a*) Standard ground plane with radials. (*b*) Ground plane with sloping radials. (*c*) Ground plane with closed ring. (*d*) Coaxial dipole. (© *C & S Antennas, Ltd.*)

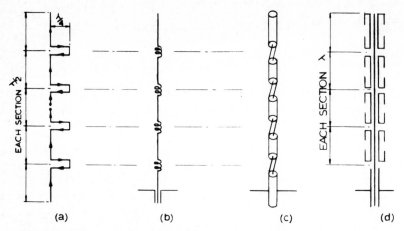

FIG. 4-8 Collinear dipole arrays. (*a*) Franklin array. (*b*) Array with meander-line phase reversal. (*c*) Array with transposed coaxial sections. (*d*) Alternative coaxial form. *(© C & S Antennas, Ltd.)*

Collinear Arrays   The ground plane and coaxial dipole have about the same gain as a half-wavelength dipole. When more gain is needed, the most popular omnidirectional antennas are simple collinear arrays of half-wave dipoles. The original array of this type is the Franklin array shown in Fig. 4-8*a*. This design is not very convenient owing to the phase-reversing stubs which project from the ends of each half-wave radiating section, but various derivatives are now widely used. The arrangement at *b* uses noninductive meander lines to provide phase reversal, while those at *c* and *d* use coaxial-line sections. Arrangements such as these may be mounted in fiberglass tubes to provide mechanical support. An input-matching section transforms the input impedance of the lower section, which may be $\lambda/2$ or $\lambda/4$ long, to 50 Ω. A set of quarter-wavelength radial elements or a quarter-wavelength choke is used to suppress currents on the outside of the feeder cable. The gain available from these arrays is limited by two factors:

1   There is mechanical instability in a very long antenna with a small vertical beamwidth.

2   The available excitation current diminishes away from the feed as a result of the power lost by radiation from the array.

The practical upper limit of useful gain is about 10 dBi.

   In the case of the coaxial-line designs, each section is shorter than a free-space half wavelength so that the correct phase shift is obtained inside the section. The examples shown would typically provide a gain of 9 dBi at the design frequency. The useful bandwidth of these antennas is inherently narrow because of the phase error between successive radiating sections which occurs when the frequency is changed from the design frequency. The typical behavior of the major lobes of the vertical radiation pattern of these arrays is shown in Fig. 4-9.

Dipoles on a Pole   Much ingenuity has been applied to the design of simple wideband high-gain omnidirectional antennas. A simple offset pole-mounted array is

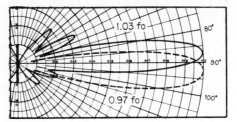

FIG. 4-9   Vertical radiation pattern of a typical end-fed collinear array.

shown in Fig. 4-10a. This will provide about 10-dBi gain in the forward direction but typically only 4 dBi rearward. An attempt to avoid this problem is shown in Fig. 4-10b, but this type of antenna has distorted vertical radiation patterns caused by the phase shifts which result from the displacement of the dipoles; gain is also reduced to about 6 dBi for the four-element array shown.

The solution in Fig. 4-10c, in which dipoles are placed in pairs and are cophased, is more satisfactory, as the phase center of each tier is concentric with the supporting pole. However, the antenna is relatively expensive, as eight dipoles provide only 6-dB gain over a single dipole.

One possibility is to use the in-line stacked array in Fig. 4-10a and place the base station toward the edge of the service area. The rearward illumination may be improved if the spacing between the dipoles and the pole is optimized for the pole size and frequency to be used.

Antennas on the Body of a Tower   Figure 4-11a shows a measured horizontal radiation pattern for a simple dipole mounted from one leg of a mast of 2-m face width. The distortion of the circular azimuthal pattern of the dipole is very typical

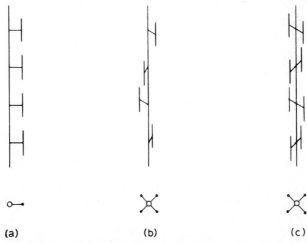

FIG. 4-10   Pole-mounted dipoles. (a) In line. (b) Four dipoles spaced around a pole. (c) Eight dipoles spaced around a pole. (© C & S Antennas, Ltd.)

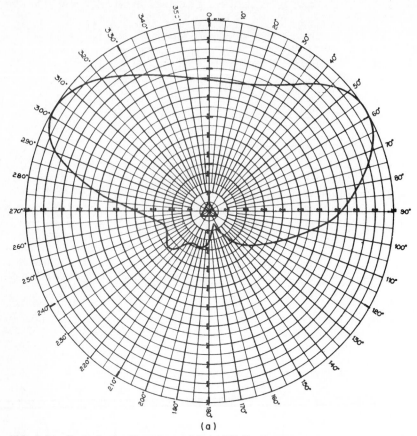

FIG. 4-11 Typical azimuth patterns of (*a*) VHF dipole mounted off one leg of a triangular mast and (*b*) three dipole panels mounted on the same structure.

and is caused by blocking and reflection from the structure. By contrast, Fig. 4-11*b* shows what can be achieved by an antenna comprising three dipole panels mounted on the same structure. The penalty of adopting this improved solution lies in the cost of the more complex antenna, so before an optimum design can be arrived at, the value of the improved service must be assessed.

The horizontal radiation pattern of a complete panel array is usually predicted from measured complex radiation-pattern data for a single panel, using a suitable computer program. For each azimuth bearing, the angle from each panel axis is found, and the relative field in that direction is obtained. The radiated phase is computed from the excitation phases and physical offsets of the phase centers of the individual panels.

Depending on the cross-sectional size of the structure, the most omnidirectional coverage may be produced with all panels driven with the same phase or by a phase rotation around the structure; for example, on a square tower the element current phases would then be 0, 90, 180, and 270°. When phase rotation is used, the individual

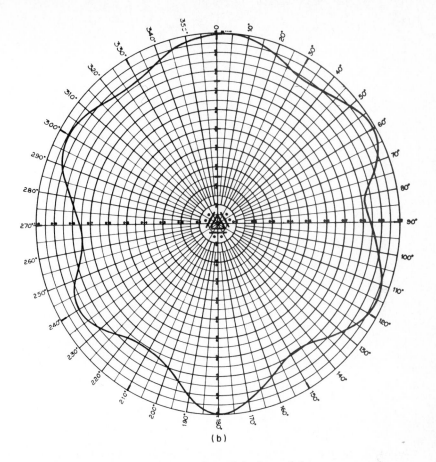

(b)

elements may be offset from the centerlines of the faces of the structure to give a more omnidirectional azimuth pattern, as in Fig. 4-12.

A well-designed panel array comprising four tiers, each of four panels, is an expensive installation, but if properly designed it can have a useful bandwidth of as much as 25 percent. This allows several user services to be combined into the same antenna, each user having access to a very omnidirectional high-gain antenna.

Special-Purpose Arrays   For applications in which the largest possible coverage must be obtained, the azimuth radiation pattern of the antenna must be shaped to concentrate the transmitted power in the area to be served, for example, an airway, harbor, or railroad track. Energy radiated in other directions is wasted and is a potential cause of interference to others.

Antennas with cardioidal azimuth radiation patterns are useful for a wide range of applications. Simple two-element arrays (dipole plus passive reflector) or dipoles mounted off the face of a tower may be adequate, but a wider range of patterns is available if two driven dipoles are mounted on a single supporting boom and excited with suitably chosen currents and phases. (See Ref. 21.)

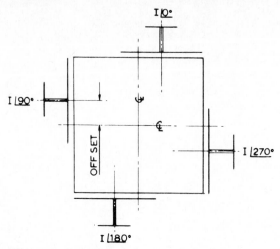

FIG. 4-12 Plan arrangement of an omnidirectional panel array on a large tower. (© C & S Antennas, Ltd.)

When a signal must be laid down over an arbitrarily shaped area of terrain, Yagi-Uda arrays may be arranged as at Fig. 4-2*f* and *g*. Due allowance must be made for the separation of the phase centers of the antennas when computing the radiation patterns. As an approximate guide, the phase center of a Yagi-Uda antenna lies one-third of the way along the director array, measured from the driven element.

Further tiers of antennas may be used to increase the gain of the system without modifying the azimuth radiation patterns.

When designing a complex array, estimation of gain can present a confusing problem. If an array contains sufficient elements to fill it, the gain of the array depends only on the size of the array aperture and not on the type of elements chosen. The gain of a filled array of identical tiers may be estimated by multiplying the vertical aperture power gain by the ratio of maximum power to mean power in the azimuth plane (the maximum-mean ratio). The vertical aperture gain may be assumed to be 1.15 times per wavelength of aperture relative to a half-wave dipole. The maximum-mean ratio may be obtained by integration of the azimuth radiation pattern, dividing the area of the pattern into the area of a circle which just encloses it (Fig. 4-13). Graphical integration is easily carried out by using a planimeter, but it is now quite simple to write integration routines for programmable pocket calculators. Once an array has been selected, its horizontal and vertical radiation patterns may be computed and used to predict the array gain more accurately.

4-5 MOBILE ANTENNAS

A road vehicle is not an ideal environment for an antenna. To make matters worse, the owner of a vehicle usually does not want antennas to be fixed in the most electrically favored positions like the center of the roof, but expects them to work when

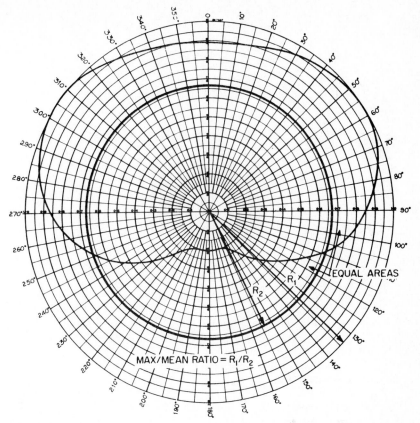

FIG. 4-13 The maximum-mean ratio of an azimuth radiation pattern.

mounted on gutters, fenders, or bumpers. The antenna is usually a severe compromise between what is ideally required and what is convenient. Figure 4-14 shows typical radiation patterns for a whip antenna measured with different mounting positions on a medium-sized automobile.

On both road vehicles and boats, antennas are subjected to severe mechanical shock, vibration, and exposure to all kinds of weather. Coaxial dipoles are widely used as VHF antennas on ships of all sizes. They can be encapsulated in a dielectric tube which protects the antenna from corrosion by seawater, and they have a vertical beamwidth which is large enough to avoid problems when the ship rolls. The most popular form of antenna for road vehicles is the simple base-fed whip. Quarter-wavelength whips require no loading, but they are inconveniently long at frequencies below about 100 MHz. Base or center loading can be used to shorten the physical length of a whip, but the efficiency of the antenna falls as the height becomes a smaller fraction of a wavelength. At higher frequencies it becomes possible to increase the gain of the antenna by using a five-eighths-wavelength whip, which has a small input coil to provide an input impedance suitably close to 50 Ω. Antennas with higher gain can be used

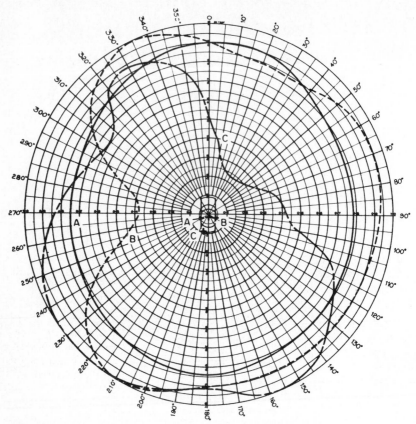

FIG. 4-14 Radiation patterns for a whip antenna on a typical automobile. Some of the data used in the preparation of this figure are from P. A. Ratliff, "VHF Mobile Radio Communications—A Study of Multipath Fading and Diversity Reception," Ph.D. thesis, University of Birmingham, 1974.

in the UHF band; they are typically short versions of the collinear arrays described in Sec. 4-4.

A variety of low-profile antennas are used on trains, buses, and security vehicles. These are usually derivatives of the inverted L or the annular slot (Fig. 4-15). Various discontinuities on a vehicle can be driven as slot radiators, although it is difficult to provide omnidirectional azimuth coverage. Antennas built into external mirrors or printed onto windows are used for applications when no conspicuous antennas must be carried.

During recent years much attention has been given to antennas which use active devices for matching or modification of radiating currents. As some of these antennas are physically small in terms of a wavelength, they are of particular interest for mobile use. Multielement antennas which provide steered beams or nulls offer promising lines of development, especially if control of the antenna is adaptively managed to optimize the received signal. Such techniques will become increasingly important with the growth of data links to vehicles.

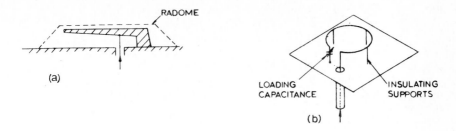

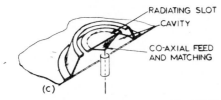

FIG. 4-15 Low-profile antennas. (*a*) Inverted L. (*b*) Hula hoop. (*c*) Annular slot. (© *C & S Antennas, Ltd.*)

Antennas for personal hand-held transceivers usually take the form of rigid telescopic whips, wires positioned in carrying straps, or short semiflexible whips. The helical whip is increasingly preferred, as rigid antennas are inconvenient and the performance of carrying-strap antennas varies greatly with their position. When designing these systems, the developing literature on biological hazards should be consulted.[14,15]

4-6 SYSTEM CONSIDERATIONS

Mounting Arrangements When mounting any antenna, it is important not to impair its performance by the influence of the supporting structure. The inevitable effect of the supporting structure on the radiation pattern of a dipole has been referred to in Sec. 4-4. This effect is accompanied by a modification of the input impedance, which may be unwelcome if a low VSWR is needed. In any critical application the change of the radiation patterns and gain must be taken into account when estimating system performance. Impedance matching of the antenna must be undertaken in the final mounting position or a simulation of it.

If Yagi-Uda arrays are mounted with their elements close to a conducting structure, they too will suffer changes of radiation patterns and impedance. The effects will be greatest if members pass through the antenna, as they do when an array is mounted on clamps fitted at the center of the cross boom. If at all possible, when an array is center-mounted, the member to which it is clamped should lie at right angles to the elements of the array.

Currents induced in diagonal members of the supporting structure will cause reradiation in polarization planes other than that intended. This will result in the cross-polar discrimination of the antenna system being reduced from that which would be measured on an isolated antenna at a test range. When polarization protection is important, the tower should be screened from the field radiated by the antenna with a cage of bars spaced not more than $\lambda/10$ apart, lying in the plane of polarization. (A square mesh is used for circular polarization.) Panel antennas are designed with an integral screen to reduce coupling to the mounting structure.

Long end-mounted antennas are subjected to large bending forces and turning moments at their support points. These forces can be reduced by staying the antenna, using nylon or polyester ropes for the purpose to avoid degrading its electrical characteristics.

In severe environments antennas may be provided with radomes or protective paints. It is very important that the antennas be tested and set up with these measures already applied, especially if the operating frequency is in the UHF band.

Coupling A further consideration when planning a new antenna installation on an existing structure is the coupling which will exist between different antennas. When a transmitting antenna is mounted close to a receiving antenna, problems which can arise include:

Radiation of spurious signals (including broadband noise) from the transmitter

Blocking or desensitization of the receiver

Generation of cross-modulation effects by the receiver

Radiation of spurious signals (intermodulation products and harmonics) due to non-linear connections in the transmitting antenna or generation of spurious signals and cross-modulation effects caused by the same mechanism in the receiving antenna

The last of these problems must be avoided by good antenna design—avoiding any rubbing, unbonded joints. The other effects depend critically on the isolation between the antennas and on parameters of the transmitters and receivers; these parameters should be specified by their manufacturers.

The isolation between two antennas may be predicted from Fig. 4-16 or from standard propagation formulas. Antenna isolations may be increased by using larger spacings between them or by using arrays of two or more antennas spaced to provide each with a radiation-pattern null in the direction of the other.

An alternative method of increasing the isolation between the antenna-system inputs is to insert filters. If a suitable filter can be constructed, the antenna isolation may be reduced until, in the limit, a single antenna is used with all equipment, transmitters and receivers, coupled to it through filters. When receivers are connected to a common antenna, the signal from the antenna is usually amplified before being divided by a hybrid network. For information on filter design, consult Refs. 17 and 18. The number of services which use a single antenna can be extended to six or more, provided adequate spacings are maintained between the frequencies allocated to different users. The whole system is expensive, but the cost may be justified if the antenna itself is large or if tower space is limited.

Coverage In free-space conditions the intensity of a radio wave diminishes as the distance from the transmitter increases in accordance with the inverse-square law. Terrestrial links are not usually in free-space conditions, and the field diminishes more rapidly with distance. Reference 16 provides a large variety of basic data and curves for planning point-to-point, aeromobile, and other services. References 19 and 20 pro-

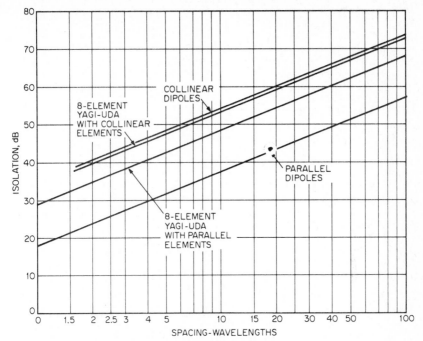

FIG. 4-16   Typical isolations between Yagi-Uda antennas.

vide further examples of the use of the data and also provide additional information on the methods to be adopted for dealing with obstructed paths.

REFERENCES

1 C. J. Richards (ed.), *Mechanical Engineering in Radar and Communications,* Van Nostrand Reinhold Company, New York, 1969.
2 V. R. Pludek, *Design and Corrosion Control,* The Macmillan Company, New York, 1977.
3 H. H. Uhlig, *Corrosion and Corrosion Control,* 2d ed., Interscience Publishers, a division of John Wiley & Sons, Inc., New York, 1971.
4 *Lightning Protection Code,* ANSI C5.1-1969, American National Standards Institute, New York, 1969.
5 *The Protection of Structures against Lightning,* British Standard Code of Practice CP 326:1965, British Standards Institution, London, 1965.
6 *Manual on Lightning Protection,* Australian Standard 1768–1975, Standards Association of Australia, Sydney, 1975.
7 H. W. Ehrenspeck and H. Poehler, "A New Method for Obtaining Maximum Gain from Yagi Antennas," *IRE Trans. Antennas Propagat.,* vol. AP-7, October 1959, pp. 379–385.
8 B. M. Thomas, "The Precise Mechanism of Radiation from Surface Wave Aerials," *J. Inst. Eng. Aust.,* September 1964, pp. 225–238.
9 M. Kosugi, N. Inasaki, and T. Sekiguchi, "Design of an Array of Circular-Loop Antennas with Optimum Directivity," *Electron. Commun. Japan,* vol. 54-B, no. 5, 1971, pp. 67–76.

10 H. V. Rumsey, *Frequency Independent Antennas,* Academic Press, Inc., New York, 1966.

11 J. D. Kraus, *Antennas,* McGraw-Hill Book Company, New York, 1950.

12 H. J. Riblet, "General Synthesis of Quarter-Wave Impedance Transformers," *IRE Trans. Microwave Theory Tech.,* vol. MTT-5, January 1957, pp. 36–43.

13 R. Neri and T. S. M. Maclean, "Receiving and Transmitting Properties of Small Grid Paraboloids by Moment Method," *IEE Proc.,* vol. 126, no. 12, December 1979, pp. 1209–1219.

14 *Safety Levels of Electromagnetic Radiation with Respect to Personnel,* C95, United States of America Standards Institute, New York, 1966.

15 J. R. Swanson, V. E. Rose, and C. H. Powell, "A Review of International Microwave Exposure Guides," in *Electronic Product Radiation and the Health Physicist,* Pub. BRH/DEP70-26, U.S. Department of Health, Education, and Welfare, Washington, 1970, pp. 95–110.

16 Texts of Thirteenth Plenary Assembly, Comité Consultatif International des Radiocommunications, Geneva, 1975.

17 A. I. Zverev, *Handbook of Filter Synthesis,* John Wiley & Sons, Inc., New York, 1967.

18 G. L. Matthaei, L. Young and E. Jones, *Microwave Filters, Impedance Matching Networks and Coupling Structures,* McGraw-Hill Book Company, New York, 1964.

19 A. Picquenard, *Radio Wave Propagation,* Philips Technical Library, John Wiley & Sons, Inc., New York, 1974.

20 M. P. M. Hall, *Effects of the Troposphere on Radio Communication,* Peter Peregrinus Ltd., London, 1979.

21 M. T. Ma, "Arrays of Discrete Elements," in R. C. Johnson and H. Jasik (eds.), *Antenna Engineering Handbook,* McGraw-Hill Book Company, New York, 1984, Chap. 3.

22 F. J. Zucker, "Surface-Wave Antennas and Surface-Wave-Excited Arrays," in R. C. Johnson and Jasik (eds.), *Antenna Engineering Handbook,* McGraw-Hill Book Company, New York, 1984, Chap. 12.

23 H. E. King and J. L. Wong, "Helical Antennas," in R. C. Johnson and H. Jasik (eds.), *Antenna Engineering Handbook,* McGraw-Hill Book Company, New York, 1984, Chap. 13.

Chapter 5

TV and FM Transmitting Antennas

Raymond H. DuHamel

Antenna Consultant

Television broadcast services are located within four bands: the lower very-high-frequency (VHF) bands of 54 to 72 and 76 to 88 MHz, the upper VHF band of 174 to 216 MHz, and the ultrahigh-frequency (UHF) band of 470 to 890 MHz. The FM band is 88 to 108 MHz. The bandwidths of the TV and FM channels are 6 and 0.2 MHz respectively. A stringent requirement for broadcast antennas is that the voltage standing-wave ratio (VSWR) should be less than 1.1:1 over the band. The bandwidth is 0.2 percent for FM and varies from 10 to about 1 percent for TV. If several channels are multiplexed into a single antenna, achieving the required VSWR is even more difficult. In many cases, another stringent requirement is that the antenna be capable of handling power inputs of 50 kW or more.

The *Technical Standards* of the Federal Communications Commission (FCC) specify the maximum effective radiated power (ERP) which TV and FM stations can radiate. The maximum power[1,2] varies with regions or zones of the United States and with the height of the antenna above ground. The height is not specified, but large heights are desired to increase coverage. Heights of 1000 to 2000 ft (305 to 610 m) are commonly used. The antennas are usually supported on guyed towers or tall buildings.

The majority of station requirements call for omnidirectional azimuth patterns. The circularity of the pattern depends upon the type of antenna when top-mounted and also on the configuration of the support structure when side-mounted. Other requirements call for various types of azimuth patterns, such as cardioid, skull-shape, peanut-shape, etc., to protect other stations or reduce radiation in low-population areas. It is desirable to have a large antenna gain so as to reduce the required transmitter power. Since the gain of the azimuth pattern is low, it is necessary to use a large vertical aperture to increase the gain. For antennas mounted at large heights, the vertical beamwidth should not be less than about 1°. This implies a vertical aperture of about 50 wavelengths, which is practical only in the UHF band. Wind loading and cost considerations limit the vertical aperture to about 120 ft (36.6 m) for FM and the other TV bands. Thus, the minimum elevation beamwidths are about 7, 4, and 2° for the lower VHF, FM, and upper VHF bands respectively. For these narrow beamwidths and certain combinations of terrain and antenna heights, it may be desirable to tilt the beam and/or fill in several nulls below the beam. Null fill is desired when the antenna is located near a residential area.

The antenna structure usually consists of a vertical array of radiating elements, such as dipoles, loops, and slots, or of one or more bays of vertical traveling-wave antennas such as helices, zigzags, and rings. It is desirable to mount the antennas on top of the tower or support structure, but in many cases it is necessary to side-mount them. Side-mounted antennas present an interesting challenge to the antenna designer to control the azimuth pattern. The elements of a vertical array are commonly referred to as *bays*. The array-feed techniques are conventional, such as a corporate feed or an end feed in a matched or resonant manner.

Beam tilt is achieved by phasing the bays of the vertical array and/or by mechanically tilting the array. Combinations of electrical and mechanical beam tilt may be used to achieve a beam tilt which varies with azimuth direction for special applications.

One simple technique for null fill is to shift the phase of one or more elements in the center of the array. The currents in these elements may be considered as the superposition of in-phase and quadrature-phase currents with respect to the rest of the array. The quadrature-phase currents produce the null fill. Another technique is to increase and decrease the current in upper and lower elements by the same amount. By using superposition again, the radiation from the difference component is in quadrature with the sum component and produces null fill. A combination of these two techniques may be used to produce null fill only below the horizon. An exponential aperture distribution in an end-fed antenna also produces null fill. Null fills of 5 percent and 20 percent reduce the gain of the array by about 0.2 and 0.6 dB respectively.

Originally, the *Technical Standards* of the FCC specified that TV and FM stations radiate horizontal polarization. Then, in the 1960s, the FCC permitted FM stations to use circular polarization. This provided improved reception, especially for vehicles with whip antennas, which are predominantly vertically polarized. In 1977 the FCC allowed TV stations to radiate right-hand circular polarization and to use the maximum ERP for horizontal polarization so as to maintain the field strength existing before conversion to circular polarization. Thus, the same ERP can be used for vertical polarization, which means doubling transmitter power. This has provided greatly improved reception for receivers with indoor antennas such as monopoles and rabbit ears. In addition, circularly polarized receiving antennas can be used to reduce "ghosting," since reflections from buildings and other objects tend to have the opposite sense of circular polarization.

With the exception of bandwidth, the requirements for TV and FM broadcast are very similar. In multiple-station FM antennas, there is little difference. Thus, in the following discussion antennas will be classified as circularly and horizontally polarized antennas.

The gain of TV and FM antennas is specified with respect to a horizontally polarized half-wave dipole, which has a gain of 1.65 with respect to an isotropic antenna. The gain of omnidirectional horizontally polarized antennas varies from about 0.9 to 1.1 times the vertical aperture in wavelengths. The gain for circularly polarized antennas is one-half of this.

5-2   PANEL-TYPE ANTENNAS

In many cases, antenna elements must be placed near the sides of a triangular or square tower because of physical restrictions. Reflecting panels are placed on the faces of the tower to prevent reflections from tower members, which lead to erratic azimuth patterns. With proper design, each panel has a unidirectional pattern with the beam direction normal to the panel. The 6-dB beamwidth should be 90° for square towers when an omnidirectional pattern is desired, since the gain should be down 6 dB in the crossover direction. For a triangular tower, a 120° 6-dB beamwidth is desired. By proper design of the antenna element and/or the addition of parasitic elements, reflecting sheets, or cavities, it is possible to vary the 6-dB beamwidth over the range of about 80 to 130°. (This is discussed in later sections.)

With equal power fed to the sides of the tower and with the antenna elements placed in the center of the sides as shown in Fig. 5-1, an omnidirectional-type pattern is obtained with a maximum-minimum ratio that increases with the face width of the tower. The short lines represent radiating elements such as dipoles, rings, and zigzags.

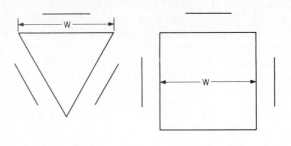

FIG. 5-1 Panel-type antennas for triangular and square towers.

Figure 5-2 shows this ratio for both square and triangular towers for optimum panel beamwidths and in-phase excitation of the panels. For good omnidirectional patterns, tower width should not be much greater than one wavelength. The null directions occur on each side of the crossover directions where the radiation from adjacent panels is not phased properly.

The elements may be displaced as shown in Fig. 5-3 to the point where quadrature phasing of adjacent panels is required to achieve in-phase radiation in the crossover direction. This requires different line length from the power splitter to the panels and provides reflection cancellation at the splitter, which in turn allows a much lower VSWR at the splitter than at the panels. An adverse effect is that the waves reflected back toward the elements distort the excitation of the elements and therefore the azimuth pattern. Azimuth patterns are about the same as for center-placed elements. The same technique may be used for triangular towers with a 120° differential phasing.

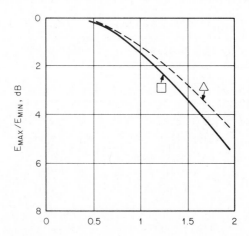

FIG. 5-2 Maximum-minimum ratio versus tower width in wavelengths for triangular and square towers.

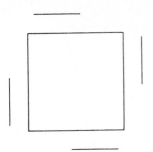

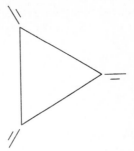

FIG. 5-3 Offset panel-type radiators on a square tower for reflection cancellation.

FIG. 5-4 Skewed panel-type radiators for triangular towers.

Skewed-panel antennas[3] may be placed on the corners of large-width towers (such as those measuring five wavelengths), as shown in Fig. 5-4, in which the antenna element is placed on a narrow panel. The panels are skewed so that the crossover direction occurs in the line of the tower face. Thus, the relative phase of the radiation from adjacent antennas varies more slowly near the crossover direction than in a non-skew arrangement. Theoretically, this greatly improves the azimuth pattern. However, reflections from the tower members and backs of the panels degrade the patterns. For large vertical arrays, it is difficult to achieve better than 6-dB nulls.

Panel-type antennas are also very useful for directional applications. A wide variety of azimuth patterns may be achieved by controlling the magnitude and phase of excitation of the panels around the tower and the orientation of the panels.

5-3 CIRCULARLY POLARIZED ANTENNAS

The FCC allowance of circularly polarized broadcast transmission has led to the introduction of a wide variety of new transmitting antennas. The antenna types usually take the form of crossed dipoles, circular arrays of slanted dipoles, helical structures, and traveling-wave ring configurations.

Helix Antennas

The multiarm helix[4,5] is a versatile antenna for radiating circularly polarized waves. Figure 5-5 shows a three-arm helix with a pitch angle ψ wrapped around a conducting cylinder, which forms the support for the antenna and allows space for a transmission-line feed network for several bays of helices. For broadside radiation the turn length of an arm is equal to M wavelengths, where M is an integer and defined as the mode number. For an N-arm helix the arms are fed with equal powers and a phase progression of 360 x $N/M°$ such that the currents in the arms along a directrix of the helical cylinder are in phase. To obtain a low axial ratio and satisfactory radiation patterns, the number of arms N should be larger than the mode number M by a factor in the range of 1.5 to 2.0.

The left-hand helix of Fig. 5-5 radiates left-handed circular polarization toward the zenith, right-handed circular polarization to the nadir, and horizontal polarization at an elevation angle of approximately $\psi°$. Figure 5-6 shows the variation of the axial ratio for right-handed elliptical polarization in the broadside direction with the pitch angle ψ. It is seen that the average pitch angle should be about 40° to achieve an axial ratio of less than 3 dB. The calculation of this curve was based on a sheath model of the helical currents and neglects the effect of the cylinder. If the cylinder circumference in wavelengths is greater than about $M - 1$ for M greater than 1, then reflections from the cylinder produce a phase error between the vertical and horizontal polarizations which degrades the axial ratio by more than that shown in Fig. 5-6. For mode 1, the cylinder circumference should be less than one-half wavelength.

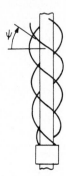

FIG. 5-5   Three-arm helical antenna.

The uniform helix is a traveling-wave antenna with an exponential attenuation rate which is a complex function of M, N, ψ, cylinder diameter, and arm diameter. Attenuation increases with arm diameter and decreases with pitch angle. Attenuation rates of up to 6 dB per axial wavelength may be achieved. To approximate a uniformly illuminated aperture, the pitch angle may be varied along the aperture (keeping the turn length constant), which leads to a spiral-type structure. If $2\,M/N$ is not an integer, the reflected wave from the end of the helix will radiate a beam in an elevation direction other than broadside. If this is not the case, the reflected wave will radiate a broadside beam which produces scallops in the azimuth pattern with $2M$ lobes. The axial ratio is not degraded because the sense of circular polarization for the reflected wave is the same as that for the incident wave in the broadside direction. This effect may be reduced by terminating the helix with radiating loads or resistors. The helix is usually designed so that one-way attenuation is about 16 dB.

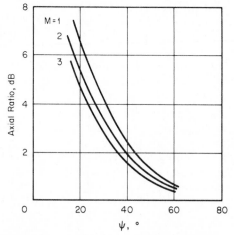

FIG. 5-6   Axial ratio of helical antennas in broadside direction versus pitch angle.

Because the helix is a traveling-wave antenna, the impedance bandwidth is large, especially if 2 M/N is not an integer, since reflection cancellation occurs at the input to the feed network. However, the pattern bandwidth is limited by beam scan with frequency since it is equivalent to an end-fed array. For desirable pitch angles, the beam of a helix bay scans about 1° for a 1 percent frequency change. For Channels 2 to 6, this limits the bay length to about two to three wavelengths. Thus, two or more bays are generally used. A three-arm mode 1 helix may be used for these channels and has less wind loading than a horizontally polarized batwing antenna. For Channels 7 to 13, bay lengths of about six wavelengths may be used. Three- or four-arm mode 2 helices are used. In the UHF band, bay lengths may be in the range of 16 wavelengths, and mode numbers of five or more are used, with the number of arms being greater than the mode number.

Because of their symmetry, helical and spiral antennas have an excellent omni-directional pattern, with a circularity of less than ±1 dB. The axial ratio is about 2 dB for the low VHF channels and even less for the other channels. The arms of a helical antenna have a characteristic impedance similar to that of a rod over a ground plane, with the height equal to the spacing of the arm from the support cylinder. Thus, the impedance is in the order of several hundred ohms for practical arm diameters. Special techniques, such as inductance-capacitance tuners or transformers, are required to match this impedance to the outputs of the power splitter in the feed network for the multiarm helix.

A novel feature of the higher-order-mode multiple-arm helical antenna is that it may be placed around triangular or square towers and still produce an omnidirectional pattern. This occurs because the waves radiated toward the support structure with a M x 360° azimuth phase variation enter a cutoff region in a manner similar to that for radial waveguides. Thus, the waves are reflected, and the support structure has little effect on the radiation pattern if the mode number is about 5 times larger than the tower diameter in wavelengths.

Slanted-Dipole Antennas

Many circularly polarized FM and TV broadcast antennas are based on the concept of a circular array of slanted dipoles. The dipoles may be linear, V-shaped, curved, or of a similar configuration. Each dipole radiates linear polarization, but the slant angles and diameter of the circular array are adjusted so that an omnidirectional circularly polarized radiation pattern is obtained. The term *circular array* is used here to include one or more dipoles placed on a circle with rotational symmetry. Actually, the array of slanted dipoles may be considered as a short length of an N-arm helical antenna with a standing-wave rather than a traveling-wave current distribution. The cross section of the helical structure need not be circular to produce circular polarization. This approach is in contrast to that in which unidirectional circularly polarized radiators (such as crossed dipoles placed in front of a reflecting screen) are placed around a mast or tower to obtain omnidirectional or other azimuth patterns.

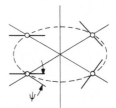

FIG. 5-7 Circular array of four slanted dipoles.

Figure 5-7 shows a circular array of four slanted dipoles[6] with a slant angle ψ (which is similar to the pitch angle of a helical antenna). If the dipoles are assumed to be fed in phase, the following is a simplified explanation of how circular polarization is achieved. In a direction in line with opposite dipoles, the phase of the radiation from the vertical and horizontal components of the current in the front dipole lead those of the rear. The vertically polarized components are vectorally added, whereas the horizontal components are subtracted, which produces a 90° phase difference between the two polarizations. The slant angle is adjusted to produce equal magnitudes of vertical and horizontal polarization, taking into account radiation from the other two dipoles, which results in circular polarization. To generalize, N slanted dipoles may be placed in a circular array and excited in mode M (an integer) to radiate omnidirectional circular polarization[7,8] in the plane of the array where the current in the nth dipole is given by

$$I_n = \exp(j2\pi nM/N) \qquad \textbf{(5-1)}$$

The circumferential spacing of the dipoles must be about one-half wavelength for mode 0 and less than that for other modes to obtain omnidirectional patterns and circular polarization. The pitch angle ψ, which depends on the shape of the dipole, is approximately

$$\psi = -\tan^{-1}\left(\frac{J'_M(\beta\rho)}{J_M(\beta\rho)}\right) \qquad \textbf{(5-2)}$$

for curved dipoles lying on the surface of a circular cylinder, where ρ is the radius of the circular array, $J_M(\beta\rho)$ is a Bessel function, and the prime represents the derivative. The effect of a vertical mast which introduces a phase error between the vertical and horizontal polarization components is not included Eq. (5-2). The slant angle for right-hand circular polarization is positive for mode 0 and negative for the other modes when the array circumference is equal to or less than M wavelengths.

Figure 5-8 illustrates a simple, compact version of this concept in which two V dipoles[9] are supported by a horizontal mast. With an included angle of about 90°, the V dipoles perform approximately as a four-dipole circular array. One-half of each dipole is shunt-excited from the center of the support mast. If the dipole length is about one-half wavelength, then the current on the parasitic arms will be about the same as the current on the shunt-driven arms of the dipoles. The antenna is matched by adjusting the positions of the shunt feeds and dipole lengths. Alternatively, one-half of each dipole may be series-fed as illustrated in Fig. 5-9, in which the two dipoles are supported in a T arrangement. The internal coaxial feeds are connected to gaps in the monopoles. Since the impedance bandwidth is on the order of 1 percent or less, these antennas are most useful for FM applications. A multiplicity of bays with wavelength spacing may be fed by and supported by a vertical transmission line.

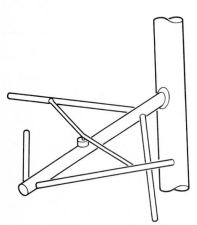

FIG. 5-8 Two shunt-fed slanted V-dipole antennas.

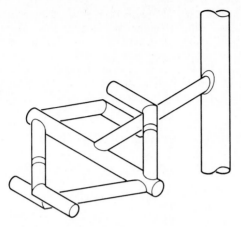

FIG. 5-9 Series-fed slanted dipoles.

The array is usually supported on the side of a mast or tower. Reflections from the support distort the azimuth patterns, especially for vertical polarization, and degrade the axial ratio. Parasitic dipoles may be added to reduce these effects.

A circular array of four curved dipoles[10] is shown in Fig. 5-10. The dipoles form a short section of a four-arm helix antenna and are approximately one-half wavelength long. The array circumference is approximately one wavelength. Thus, the overlap of the dipoles provides approximately the equivalent of a constant circular current distribution for both the horizontal and the vertical components. The four dipoles are shunt-fed asymmetrically by four rods emanating from the center of the array. The rods are connected to the center conductor of a coaxial feed enclosed in the horizontal support structure. The same approach may be used for circular arrays of two or three curved dipoles when the array circumference is approximately one-half and three-fourths of a wavelength respectively. The pattern circularity in free space is ± 1 dB,

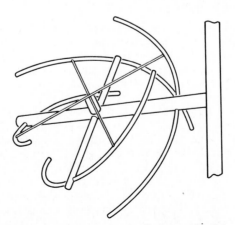

FIG. 5-10 Four shunt-fed helical-type dipoles.

and the axial ratio is about 3 dB. The support mast or tower degrades these values by several decibels. The power rating and bandwidth increase with the number of dipoles in the array. An 11 percent bandwidth has been achieved for a four-element FM array with 2-in-diameter arms.

A single dipole may be bent in the form of a one-turn helical antenna to produce circular polarization. It may be fed by a slotted coaxial-line balun[11] or inductive loop coupling. A disadvantage is that it radiates up and down so that bay spacings of less than one wavelength should be used. For two or more dipoles in each bay there are nulls up and down.

The degradation of azimuth patterns and axial ratio for side-mounted antennas may be eliminated by placing the circular array of slanted dipoles around the mast and adding shunt dipoles[8,12] to compensate for the vertical currents flowing on the

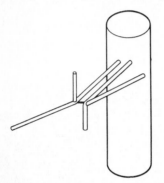

FIG. 5-11 Slant dipole with a parallel-connected short vertical dipole.

mast. Figure 5-11 shows one element of a circular array which consists of a slant half-wave dipole and a short vertical dipole that are fed and supported by a balun structure. For mode 0 and an array diameter of about one-half wavelength, only three elements are needed in the circular array to produce a circularity of $\pm$ 1.5 dB and an axial ratio of less than 3 dB. The elements are fed in phase with equal power. The slant angle is approximately that given by Eq. (5-2) (without mast reflections), and the length of the vertical shunt dipole is adjusted to achieve a low axial ratio. A bandwidth greater than 10 percent may be achieved with a bay spacing of 0.8 wavelength. Thus, the antenna may be used for both TV and FM applications. Since the wind loading is equal to or less than that for the batwing antenna, it may be used to replace the batwing on existing towers for conversion to circular polarization.

Crossed-Dipole Antennas

A common technique for producing circular polarization has been to place two linear dipoles at right angles in front of a reflecting screen and to feed them with equal voltage magnitudes and a 90° phase difference. However, the azimuth beamwidth for horizontal and vertical polarization is about 60° and 120° respectively. Thus, the axial ratio is low only for directions near the normal to the screen. This deficiency may be corrected in several ways.

V dipoles as illustrated in Fig. 5-12 may be used to increase the azimuth beamwidth for horizontal polarization. Three crossed V-dipole panels may be placed around a triangular tower to obtain a circularity of ± 2 dB and a maximum axial ratio of 4 dB. The crossed dipoles may be identical and fed in phase quadrature or be unequal in length and fed in phase with the lengths adjusted to produce quadrature currents in the dipoles. Another version of this type of antenna has three reflecting panels placed in a Y configuration and supported by a central mast. Three crossed V dipoles are phased in the 120° sectors formed by the panels. This provides a more compact structure than the triangular tower.

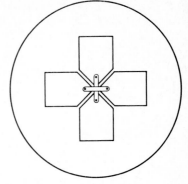

FIG. 5-12 Panel antenna with crossed V dipoles.

FIG. 5-13 Crossed broadband dipoles in a cylindrical cavity.

A better technique for equalizing the azimuth beamwidths for vertical and horizontal polarization is to enclose crossed planar dipoles in a cylindrical cavity[13] as shown in Fig. 5-13. The length-to-width ratio of the planar dipoles is about 3 and provides a bandwidth of 10 percent with a VSWR less than 1.1:1. The cavity depth is 0.25 wavelength. Cavity diameters of 0.65 and 0.8 wavelength produce azimuth 6-dB beamwidths of 120° and 90° respectively, which are desired for triangular and square towers respectively. The dipoles are fed in phase quadrature by two baluns forming a four-tube support structure. The circularity is ±2 dB, and the axial ratio is less than 2 dB.

Another approach is to place four half-wavelength dipoles in a square arrangement, with the side of the square being somewhat larger than a half wavelength. The vertical and horizontal dipoles are fed in phase quadrature. Since the 6-dB azimuth beamwidths are about 90°, four panels around a square tower may be used for omnidirectional applications.

Ring-Panel Antennas

The ring-panel antenna[14] consists of a multiplicity of ring radiators fed in series by a transmission line. Figure 5-14 shows two circular rings formed by strips over a panel and connected by rods over the panel which provide simple low-radiation transmission lines. The ring circumference is approximately one wavelength, as is the spacing between rings. The antenna is designed so that the characteristic impedance of the ring strip over ground is equal to the characteristic impedance of the transmission-line rod over ground. A practical value of the characteristic impedance is 140 Ω. By using a resistive termination on the last ring and/or special tuning techniques, it is possible to achieve a traveling-wave type of antenna. A traveling wave on a ring of one-wavelength circumference as shown in Fig. 5-14 radiates right-hand circular polarization. The ring is equivalent to four quarter-wave dipoles placed on a square with −90° progressive phasing. The azimuth beamwidth for horizontal polarization is usually about 10° less than that for vertical polarization. The beamwidths may be equalized and changed by means of parasitic elements such as monopoles on each side of the rings and dipoles in front of the rings. A cavity is not required.

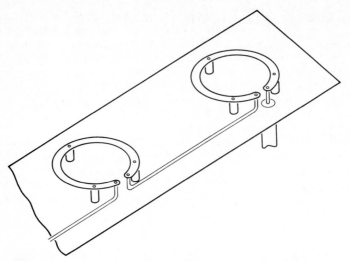

FIG. 5-14 Two elements of a series-fed traveling-wave ring-panel antenna.

The radiation from, or the attenuation through, a traveling-wave ring increases with the height of the ring above the panel and decreases with the characteristic impedance of the strip. The attenuation ranges from about 0.5 to 8 dB for ring heights of 0.05 to 0.2 wavelengths, respectively. Because of this attenuation, it is necessary to increase the height of the rings as one progresses from the feed point in order to approximate a uniform array.

Since the distance along the transmission line and ring between similar points on adjacent rings is two wavelengths for broadside radiation, the beam direction will scan 1.15° for a 1 percent change in frequency. This limits the number of end-fed rings to 3 for Channel 2 and to about 10 for the UHF channels.

The axial ratio may be reduced to a very low level by introducing reflections on the transmission-line rods which radiate left-hand circularly polarized waves. The magnitude and phase of the reflections may be controlled by the size and position of the reflecting device to cancel undesired left-hand circular polarization from other parts of the antenna.

Several panels may be stacked vertically to achieve the desired gain and beamwidth. Beam tilt and null fill may be achieved by control of the excitation of the panels and/or control of the amplitude and phase of the radiation from each ring by the height of the ring and the transmission line-rod length respectively.

Axial-Ratio Measurement

The axial ratio of elliptically polarized antennas may be measured by rotating a linearly polarized transmit antenna and receiving with the test antenna. If there are negligible multipath reflections, as in an anechoic chamber, then the axial ratio is simply the ratio of the maximum to the minimum field received by the test antenna. For TV and FM antennas it is not practical to eliminate ground reflections. However, it is possible to design some long ranges so that the reflected wave is at an angle much smaller than the Brewster angle. In this case the reflection coefficients for vertical and horizontal polarization are nearly identical, and the rotating linearly polarized antenna technique[15] can be used. In many other cases, such as a shorter range and/ or variable moisture conditions, the reflection coefficients for the two polarizations will not be the same. However, the range can be calibrated and the axial ratio can be computed as follows. A linearly polarized antenna with a clockwise 45° slant angle, as viewed from the transmit site, is placed at the test site. The linearly polarized transmit antenna is then rotated, and the maximum–minimum ratio S_0 and the null angle of the transmit antenna ψ_0 are measured.

A convenient form of these antennas is a rotatable dipole in a cylindrical cavity. The process is then repeated by using the test antenna to measure S_1 and ψ_1, the maximum–minimum ratio and null angle for the transmit antenna. The following calculations are performed:

$$K = j\frac{1 + \Gamma}{1 - \Gamma} \tag{5-3}$$

where

$$\Gamma = \frac{S_0 - 1}{S_0 + 1} \exp{(j2\psi_0)} \tag{5-4}$$

and

$$\rho = \left|\frac{K(1 - P) - (1 + P)}{K(1 - P) + (1 + P)}\right| \tag{5-5}$$

where

$$P = \frac{S_1 - 1}{S_1 + 1} \exp{(j2\psi_1)} \tag{5-6}$$

and finally the axial ratio AR is

$$AR = \frac{1 + \rho}{1 - \rho} \tag{5-7}$$

Another approach is to use a crossed-dipole transmit antenna with a switchable and adjustable feed network so as to produce right circular and left circular polarization at the test site. The feed circuit is adjusted so that the received signal of a rotating linearly polarized antenna is uniform for each of the transmit polarizations. The axial ratio is given by Eq. (5-7), where ρ is the ratio of the right circular to left circular polarization signals.

5-4   HORIZONTALLY POLARIZED ANTENNAS

Reference may be made to the first edition of the *Antenna Engineering Handbook* for descriptions of loop, cloverleaf, V, and other types of horizontally polarized antennas which were popular before the widespread use of circular polarization.

Dipole Antennas

A top view of a dipole panel-type antenna is illustrated in Fig. 5-15. The support arms are electrically connected to the dipole arms and form a part of the radiating structure and also aid in the matching of the dipole. The dipole is usually about a half wavelength long and spaced about a quarter wavelength from the panel. The dipole may be fed from a balanced transmission line connected to the center of the dipole or from

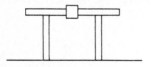

a coaxial line entering one of the support arms and extending to the central feed gap, forming a balun. Bandwidths of 10 percent may be achieved with shunt stub-matching techniques. The panels may be placed on square or triangular towers to obtain omnidirectional or directional patterns.

FIG. 5-15 Dipole panel antenna.

Batwing Antennas

The batwing antenna[16,17,18] illustrated in Fig. 5-16 is the most popular horizontally polarized VHF TV antenna. It consists of several bays of turnstile configurations of broadband planar dipoles. The dipoles are formed by a grid of rods and have a length and width of about one-half wavelength. Each half of a dipole is supported by a spacer rod which is shorted to the supporting mast at the top and bottom of the dipole. Opposite halves of each dipole are fed out of

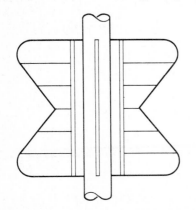

FIG. 5-16 Batwing, or superturnstile, antenna.

phase from a power divider with equal-length coaxial transmission lines by feeding one-half from a coaxial line grounded to the mast and the center conductor connected to the center of the spacer and the other half from a coaxial line running along and connected to the spacer with the center conductor connected to the mast at the center of the spacer. The quadrature dipoles are fed 90° out of phase to provide an azimuth pattern with a circularity of ± 2 dB. The 90° phasing may be obtained by different line lengths for the quadrature dipoles or by a quadrature hybrid. In the latter case, the visual and aural transmitters may be diplexed into the antenna system through the hybrid. The isolation between the transmitters is about the same as the return loss from the dipoles, which is greater than 26 dB. Each half dipole has an impedance of 75 Ω, and bandwidths of 20 percent with a return loss greater than 26 dB have been achieved. The dipoles have nulls in the nadir and zenith directions because of their width or height when viewed from the horizontal plane. Thus the bays are spaced by one wavelength to achieve maximum gain. Two to six bays are normally used for Channels 2 to 6 and up to 18 bays for Channels 7 to 13. Peanut-shaped azimuth patterns may be achieved by an unequal power split between the orthogonal dipoles.

Slot Antennas

Both resonant and nonresonant (or standing-wave and traveling-wave respectively) end-fed arrays of slots are used for TV broadcasting. The resonant arrays are restricted to UHF applications because of their limited bandwidth.

The traveling-wave slot antenna[19] illustrated in Fig. 5-17 is a large end-fed coaxial transmission line with a slotted outer conductor. The slots are arranged in pairs at each layer, with the pairs separated by a quarter wavelength along the length of the antenna. Adjacent pairs occupy planes at right angles to each other. The slot pairs, which are approximately one-half wavelength long, are fed out of phase by the coaxial line by capacitive probes projecting radially inward from one side of each slot so as to produce a figure-eight pattern. The probes are placed on opposite sides of adjacent in-line slots which are spaced one-half wavelength to provide in-phase excitation. The quarter-wavelength separation of layers in conjunction with the space-quadrature arrangement of successive layers of slots effects a turnstile-type feed which produces a horizontally polarized azimuth pattern with a circularity of ±1 dB for VHF applications. An equal percentage of the power in the coaxial line is fed to each layer of slots, which results in an exponential aperture distribution that provides null fill.

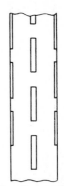

FIG. 5-17   Traveling-wave slot antenna.

Reflections from adjacent layers tend to cancel, which allows the traveling-wave operation. The top slots are strongly coupled to the line to reduce reflections. For high-gain applications, one-half of the slots may be eliminated, resulting in a one-wavelength spacing of in-line slots. The pipe diameter is 10 to 20 in for Channel 7–13 applications.

The standing-wave antenna consists of layers or bays of one or more axial slots spaced by one wavelength and fed by a coaxial line with the slotted pipe forming the outer conductor. For UHF applications, the pipe diameter may vary from 6 to 18 in, depending on the gain and frequency. Azimuth patterns are controlled by the number of slots per bay. One slot per bay produces a skull-shaped pattern, two slots a peanut-shaped pattern, and three slots a trilobe pattern. Four or more slots per bay are usually required for an omnidirectional pattern with a circularity of ±1 dB. The slots may be coupled to the coaxial line by several means such as capacitive probes or bars[20] connected to one side of the slot, rods connecting one side of the slot to the center conductor, or balanced loops connected to the slot with a variable orientation to control the magnetic coupling. The symmetry of the slots must be controlled to prevent excitation of propagating higher-order modes in the coaxial line. The length of the slot and the coupling mechanism are usually adjusted so that the shunt reactance of each bay of slots is zero and the shunt resistance is $NZ_0/2$, where N is the number of end-fed bays and Z_0 is the characteristic impedance of the coaxial line. This produces an overloaded resonant array[21] which provides larger bandwidth with the proper input-matching device. For antennas with a gain of more than 20, it is necessary to center-feed the slotted array. This may be accomplished by using a triaxial line in the lower half of the antenna.

Parasitic elements such as monopoles may be added to the cylinder to shape the azimuth pattern.

Zigzag Antennas

The panel type of zigzag antenna[22] shown in Fig. 5-18 is a simple type of traveling-wave antenna which may be placed around triangular or square towers to produce a wide variety of azimuth patterns. The antenna consists of a wire or rod that is bent at

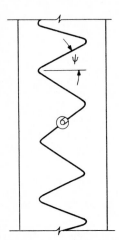

half-wavelength intervals to form the zigzag structure. This provides a broadside beam with horizontal polarization since the radiation from the vertical components of the currents in adjacent half-wavelength segments cancels, whereas that for the horizontal components adds. The azimuth 6-dB beamwidth is about 90°, which is desirable for a square tower. The beamwidth may be increased for triangular-tower configurations by adding arrays of parasitic monopoles along each side of the zigzag or bending the panel and zigzag in a V configuration. It is preferable to feed the zigzag in a balanced manner at the center of one rod as shown, which eliminates radiation from the feed structure. If an unbalanced feed is used, e.g., at the bend, the feed radiation will distort the azimuth pattern.

FIG. 5-18 Zigzag antenna with balanced feed.

The zigzag may be designed on the basis of a leaky-wave antenna, in which the attenuation per wavelength due to radiation increases with distance from the feed point. Figure 5-19 shows the attenuation per axial wavelength versus the height of the zigzag above the panel for several pitch angles ψ. The rod diameter and band radius are 0.01 and 0.03 wavelength respectively. To simulate a uniform aperture distribu-

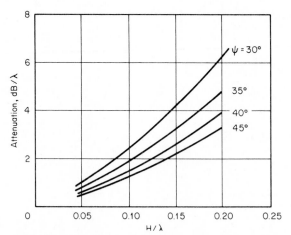

FIG. 5-19 Attenuation of zigzag antenna versus height. *(Courtesy of Cetec Antennas.)*

tion, the pitch angle decreases with distance from the feed point and is adjusted along with the height of the zigzag to give a one-way attenuation of about 15 dB for the current on the zigzag. It is preferable and simpler to use a constant-height zigzag. Since the beam direction of the incident and reflected waves on each half of the zigzag scan with frequency but in opposite directions, the length of a zigzag panel is limited by the bandwidth of the channel. The average pitch angle is usually about 35°, which produces a beam scan of 1° per 1 percent change in frequency. For UHF, panel lengths of 16 wavelengths may be used.

Since reflections occur at each bend, it is necessary to compensate for these in order to achieve a traveling-wave antenna. For a bend radius of 0.05 wavelength, the reflection coefficient varies over the range of 0.1 to 0.2 for practical pitch angles and heights of the zigzag. These reflections may be reduced considerably by placing the support insulators one-eighth wavelength before each bend as viewed from the feed point. It is usually necessary to add shunt capacitive tuners along the zigzag to achieve a traveling-wave condition, which is required for a wide impedance bandwidth.

Beam tilt may be achieved by displacing the feed from the center of the middle zigzag element or changing the lengths of the rods from the feed to the upper and lower zigzags.

Helix Antennas

Figure 5-20 shows a single bay of a single-arm right-hand and left-hand helix fed in phase at the center so that the vertically polarized components of the two helices cancel in the broadside direction. The pitch angle is about 12° so that the vertically polarized radiation from each helix is about 10 dB down from the horizontally polarized radiation, which produces about 0.5-dB loss in gain due to cross-polarization radiation. Since the beam of each helix scans about 2.7° per 1 percent change in frequency, the bay length is limited to about six wavelengths for Channels 7 to 13. For these channels the mechanical requirements of the supporting mast usually result in a mode 2 helix with a turn length of two wave-

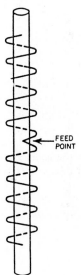

FIG. 5-20 Horizontally polarized helical antenna.

lengths for broadside radiation. The mast diameter is chosen so that the one-way attenuation through each helix is 24 dB. Because of this, the circularity of the horizontal pattern is less than ±1.5 dB.

5-5   MULTIPLE-ANTENNA INSTALLATIONS

It is desirable to have a number of TV and FM stations share the same supporting structure for their antennas. This reduces the costs of the supporting structures and reduces or eliminates the need for rotatable receiving antennas.

FIG. 5-21 Candelabra antenna in San Francisco. *(Courtesy of RCA Corp.)*

It is most economical to stack the antennas vertically. Unfortunately, each broadcaster wants to be "king of the hill," i.e., to have the top antenna, even though there is only a fraction of a dB difference in coverage for tall towers. This has resulted in candelabra installations in which antennas are mounted on the corners of a triangular support structure with a separation of 50 to 100 ft (15 to 30 m). Figure 5-21 shows the San Francisco Mount Sutro structure,[23] which supports five VHF, three UHF, and four FM antennas. A problem[24,25] with candelabra installations is that reflections from the other antennas produce ripples of several decibels in the azimuth pattern with the ripple directions being frequency-sensitive, which can lead to distortions in the received picture, especially for color-TV stations. The problem is tolerable for horizontally polarized antennas but more difficult for circularly polarized antennas, where reflections for vertical polarization from the antenna support structures are much stronger.

REFERENCES

1 Henry Jasik (ed.), *Antenna Engineering Handbook,* 1st ed., McGraw-Hill Book Company, New York, 1961, chap. 23.

2 *Reference Data for Radio Engineers,* 6th ed., Howard W. Sams & Co., Inc., Indianapolis, 1975, chap. 30.

3 J. Perini, "A Method of Obtaining a Smooth Pattern on Circular Arrays of Large Diameters with $\cos^n \phi$ Elements," *IEEE Trans. Broadcast.,* vol. BC-14, no. 3, September 1968, pp. 126–136.

4 R. H. DuHamel, "Circularly Polarized Helix and Spiral Antennas," U.S. Patent 3,906,509, Sept. 16, 1975.

5 O. Ben-Dov, "Circularly Polarized, Broadside Firing Tetrahelical Antenna," U.S. Patent 3,940,722, Feb. 24, 1976.

6 G. H. Brown and O. M. Woodward, Jr., "Circularly Polarized Omnidirectional Antenna," *RCA Rev.,* vol. 8, no. 2, June 1947, pp. 259–269.

7 O. M. Woodward, Jr., "Circularly Polarized Antenna Systems Using Tilted Dipoles," U.S. Patent 4,083,051, Apr. 4, 1978.

8 R. H. DuHamel, "Circularly Polarized Antenna with Circular Arrays of Slanted Dipoles Mounted around a Conductive Mast," U.S. Patent 4,315,264, Feb. 9, 1982.

9 Peter K. Onnigian, "Circularly Polarized Antenna," U.S. Patent 3,541,570, Nov. 17, 1970.

10 M. S. Suikola, "New Multi-Station Top Mounted FM Antenna," *IEEE Trans. Broadcast.,* vol. BC-23, no. 2, June 1977, p. 56.

11 R. D. Bogner, "Improve Design of CP FM Broadcast Antenna," *Communications News,* vol. 13, no. 6, June 1976.

12 M. S. Suikola, "A Circularly Polarized Antenna Replacement for Channel 2–6 Superturnstiles," *Broadcast Eng.,* February 1981.

13 R. E. Fisk and J. A. Donovan, "A New CP Antenna for Television Broadcast Service," *IEEE Trans. Broadcast.,* vol. BC-22, no. 3, September 1976, p. 91.

14 R. H. DuHamel, "Circularly Polarized Loop and Helix Panel Antennas," U.S. Patent 4,160,978, July 10, 1979.

15 J. A. Donovan, "Range Measurement Techniques for CP Television Antennas," *IEEE Trans. Broadcast.,* vol. BC-24, no. 1, March 1978, p. 4.

16 G. H. Brown, "A Turnstile Antenna for Use at Ultra-High Frequencies," *Electronics,* March 1936; "The Turnstile Antenna," *Electronics,* April 1936.

17 R. F. Holtz, "Super Turnstile Antenna," *Communications,* April 1946.

18 H. E. Gihring, "Practical Considerations in the Use of Television Super Turnstile and Super-Gain Antennas," *RCA Rev.,* June 1951.

19 M. S. Siukola, "The Traveling-Wave VHF Television Transmitting Antenna," *IRE Trans. Broadcast Transmission Syst.,* October 1957.

20 O. O. Fiet, "New UHF-TV Antenna," part 1: "Construction and Performance Details of TFU-24B UHF Antennas," *FM & Television,* July 1952; part 2: "TFU-24B Horizontal and Vertical Radiation Characteristics," *FM & Television,* August 1952.

21 S. Silver, *Microwave Antenna Theory and Design,* McGraw-Hill Book Company, New York, 1949, sec. 9.20.

22 O. M. Woodward, Jr., U.S. Patent 2,759,183, August 1956.

23 "Bay Area TV Viewers Turn to Sutro Tower," *RCA Broadcast News,* vol. 151, August 1973, pp. 19–31; H. H. Westcott, "A Closer Look at the Sutro Tower Antenna Systems," *RCA Broadcast News,* vol. 152, February 1974, pp. 35–41.

24 M. S. Siukola, "Predicting Characteristics of Multiple Antenna Arrays," *RCA Broadcast News,* vol. 97, October 1957, pp. 63–68.

25 M. S. Siukola, "Size and Performance Tradeoff Characteristics in Multiple Arrays of Horizontally and Circularly Polarized TV Antennas," *IEEE Trans. Broadcast.,* vol. BC-22, no. 1, March 1976, pp. 5–12.

Chapter 6

TV Receiving Antennas

Edward B. Joy

Georgia Institute of Technology

The Federal Communications Commission (FCC) frequency allocations for commercial television consist of 82 channels, each having a 6-MHz width.[1] Channels 2 through 6 are known as the low-band very high frequency (VHF) and span 54 to 88 MHz, Channels 7 through 13 are known as the high-band VHF and span 174 to 216 MHz, and Channels 14 through 83 are known as the ultrahigh-frequency (UHF) band and span 470 to 890 MHz. Channels 70 through 83 are designated for translator service. Channel designations and frequency limits are given in Table 6-1.

TABLE 6-1 Designations and Frequency Limits of Television Channels in the United States

Channel designation	Frequency band, MHz	Channel designation	Frequency band, MHz	Channel designation	Frequency band, MHz
2	54–60	30	566–572	57	728–734
3	60–66	31	572–578	58	734–740
4	66–72	32	578–584	59	740–746
5	76–82	33	584–590	60	746–752
6	82–88	34	590–596	61	752–758
		35	596–602	62	758–764
7	174–180	36	602–608	63	764–770
8	180–186	37	608–614	64	770–776
9	186–192	38	614–620	65	776–782
10	192–198	39	620–626	66	782–788
11	198–204	40	626–632	67	788–794
12	204–210	41	632–638	68	794–800
13	210–216	42	638–644	69	800–806
		43	644–650		
14	470–476	44	650–656	70	806–812
15	476–482	45	656–662	71	812–818
16	482–488	46	662–668	72	818–824
17	488–494	47	668–674	73	824–830
18	494–500	48	674–680	74	830–836
19	500–506	49	680–686	75	836–842
20	506–512	50	686–692	76	842–848
21	512–518	51	692–698	77	848–854
22	518–524	52	698–704	78	854–860
23	524–530	53	704–710	79	860–866
24	530–536	54	710–716	80	866–872
25	536–542	55	716–722	81	872–878
26	542–548	56	722–728	82	878–884
27	548–554			83	884–890
28	554–560				
29	560–566				

6-2 TV SIGNAL-STRENGTH ESTIMATION

The FCC limits the maximum effective radiated power (antenna input power times antenna gain) of commercial television stations to 100 kW for low-band VHF, 316 kW for high-band VHF, and 5 MW for the UHF band. Maximum TV transmitting-antenna heights are limited to 2000 ft (609.6 m). Figure 6-1 presents predicted field-strength levels for Channels 7 to 13 versus transmitting-antenna height and distance from the transmitting antenna for 1 kW radiated from a half-wavelength dipole antenna in free space. The figure predicts field strength 30 ft (9.14 m) above ground that is exceeded 50 percent of the time at 50 percent of the receiving locations at the specified distance.

Table 6-2 shows correction factors to be used with Fig. 6-1 to determine predicted field-strength levels for Channels 2 through 6 and Channels 14 through 69. The table presents correction factors for a 1000-ft transmitting antenna, but it is reasonable to use these factors for all transmitting-antenna heights. The table shows that field-strength levels for the low-VHF band and the UHF-band channels are typically lower than those for the high-VHF-band channels. Ground roughness between the transmitting and receiving antennas is also important in predicting field-strength levels for receiving sites more than 6 mi (9.6 km) from the transmitting antenna. It is measured along a line connecting the transmitting antenna and the receiving antenna that begins 6 mi from the transmitting antenna and terminates either at the receiving-antenna location or 31 mi (49.9 km) from the transmitting antenna, whichever is least. The roughness ΔH, measured in meters, is the difference between the elevation level exceeded by 10 percent of the elevations along the line and the elevation level not reached by 10 percent of these elevations. The value of roughness assumed in the formulation of Fig. 6-1 is 50 m. The correction factor ΔF, given in decibels, to be applied to the field-strength value of Fig. 6-1 for a frequency f in megahertz is

$$\Delta F = C - 0.03\Delta H (1 + f/300) \qquad \textbf{(6-1)}$$

where $C = 1.9$ for Channels 2 to 6

TABLE 6-2 Correction Factors for Fig. 6-1 for an Antenna Height of 1000 Ft

Distance, mi	Channels 2–6, correction, dB	Channels 14–69, correction, dB
1	0	0
5	−1	−1.5
10	−3	−3
20	−3	−3.5
30	−3	−6
40	−2.5	−9
50	−1.5	−10
60	−1	−9
80	0	−6.5
100	0	−4
200	0	−2.5

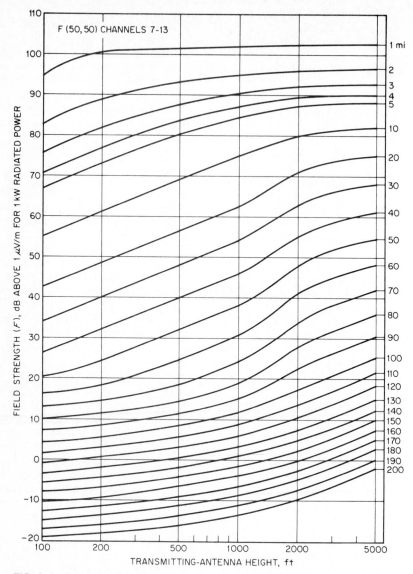

FIG. 6-1 Television Channels 7 to 13; estimated field strength exceeded at 50 percent of the potential receiver locations for at least 50 percent of the time at a receiving antenna height of 30 ft (9.14 m).

$C = 2.5$ for Channels 7 to 13
$C = 4.8$ for Channels 14 to 69

It can be seen that the correction factor lowers predicted field strength with increasing roughness and increasing frequency, reaching -39.2 dB for a roughness of 400 m and a frequency of 800 MHz.

Ground reflections cause the field strength to increase approximately linearly with height; i.e., the field strength is 6 dB less at a height of 15 ft (4.6 m) and 6 dB more at 60 ft (18.3 m). The rms signal voltage delivered to a 75-Ω load at the terminals of an antenna with gain G_A relative to a half-wavelength dipole immersed in a field with an rms field strength of E V/m at a frequency of f MHz is given by

$$V = \frac{48.5E\sqrt{1.64G_A}}{f} \tag{6-2}$$

As an example, f = 689 MHz (Channel 50), receiving-antenna height = 10 ft (3 m), receiving-antenna gain = 12 dB (G_A = 15.8), transmitting-antenna height = 1000 ft (305 m), transmitting effective radiated power = 5 MW, distance = 40 mi (64.4 km), and surface roughness = 50 m. Let us find the rms received voltage. From Fig. 6-1, E at a 30-ft (9.14-m) height is 83 dB above 1 μV/m = 14,125 μV/m (it is noted that surface roughness is negligible); E at 10 ft is 4708 μV/m. The voltage received 50 percent of the time at 50 percent of such locations is greater than 1687 μV.

The received-signal voltage at a given receiving location can be increased by increasing the height and gain of the receiving antenna. In addition, the effect of building attenuation must be considered for TV receiving antennas installed indoors.[2] Measured attenuation values for a single wall between the receiving and the transmitting antennas range from 8 to 14 dB.

6-3 TV RECEIVING SYSTEMS

A TV receiving system is composed of an antenna, a transmission line, and a TV receiver. Baluns and splitters may be employed at the antenna or antennas and the TV receiver for impedance matching and for the separation or the combination of signals. Two TV receiving systems are typical: (1) an indoor antenna system using a small indoor antenna mounted on or near the TV receiver and (2) an outdoor antenna system using a larger antenna mounted high on a mast outdoors and connected to the TV receiver by an extensive transmission line. Indoor antenna systems typically are employed in high-signal-strength areas where multipath propagation is not severe. Outdoor antenna systems become necessary at a large distance from the transmitting antenna.

Overview of TV Antenna Types

A TV receiving antenna should have sufficient gain and a good impedance match to deliver a signal to a transmission line and subsequently to a TV receiver to produce a single clear TV picture and sound. Depending on its location, the antenna must reject reflected signals and other extraneous signals arriving from directions well off the direction to the transmitting antenna. Signal suppression should be particularly effective in the hemisphere directly opposite the transmitting antenna, as even small reflecting surfaces in this region can produce undesirable "ghost" images. TV receiving antennas should provide these properties over all TV channels of interest, which may include the complete 54- to 890-MHz range. They can be designed to receive all TV channels, or groups of channels such as all low-band VHF channels, all high-band

VHF channels, all VHF channels, and all UHF channels, or a single VHF or UHF channel.

TV receiving antennas fall into two major categories, indoor antennas and outdoor antennas. The most common configuration of an indoor antenna consists of two antennas, one for all VHF channels and one for all UHF channels. The most popular indoor VHF antennas are extendable monopole and dipole rods (rabbit-ear antennas). These antennas have measured average VHF gains of −4 dB with respect to a half-wavelength dipole and normally must be adjusted in length and orientation for best signal strength and minimum ghosting for each channel.[3] Rabbit-ear antennas are available with 75- or 300-Ω impedance. There are several popular indoor UHF antennas, including the circular loop, triangular dipole, and triangular dipole with reflecting screen. The loop and triangular-dipole antennas have low gain; the triangular dipole with reflecting screen has increased gain and a greatly improved ability to reject signals arriving from behind the reflecting screen. Indoor UHF antennas are most commonly designed with a balanced 300-Ω impedance. A popular indoor VHF-UHF combination antenna consists of a VHF rabbit-ear dipole antenna and a UHF loop antenna mounted on a fixture containing a switchable impedance-matching network. If a preamplifier is included as an integral part, the antenna is known as an *active antenna*.

Outdoor antenna systems can provide up to a 15-dB increase in antenna gain (including transmission-line loss), typically 15- to 20-dB greater rejection of ghost signals, and greater immunity to electrical interference over indoor antenna systems. This advantage, combined with a typical signal-strength increase due to antenna height of 14 dB [6 to 30 ft (1.8 to 9.1 m)] plus the removal of a typical 11-dB building attenuation, could result in an overall signal increase at the TV receiver of up to 40 dB. The most common outdoor antenna configuration is a combined VHF and UHF antenna. The combined antenna usually consists of two separate antennas mounted together to form a single structure. The most common VHF antenna is some variation of the log-periodic dipole array (LPDA). This antenna may be designed for a 300-Ω balanced or a 75-Ω unbalanced input impedance. There are several common types of UHF antennas including the LPDA, the broadband Yagi-Uda parasitic dipole array, the corner reflector, the parabolic reflector, and an array of triangular dipoles with a flat reflecting screen. Most UHF antenna types are designed with a balanced 300-Ω input impedance. The UHF LPDA may also be designed for an unbalanced 75-Ω impedance. Later sections treat these antenna types.

Transmission Lines for TV Receiving Systems

The standard impedances for the transmission lines used for TV and FM receiving systems are 75 Ω unbalanced and 300 Ω balanced. RG-59/U and RG-6/U types are the most commonly used 75-Ω lines, and twin-lead flat, foam, tubular, and shielded types are the most common 300-Ω lines. Table 6-3 shows average measured transmission-line losses for 100 ft (30.5 m) of the various types of lines.[4] All measurements were performed on new dry lines suspended in free space. The standard deviation of the measurements within the types was approximately 0.5 dB at VHF frequencies and 1.5 dB at UHF frequencies except for the flat 300-Ω twin-lead type, which was 0 dB at all frequencies, and for the 300-Ω shielded twin-lead type, which was 0.5 dB at VHF and increased to 6 dB at the high UHF frequencies.

The voltage standing-wave ratio (VSWR) measured on 100-ft sections (with

TABLE 6-3 Average Transmission-Line Insertion Loss, dB, for a 100-Ft Length of New Dry Line

Cable \ Channel	4	9	14	24	34	44	54	64	74	83
RG-59/U types	2.4	4.1	6.8	7.2	8.0	8.5	8.8	8.8	9.2	9.5
RG-6/U types	1.7	3.0	5.1	5.5	5.8	6.1	6.5	6.8	7.1	7.4
Flat twin-lead types	0.9	1.6	3.0	3.3	3.5	3.8	4.0	4.2	4.4	4.6
Foam twin-lead types	1.0	1.9	3.5	3.8	4.1	4.3	4.6	4.8	5.0	5.3
Tubular twin-lead types	0.9	1.7	3.2	3.5	3.7	4.0	4.2	4.5	4.7	5.0
Shielded twin-lead types	2.7	4.9	8.3	9.0	9.6	10.5	11.5	12.8	13.5	15.3

appropriate terminations) of the various types of lines is fairly independent of frequency on the VHF and UHF bands; it is shown in Table 6-4. Additional line losses and increased VSWR are reported for 50-ft (15.2-m) lengths of unshielded twin-lead transmission line because of proximity to metal and wetness. Table 6-5 presents these data for three cases: (1) 10 ft (3 m) of the 50-ft line parallel to and touching a metal mast, (2) 10 ft of the 50-ft line wrapped around a metal mast, and (3) a 50-ft section which is spray-wet.

Impedance-mismatch VSWR increases transmission loss. Such losses may occur at all junction points from the antenna to the TV receiver, including junctions at baluns, splitters, couplers, connectors, and preamplifiers. Impedance-mismatch loss is given by

$$\text{Mismatch loss (dB)} = -10 \log_{10} \left[1 - \left(\frac{\text{VSWR} - 1}{\text{VSWR} + 1} \right)^2 \right] \qquad \textbf{(6-3)}$$

Mismatch loss is 0.2 dB for a VSWR of 1.5, 0.5 dB for a VSWR of 2.0, 1.9 dB for a VSWR of 4.0, and 4.0 dB for a VSWR of 8.0.

Baluns are used to match a 75-Ω unbalanced transmission line to a 300-Ω balanced line. Most commercially available baluns used for TV receiving systems employ bifilar transmission-line windings around a ferrite core to achieve broadband operation. Desirable features of baluns are minimum insertion loss and VSWR. Table 6-6 shows averaged measured balun insertion loss and VSWR for several TV channels. The standard deviation of these measurements is approximately 0.3 dB for insertion loss; it is 0.1 for VSWR at Channel 4, increasing to 0.5 at Channel 64.

TABLE 6-4 Average Measured Transmission-Line VSWR

Line	Average VSWR
RG-59/U types	1.2
RG-6/U types	1.2
Flat 300-Ω twin-lead types	2.0
Foam 300-Ω twin-lead types	1.8
Tubular 300-Ω twin-lead types	1.6
Shielded 300-Ω twin-lead types	2.4

TABLE 6-5 Average Additional Transmission-Line Loss and VSWR Because of Proximity to Metal and Wetness

Line	Additional loss, dB			VSWR increase		
	10 ft parallel to mast	10 ft wrapped on mast	50 ft spray-wet	10 ft parallel to mast	10 ft wrapped on mast	50 ft spray-wet
Flat 300-Ω twin-lead types	5.8	5.8	8.4	2.1	1.2	1.7
Foam 300-Ω twin-lead types	0.8	2.0	3.4	0.1	0.8	0.7
Tubular 300-Ω twin-lead types	0.8	3.0	0.8	0.1	0.4	0.3

Signal splitters are used to separate the VHF, FM, and UHF signals present on a transmission line at the receiver. They are also used as combiners to combine the signals from separate VHF, FM, and UHF antennas onto a single transmission line. Most commercially available splitters use three- to five-element high-pass and low-pass LC filters to accomplish the frequency separation. Some splitters contain input and/or output baluns to achieve 75- to 300-Ω impedance transformations. Desirable features of signal splitters, in addition to frequency selectivity, are low insertion loss and VSWR. Table 6-7 shows the average measured insertion loss and VSWR of a selection of commercially available signal splitters. The standard deviation of these measurements is approximately 1.0 dB for the insertion loss and 0.5 for the VSWR.

Transmission-line connectors can cause significant VSWR on a transmission line. Properly installed F-type connectors provide reasonably low VSWR for RG-59/U and RG-6/U coaxial lines for a limited number of matings. The connector problem is particularly severe for 300-Ω-twin-lead transmission lines. Standard twin-lead plugs and sockets and screw-connector terminal boards were found to have VSWRs near 1.5. Lower VSWR for twin-lead lines can be achieved by direct soldering of the lines in a smooth and geometrically continuous manner.

The noise figure for a lossy transmission-line system is equal to its total attenuation from input to output; 5 dB of attenuation means a noise figure of 5 dB. The total transmission-system loss has a significant impact on the noise figure of the TV receiving system.

TABLE 6-6 Average Measured Balun Insertion Loss and VSWR

Channel	4	9	24	44	64
Insertion loss, dB	0.8	0.6	1.1	1.2	1.7
VSWR	1.4	1.6	1.5	1.8	2.0

TABLE 6-7 Average Measured Splitter Insertion Loss and VSWR

Channel	4	9	24	44	64
Insertion loss, dB	1.3	2.5	1.5	1.5	2.1
VSWR	1.8	2.1	1.7	1.7	2.5

TV Receivers

Typically, TV receivers have separate UHF and VHF inputs which are directly connected to the respective UHF and VHF tuners.[5] These inputs are either 75-Ω coaxial with a Type F connector or 300-Ω twin lead with a twin-lead screw-connector terminal board. The trend is toward the 75-Ω coaxial input, as it is more compatible with cable and other video systems which use the TV receiver as a display unit. The noise figure of a typical TV receiver is 6 dB for the low-VHF band and 8 dB for the high-VHF band; for the UHF band, the noise figure increases from 11 to 13 dB with increasing frequency. The input VSWR is approximately 2.5 for both the VHF and the UHF inputs.

6-4 TV RECEIVER-SYSTEM NOISE

The required signal strength for TV reception must be measured with respect to the TV receiver-system noise, including antenna noise. Figure 6-2 shows average observer picture-quality ratings versus the signal-to-random-noise ratio.[6] From this figure it is seen that a signal-to-noise ratio of approximately 42 dB is required to produce a rating of excellent by the average observer. The figure also shows that a signal-to-noise ratio of 10 dB produces an unusable picture.

The equivalent noise voltage at the terminals of a TV receiving antenna when

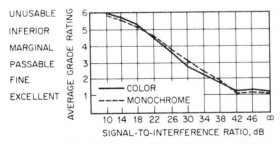

FIG. 6-2 Average observer rating versus signal-to-interference ratio. The solid line indicates color; the dashed line, monochrome. [*Source: "Engineering Aspects of Television Allocations," Television Allocation Study Organization (TASO), report to FCC, Mar. 16, 1959.*]

connected to a transmission-line–TV-receiving system is the rms sum of the antenna noise voltage and the equivalent noise voltage of the transmission-line–TV-receiving system.

Antenna noise voltage can be estimated by assuming that one-half of the gain of the antenna is receiving cosmic noise and the other half is receiving blackbody radiation from the earth, which is assumed to be at a temperature of 290 K.

The rms antenna noise voltage delivered to a matched impedance is given by

$$V_n = \sqrt{0.82 G_A kRB(T_g + 290°)} \tag{6-4}$$

where G_A = gain of antenna with respect to a half-wavelength dipole
k = Boltzmann's constant = 1.38×10^{-23} W/Hz·K
T_g = equivalent noise temperature of sky, K
R = matched input resistance of antenna, Ω
B = bandwidth, Hz = 6×10^6 Hz

T_g is a function of frequency and varies primarily because of solar activity. Figure 29-3 gives measured maximum and minimum sky temperatures versus frequency.[7] It shows that sky temperature is larger than ambient temperature for VHF frequencies and smaller than ambient temperature for UHF frequencies. The rms noise voltage for Channel 2 (using a sky temperature of 10,000 K) impressed on a 75-Ω transmission line by a 6-dB-gain ($G_A = 4$) antenna is 14.5 μV.

The equivalent rms noise voltage of a transmission-line–TV-receiver system with an overall noise figure N is given by

$$V_n = \sqrt{NBRTk} \tag{6-5}$$

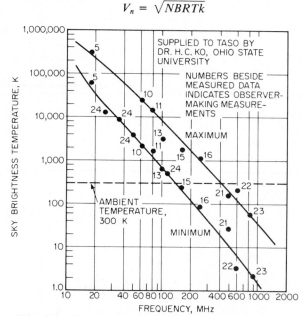

FIG. 6-3 Sky temperature as a function of frequency. [*Source: "Engineering Aspects of Television Allocations," Television Allocations Study Organization (TASO), report to FCC, Mar. 16, 1959.*]

where T is the temperature of the system in degrees Kelvin and is assumed to be 290 K. The parameters B, R, and k are as defined for Eq. (6-4). The overall noise figure of two systems connected in cascade can be calculated if the noise figures of both systems are known and the gain of the first system is known. The transmission-line system and the TV receiver form two systems in cascade as viewed from the terminals of the antenna. Let N_T represent the noise figure of the transmission line, G_T the gain of the transmission line (actually $G_T < 1$ for transmission lines), and N_{TV} be the noise figure of the TV receiver. The overall noise figure N_{T-TV} of this cascade system is given by

$$N_{T-TV} = N_T + \frac{N_{TV} - 1}{G_T} \qquad (6\text{-}6)$$

A Channel 2 outdoor antenna system might have a transmission-line system with an attenuation of 4 dB ($G_T = 0.4$) and thus a noise figure of 4 dB ($N_T = 2.5$). By using a noise figure for the TV receiver of 6 dB, the overall noise for the transmission-line–TV-receiver system is 9.95, or approximately 10 dB. From Eq. (6-5) the equivalent rms noise voltage is therefore 4.24 μV. The rms sum of this noise voltage and the antenna noise voltage for this Channel 2 example is 15.1 μV. The required signal voltage for an excellent picture as defined above is therefore 1901 μV at the terminals of the antenna. The Channel 2 example showed that the noise voltage is due primarily to the high sky temperature and the low transmission-line-system loss and low TV-receiver-noise figure. A Channel 69 example is quite different. Figure 6-3 shows a sky temperature of approximately 10 K, yielding an rms noise voltage for a 10-dB-gain, 75-Ω antenna of 3.96 μV. If transmission-line-system losses are 10 dB and the noise figure of the UHF tuner is 13 dB, the total-system noise figure is 23 dB, and the equivalent rms noise voltage for the voltage-line–TV-receiver system is 18.9 μV. The total equivalent system noise voltage at the terminals of the antenna is 19.3 μV. The Channel 69 example shows that antenna noise is small compared with TV receiving-system noise. An excellent-quality picture would therefore require a 2430-μV signal at the terminals of the antenna.

A preamplifier mounted at the terminals of the antenna with a noise figure of 5 dB ($NF = 3.16$) and a gain of 25 dB ($G = 316$) can significantly improve the noise figure of the UHF TV receiving system. Consider a cascade system composed of two sections. The first section is the preamplifier, and the second section is the above-mentioned UHF transmission-line–TV-receiver system with a noise figure of 23 dB. The overall noise figure of this cascade system, using Eq. (6-6) with the variables relabeled for the new cascade, is

$$N = 3.16 + \frac{199.5 - 1}{316} = 3.79 = 5.8 \text{ dB}$$

The equivalent noise voltage is 2.6 μV and combined with the antenna noise is 4.7 μV. The required signal at the terminals of the antenna for an excellent picture is now 592 μV, or 12.3 dB less than without the preamplifier. It is noted that greater preamplifier gain would not significantly reduce the required signal strength and might impair the large signal performance of the preamplifier.

6-5   TRIANGULAR-DIPOLE ANTENNAS

A dipole formed from two triangular sheets of metal is sufficiently broadband with respect to gain and VSWR (with respect to 300 Ω) for all-UHF-channel reception. The flare angle α and the length $2A$ of the triangular dipole (also known as the bowtie antenna) are shown in Figure 6-4. The triangular dipole has many of the same

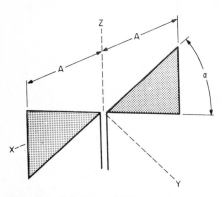

FIG. 6-4   Triangular dipole.

pattern and input-impedance characteristics of the biconical dipole discussed in Reference 12, but it is lighter in weight and simpler to construct. Further simplification of the triangular dipole to a wire outline of the two triangles results in significant degradation of its broadband performance. However, the metal triangles can be approximated with wire mesh, provided the mesh spacing is less than one-tenth wavelength. The input impedance of a triangular monopole versus length A for a variety of flare angles α is shown in Ref. 8. The larger the values of A and α, the further the pattern in the XZ plane deviates from those in the

XY plane and the more predominant are the sidelobes. On the basis of pattern and impedance characteristics, compromise values of α between 60 and 80° may be used for values of A up to 210° (or 0.58λ). Figure 6-5 shows the measured-gain characteristics of a triangular dipole over the UHF band.[9] The triangular dipole can also be mounted approximately one-quarter wavelength in front of a flat reflecting screen to increase gain and decrease back radiation; typically, it is stacked vertically in two- or four-bay configurations for increased gain. The measured gain for several triangular dipoles with flat-screen-reflector configurations is also shown in Fig. 6-5. Measurements of several commercially available, vertically stacked multibay triangular dipoles with flat-screen-reflector arrays show that the VSWR is typically less than 2.0, the front-to-back ratio is greater than 15 dB, and sidelobe levels are less than 13 dB below the peak gain over 90 percent of the UHF band.[4]

6-6   LOOP ANTENNAS

The single-turn circular-loop antenna is a popular indoor UHF antenna primarily because of its low cost. The single-turn loop in free space is discussed in detail in Sec. 5-3 in Ref. 13. Figure 5-10a* shows the typical configuration used for television reception when the impedance of the balanced feed line is 300 Ω. Although the loop is a resonant structure, entire-UHF-band operation is possible by using a 20.3-cm-diameter single loop where the circumference varies across the band from 1.0 wavelength at 470 MHz to 1.7 wavelengths at 806 MHz. Figure 5-16* shows the directivity of a single-turn loop versus circumference and thickness. This figure shows that the directivity is above 3.5 dB for a loop circumference between 1.0 and 1.7 wavelengths. Fig-

---

*This figure is reproduced at the end of this chapter.

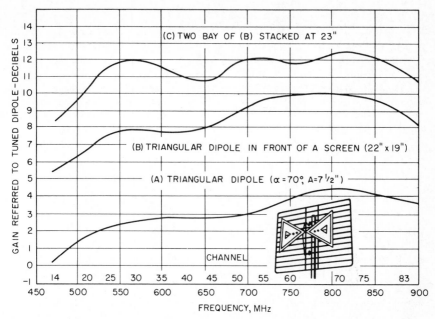

FIG. 6-5 Measured-gain characteristics. (*a*) Triangular dipole (α = 70°, *A* = 7½ in). (*b*) Triangular dipole in front of screen (22 by 19 in). (*c*) Two bays of antenna *b* stacked 23 in apart.

ures 5-11† and 5-12† show the input resistance and reactance, respectively, of a single-turn circular loop. These figures show that the input resistance of a loop with a conductor thickness parameter* equal to 10 ranges from 100 to 520 Ω over the circumference range of 1.0 to 1.7 wavelengths, and the reactance is less than 100 Ω for a circumference between 1.0 and 1.45 wavelengths but increases to 210 Ω between 1.45 and 1.7 wavelengths. Thus, across the 1.0- to 1.7-wavelength band the input VSWR on a 300-Ω feed line is less than 3 and is near 1 for a circumference near 1.3 wavelengths. The far-zone patterns of a one-wavelength loop with the thickness parameter equal to 10 are shown in Figs. 5-14† and 5-15*b*†. These figures show that a one-wavelength loop has a bidirectional pattern with maximum directivity along the loop axis and that a vertical loop fed at the bottom is horizontally polarized. Measurements made on commercially available single-turn loops for UHF TV reception show that midband gain with respect to a half-wavelength dipole including mismatch losses is near 3 dB and falls off to −1 dB at both ends of the band.[4] The measured VSWR is close to 1 near the center of the band, increasing to 4 at both ends of the band. Measurements also show that concentric groupings of loops, sometimes configured so that the loops can be turned, have lower gain but better VSWR characteristics than the single loop.

A planar reflector mounted parallel to the loop can be used to increase directivity and greatly reduce the sensitivity of the loop to signals arriving from behind the reflec-

---

*The thickness parameter is 2 ln $(2\pi b/a)$, where *b* is the loop radius and *a* is the conductor radius.

†This figure is reproduced at the end of this chapter.

tor. Figure 5-17* shows the geometry and directivity of a loop with a reflector. This figure shows that directivities greater than 8 dB are possible for a one-wavelength-circumference loop when the separation between the loop and the reflector is in a range from 0.1 to 0.2 wavelength. Figure 5-18* shows the input resistance and reactance for a one-wavelength-circumference loop versus separation from the planar reflector. The impedance is seen to be primarily real, increasing from 50 to 130 Ω for separations increasing from 0.1 to 0.2 wavelength. Figure 5-19* shows measured far-zone patterns for a loop with a reflector. This figure and Fig. 5-18* show the measured effects of variations in reflector size, suggesting a reflector size in the range of 0.6 to 1.2 wavelength on a side. Figure 5-23b* shows a method of feeding the antenna with coaxial line.

Yagi-Uda arrays of loops can be used for both single-channel and whole-band reception. Figure 5-21* shows the configuration and directivity achievable for single-frequency operation.

6-7   LOG-PERIODIC DIPOLE ARRAY ANTENNAS

The log-periodic dipole array (LPDA) antenna is the most commonly used all-VHF-channel antenna; it also is becoming a popular all-UHF-channel antenna. The LPDA is a broadband antenna capable of a 30:1 constant gain and input-impedance bandwidth. It has a gain of 6.5 to 10.5 dB with respect to a half-wavelength dipole, with most practical designs being limited to gains of 6.5 to 7.5 dB. The LPDA evolved from a broadband spiral antenna and is discussed more fully in Ref. 14. Figure 6-6 shows a schematic diagram of the LPDA and defines the angle α, dipole lengths ℓ_n, element diameters d_n, and element locations with respect to the apex of the triangle x_n. τ is shown as the ratio of a dipole's length or location to the length or location of the next larger dipole. The ratio τ is constant throughout the array. This figure also shows that all dipoles are connected to a central transmission line with a phase reversal between dipoles. In practice, the central transmission line takes two forms, a high-impedance

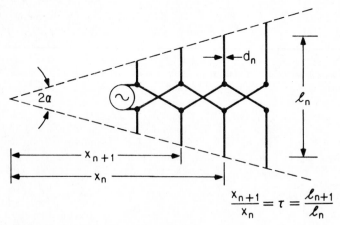

FIG. 6-6   Log-periodic dipole array.

$$\frac{x_{n+1}}{x_n} = \tau = \frac{\ell_{n+1}}{\ell_n}$$

---

*This figure is reproduced at the end of this chapter.

form and a low-impedance form. The high-impedance form designed for a 300-Ω balanced input inpedance consists of a single boom insulated from all array dipoles. The transmission line in this case is a high-impedance open two-wire line extending the length of the antenna with crisscross connections from dipole to dipole to accomplish the phase reversal. The high-impedance form should be limited to lower frequencies, where the required spacing between the two wires of the line is a small fraction of a wavelength. The low-impedance form designed for an unbalanced 75-Ω input impedance is composed of two parallel conducting booms extending the length of the antenna, forming a low-impedance two-conductor transmission line. Phase reversal from dipole to dipole is accomplished by alternating the attachment of the dipoles along the two-boom transmission line. A coaxial feed line may be placed within one of the booms and connected to the short-dipole end of the array. In this configuration the array acts as a balun.

Figure 14-26* in Ref. 14 is a graph of equal-gain contours for the LPDA with a line showing optimal design conditions. This figure determines values of σ and τ on the basis of the desired gain. The angle α is calculated from σ and τ as

$$\alpha = \tan^{-1}\left[\frac{1-\tau}{4\sigma}\right] \tag{6-7}$$

The desired bandwidth B is given as the ratio of the highest frequency to the lowest frequency of operation of the array. A larger design bandwidth is determined to accommodate the active region of the array at the high-frequency end of the array. The design bandwidth B_S is given by

$$B_S = B[1.1 + 7.7(1 - \tau)^2 \cot \alpha] \tag{6-8}$$

The required number of dipoles N in the array is given by

$$N = 1 + \frac{\ln (B_S)}{\ln (1/\tau)} \tag{6-9}$$

The length of the longest dipole ℓ_1 of the array is equal to the half wavelength at the lowest frequency in the desired bandwidth. The location of the dipole x_1 from the apex of the triangle is given by

$$x_1 = \frac{\ell_1}{2} \cot \alpha \tag{6-10}$$

The other dipole lengths ℓ_n and locations x_n are given by

$$\ell_n = \ell_1 \tau^{(n-1)} \quad 2 \leq n \leq N \tag{6-11}$$

$$x_n = x_1 \tau^{(n-1)} \quad 2 \leq n \leq N$$

Ideally, the length-to-diameter ratio of each dipole K_n should be identical. In practice, this is not usually the case. The primary effect of the variation of the length-to-diameter ratio is variation of the input impedance versus frequency. The impedance of the central transmission line Z_0 is next designed to transform the impedance of the active-region dipoles Z_A to the desired input resistance R_0. The impedance of the active-region dipoles is a function of the average K_n values of the elements K_{AVG} and is given by

$$Z_A = 120[\ln (K_{AVG}) - 2.25] \tag{6-12}$$

---

*This figure is reproduced at the end of this chapter.

The characteristic impedance of the central transmission line Z_0 is given by[10]

$$Z_0 = R_0 \left[\frac{\sqrt{\tau} R_0}{8\sigma Z_A} + \sqrt{\left(\frac{\sqrt{\tau} R_0}{8\sigma Z_A}\right)^2 + 1} \right] \qquad (6\text{-}13)$$

The characteristic impedance Z_0 may be achieved with a two-conductor transmission line in which each conductor has a diameter D and the center-to-center spacing S of the conductors is given by

$$S = D \cosh(Z_0/120) \qquad (6\text{-}14)$$

Because of the wide bandwidth of the antenna, the accuracy of construction of an LPDA is not critical.

Several variations of the basic LPDA are also common. A large percentage of commercially available VHF LPDA receiving antennas are designed with reduced gain at the lower VHF because of the cost of the longer elements and the high signal strengths at the lower VHF frequencies. The low-frequency gain is normally reduced by elimination of the lower-frequency dipoles. A VHF LPDA can also be designed by using dipoles forming a V. This configuration allows the operation of the dipoles in their half-wavelength and ¾-wavelength modes and therefore eliminates the need for the higher-frequency dipoles. An LPDA array designed with V dipoles for the low-VHF band will operate in the ¾-wavelength mode for the high-VHF band, as the frequency ratio of these two bands is approximately 3.

The high-frequency gain of an LPDA can be enhanced with the addition of parasitic directors at the short-dipole end of the array. Most home-use LPDA antennas are designed for a balanced 300-Ω input impedance, while most master-antenna-television (MATV) and community-antenna-television (CATV) applications use the 75-Ω unbalanced input-impedance configuration.

Measured performance[4] of 19 home-use VHF LPDA antennas indicated an average low-VHF-band gain of 4.5 dB with a standard deviation of 1.5 dB with respect to a half-wavelength dipole at the same height above ground. The average high-VHF-band gain is 7.2 dB with a standard deviation of 1.4 dB. VSWR for the VHF band averages 1.9 with a standard deviation of 0.7. Three measured 300-Ω UHF LPDA antennas show an average gain of 7.1 dB with a standard deviation of 2.7 dB and an average VSWR of 2.7 with a standard deviation of 1.3. These averages are taken over frequency within the respective bands and over the antennas tested.

A circularly polarized LPDA can be formed by combining two linearly polarized LPDAs. The low-impedance configuration requires four booms, two for the horizontal LPDA and two for the vertical LPDA. All dipole lengths and locations of the second LPDA are obtained by multiplication of the respective dipole lengths and locations by $\tau^{1/2}$. The second LPDA should therefore have the same apex location, τ, α, and gain as the first. Multiplication by $\tau^{1/2}$ accomplishes the required frequency-independent 90° phase shift of the second LPDA with respect to the first.

6-8 SINGLE-CHANNEL YAGI-UDA DIPOLE ARRAY ANTENNAS

The Yagi-Uda dipole array antenna is a high-gain, low-cost, low-wind-resistance narrowband antenna suitable for single-TV-channel reception. Such antennas are popular in remote locations where high gain is required or where only a few channels are to be received. The Yagi-Uda dipole array is also discussed in Chap. 4.

An empirically verified design procedure for the design of Yagi-Uda dipole arrays which includes compensation of dipole element lengths for element and metallic-boom diameters is presented.[11] Optimum Yagi-Uda arrays have one driven dipole, one reflecting parasitic dipole, and one or more directing parasitic dipoles as shown in Fig. 6-7. The design procedure is given for six different arrays having gains of 7.1, 9.2, 10.2, 12.25, 13.4, and 14.2 dB with respect to a half-wavelength dipole at the same height above ground. The driven dipole in all cases is a half-wavelength folded dipole which is empirically adjusted in length to achieve minimum VSWR at the design frequency. The length of the driven dipole has little impact on the gain of the array. Table 6-8 lists the optimum lengths and spacings for all the parasitic dipoles for each of the six arrays. The dipole lengths in this table are for elements which have a diameter of 0.0085 wavelength. Figures 6-8 and 6-9 are used to adjust these lengths for other dipole diameters and the diameter of a metallic boom, respectively. The frequency used in the design of single-channel Yagi-Uda arrays should be 1 percent

TABLE 6-8 Optimized Lengths of Parasitic Dipoles for Yagi-Uda Array Antennas of Six Different Lengths

$d/\lambda = 0.0085$ $s_{12} = 0.2\lambda$	Length of Yagi-Uda array, λ					
	0.4	**0.8**	**1.20**	**2.2**	**3.2**	**4.2**
Length of reflector, ℓ_1/λ	0.482	0.482	0.482	0.482	0.482	0.475
ℓ_3	0.424	0.428	0.428	0.432	0.428	0.424
ℓ_4		0.424	0.420	0.415	0.420	0.424
ℓ_5		0.428	0.420	0.407	0.407	0.420
ℓ_6			0.428	0.398	0.398	0.407
ℓ_7				0.390	0.394	0.403
ℓ_8				0.390	0.390	0.398
ℓ_9				0.390	0.386	0.394
ℓ_{10}				0.390	0.386	0.390
ℓ_{11}				0.398	0.386	0.390
ℓ_{12}				0.407	0.386	0.390
ℓ_{13}					0.386	0.390
ℓ_{14}					0.386	0.390
ℓ_{15}					0.386	0.390
ℓ_{16}					0.386	
ℓ_{17}					0.386	
Spacing between directors (s/λ)	0.20	0.20	0.25	0.20	0.20	0.308
Gain relative to half-wave dipole, dB	7.1	9.2	10.2	12.25	13.4	14.2
Design curve	(*A*)	(*B*)	(*B*)	(*C*)	(*B*)	(*D*)
Front-to-back ratio, dB	8	15	19	23	22	20

Length of director, ℓ_i/λ (row label spanning ℓ_3 through ℓ_{17})

*source: P. P. Viezbicke, "Yagi Antenna Design," NBS Tech. Note 688, National Bureau of Standards, Washington, December 1968.

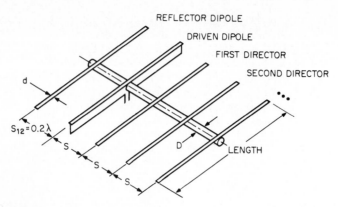

FIG. 6-7 Yagi-Uda array antenna.

below the upper frequency limit of the channel for VHF channels because the band-width of the Yagi-Uda is not symmetrical and gain falls rapidly on the high-frequency side of the band. The design frequency for UHF channels should be the center fre-quency of the channel. On the basis of desired gain, the lengths of the parasitic ele-ments are found in one of the columns of Table 6-8. The reflector spacing from the driven element for all six configurations is 0.2 wavelength, and all directors are equally spaced by the spacing shown in the table. The table also specifies the design curve A, B, C, or D to be used in Fig. 6-8. The parasitic-dipole lengths are next adjusted for the desired dipole diameter other than the diameter of 0.0085 wavelength. The reflec-tor length for the desired diameter can be read directly from one of the two upper curves in Fig. 6-8. Likewise, the length of the first director can be read directly from the designated director design curve for the desired diameter. This first director length is marked on the curve for subsequent use. The length of the first director for a dia-

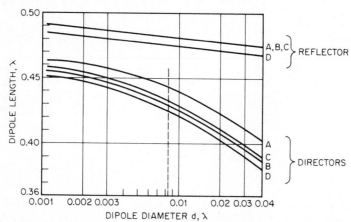

FIG. 6-8 Design curves to determine dipole lengths of Yagi-Uda arrays. *(Source: P. P. Viezbicke, "Yagi Antenna Design," NBS Tech. Note 688, U.S. Department of Commerce–National Bureau of Standards, October 1968.)*

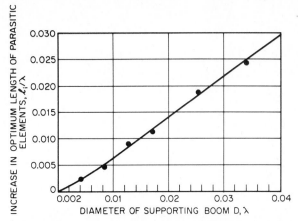

FIG. 6-9 Increase in optimum length of parasitic elements as a function of metal-boom diameter. *(Source: P. P. Viezbicke, "Yagi Antenna Design," NBS Tech. Note 688, U.S. Department of Commerce–National Bureau of Standards, December 1968.)*

meter of 0.0085 wavelength is also marked on the designated design curve. The arc length along the design curve between the two marks for the first director length is measured and used to adjust the remaining director lengths as follows. The other director lengths are plotted on the designated design curve without regard to their 0.0085-wavelength diameter. Each of the points is moved in the same direction and arc length as required for the first director. The new length is the diameter-compensated length for each of the remaining directors. If a metallic boom is used to support the reflector and directors, Fig. 6-9 is used to increase the length of each by the amount shown in the figure based on the diameter of the boom.

A circularly polarized Yagi-Uda dipole array can be formed by combining two linearly polarized Yagi-Uda dipole arrays on the same boom. The dipoles of one array might be horizontal and those of the other vertical. One array is positioned a quarter wavelength ahead of the other along the boom to accomplish the required 90° phase shift, and the feed position is midway between the two driven dipoles. Such a circularly polarized Yagi-Uda dipole array for Channel 13 is shown in Fig. 6-10.

6-9 BROADBAND YAGI-UDA DIPOLE ARRAY ANTENNAS

Several approaches to widening the inherently narrow bandwidth of the Yagi-Uda array are discussed. Yagi-Uda arrays with no more than five or six elements can be broadbanded by shortening the directors for high-frequency operation, lengthening the reflector for low-frequency operation, and selecting the driven dipole for midband operation. Figure 6-11 shows the measured gain versus frequency for a five-element single-channel and a five-element broadband Yagi-Uda array.[9]

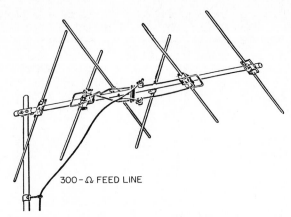

FIG. 6-10   Circularly polarized Yagi-Uda dipole array. (*Channel Master Corp.*)

The driven dipole of a Yagi-Uda array can be replaced with a two-element ordinary end-fire driven array. The driven array is designed for midband operation, the directors for high-frequency operation, and the reflector for low-frequency operation. An extension of this concept is the replacement of the driven element with a log-periodic dipole array. The parasitic elements are then used primarily to increase gain at the upper and lower frequency limits of the log-periodic dipole array.

Two Yagi-Uda dipole arrays of different design frequencies may be designed on the same boom and use the same driven element. This is possible when the design frequency ratio of the two Yagi-Uda arrays is 3:1 as in the high-VHF and low-VHF bands. The interlaced parasitic elements have only a small impact on one another, and the driven element is used simultaneously in the first and third half-wavelength modes.

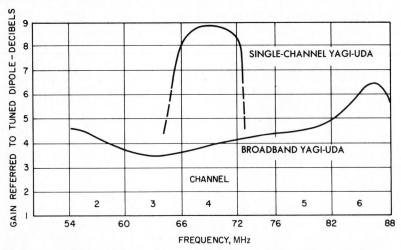

FIG. 6-11   Measured gain of five-element Yagi-Uda. (*a*) Single-element Yagi-Uda. (*b*) Broadband Yagi-Uda.

The reflector dipole of a Yagi-Uda array may be replaced with a corner reflector or a paraboloidal reflector. The reflector and driven element are designed for low-frequency and midfrequency operation, and the directors are designed for high-frequency operation.

The broadband Yagi-Uda is not commonly used for all-VHF-channel operation, but it is commonly used, generally with a corner reflector, for all-UHF-channel reception. Measurements of commercially available all-UHF-channel Yagi-Uda antennas typically show large variation in gain and input impedance as the frequency is varied across the UHF band.[4] Typical of such an antenna is an average gain of 7 dB with respect to a half-wavelength dipole at the same height above ground with a variation of ± 8 dB across the UHF band. The VSWR of such an antenna is found to vary within the range of 1.1 to 3.5 across the band.

6-10   CORNER-REFLECTOR ANTENNAS

The corner-reflector antenna is very useful for UHF reception because of its high gain, large bandwidth, low sidelobes, and high front-to-back ratio. (Corner-reflector antennas are discussed further in Sec. 17-1 in Ref. 15.) A 90° corner reflector, constructed in grid fashion, is generally used. One typical design is shown in Fig. 6-12. A wide-band triangular dipole of flare angle 40°, which is bent 90° along its axis as shown in Fig. 6-13 so that the dipole is parallel to both sides of the reflector, is a good choice for the feed element.[9] Experimental results indicate that an overall length of 14¼ in (362 mm) for the dipole gives the optimum value of average gain over the band. The dipole has a spacing of about $\lambda/2$ at midband from the vertex of the corner reflector. A larger spacing would cause a split main lobe at the higher edge of the band, where the spacing from the vertex becomes nearly a wavelength.

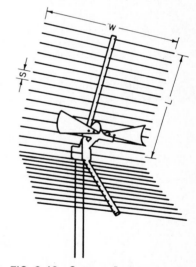

Figures 6-13 and 6-14 are experimental results useful in designing grid-type corner reflectors. Figure 6-13 shows the relation between grid length L and gain at three different frequencies. It is seen that beyond 20 in (508 mm) of grid length very little is gained. Figure 6-14 shows the relation between gain and grid width W. At 700 and 900 MHz, the improvement in gain for a grid over 20 in wide

FIG. 6-12   Corner-reflector antenna.

is rather insignificant. To reduce the fabrication cost, a width of 25 in (635 mm) is considered a good compromise. In Fig. 6-15 the spacing for grid tubing of ¼-in (6.35-mm) diameter can be determined from the allowable level of the rear lobe in percentage of the forward lobe at the highest frequency, 900 MHz. If 10 percent is the allowable value, the spacing should be slightly under 1½ in (38 mm). Below 900 MHz, the rear pickup will be less than 10 percent provided that the grid screen is wide

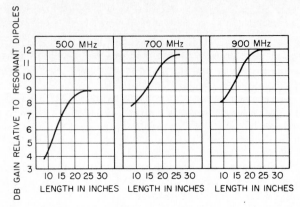

FIG. 6-13 Measured-gain characteristics versus grid length *L*.

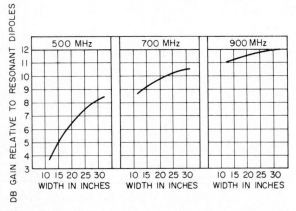

FIG. 6-14 Measured-gain characteristics versus grid width *W*.

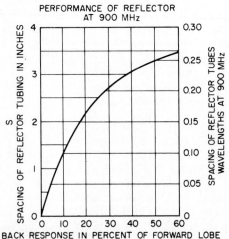

FIG. 6-15 Relation between the rear lobe in percentage of the forward lobe and the grid spacing *S* at 900 MHz.

$Z_0 = 280$

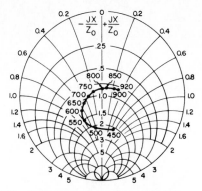

FIG. 6-16 Impedance characteristics of an ultrahigh-frequency corner reflector.

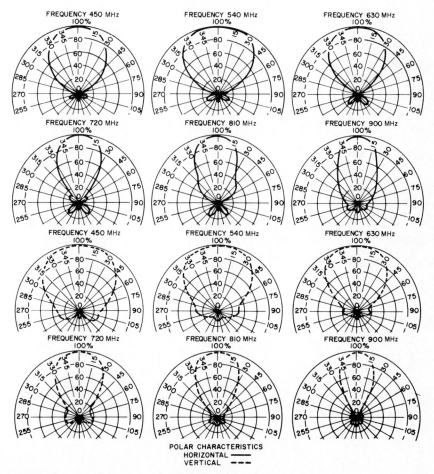

FIG. 6-17 Field patterns of an ultrahigh-frequency corner-reflector antenna.

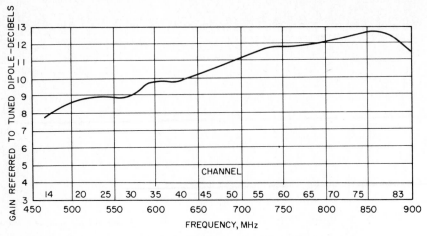

FIG. 6-18 Measured gain of an ultrahigh-frequency corner-reflector antenna.

enough. Thus, the width is determined by the lowest frequency, while the spacing of the grid tubing is determined by the highest frequency.

Figures 6-16 and 6-17 show impedance characteristics and field patterns respectively. Figure 6-18 shows the measured gain over the band.

REFERENCES

1 Federal Communications Commission, "Broadcast Television," *Rules and Regulations,* September 1972, sec. 73.6.

2 J. B. Snider, "A Statistical Approach to Measurement of RF Attenuation by Building Materials," NBS Rep. 8863, July 1965.

3 R. G. Fitzgerrell, "Indoor Television Antenna Performance," NTIA Rep. 79/28, NBS-9104386 Rep., 1979.

4 W. R. Free, J. A. Woody, and J. K. Daher, "Program to Improve UHF Television Reception," final rep. on FCC Proj. A-2475, Georgia Institute of Technology, Atlanta, September 1980.

5 D. G. Fink (ed.), *Electronics Engineers' Handbook,* McGraw-Hill Book Company, 1975. See sec. 21, "Television Broadcasting Practice," by J. L. Stern, and "Television Broadcasting Receivers," by N. W. Parker; sec. 18, "Antennas and Wave Propagation," by W. F. Croswell and R. C. Kirby.

6 "Engineering Aspects of Television Allocations," Television Allocations Study Organization (TASO) rep. to FCC, Mar. 16, 1959.

7 H. C. Ko, "The Distribution of Cosmic Radio Background Radiation," *IRE Proc.,* January 1959, p. 208.

8 G. H. Brown and D. M. Woodward, Jr., "Experimentally Determined Radiation Characteristics of Conical and Triangular Antennas," *RCA Rev.,* vol. 13, December 1952, p. 425.

9 H. T. Lo, "TV Receiving Antennas," in H. Jasik (ed.), *Antenna Engineering Handbook,* 1st ed., McGraw-Hill Book Company, New York, 1961, chap. 24.

10 G. L. Hall (ed.), *The ARRL Antenna Book,* American Radio Relay League, Newington, Conn., 1982.

11 P. P. Viezbicke, "Yagi Antenna Design," NBS Tech. Note 688, National Bureau of Standards, Washington, December 1976.
12. C. T. Tai, "Dipoles and Monopoles," in R. C. Johnson and H. Jasik (eds.), *Antenna Engineering Handbook,* McGraw-Hill Book Company, New York, 1984, Chapter 4.
13. G. S. Smith, "Loop Antennas," in R. C. Johnson and H. Jasik (eds.), *Antenna Engineering Handbook,* McGraw-Hill Book Company, New York, 1984, Chapter 5.
14. R. H. DuHamel and G. G. Chadwick, "Frequency-Independent Antennas," in R. C. Johnson and H. Jasik (eds.), *Antenna Engineering Handbook,* McGraw-Hill Book Company, New York, 1984, Chapter 17.
15. K. S. Kelleher and G. Hyde, "Reflector Antennas," in R. C. Johnson and H. Jasik (eds.), *Antenna Engineering Handbook,* McGraw-Hill Book Company, New York, 1984, Chapter 17.

BIBLIOGRAPHY

Balinis, C. A.: *Antenna Theory Analysis and Design,* Harper & Row, Publishers, Incorporated, New York, 1982.
Elliott, R. S.: *Antenna Theory and Design,* Prentice-Hall, Inc., Englewood Cliffs, N.J., 1981.
Fink, D. G. (ed.): *Television Engineering Handbook,* McGraw-Hill Book Company, New York, 1957.
Fitzgerrell, R. G., R. D. Jennings, and J. R. Juroshek: "Television Receiving Antenna System Component Measurements," NTIA Rep. 79/22, NTIA-9101113 Rep., June 1979.
Free, W. R., and R. S. Smith: "Measurement of UHF Television Receiving Antennas," final rep. on Proj. A-2066 to Public Broadcasting Service, Engineering Experiment Station, Georgia Institute of Technology, Atlanta, February 1978.
Kiver, M. S., and M. Kaufman: *Television Simplified,* Van Nostrand Reinhold Company, New York, 1973.
Martin, A. V. J.: *Technical Television,* Prentice-Hall, Inc., Englewood Cliffs, N.J., 1962.
Stutzman, W. L., and G. A. Thiele: *Antenna Theory and Design,* John Wiley & Sons, Inc., New York, 1981.
Television Allocations Study Organization: *Proc. IRE* (special issue), vol. 48, no. 6, June 1960, pp. 989–1121.
UHF Television: *Proc. IEEE* (special issue), vol. 70, no. 11, November 1982, pp. 1251–1360.
Weeks, W. L.: *Antenna Engineering,* McGraw-Hill Book Company, New York, 1968.
Wells, P. I., and P. V. Tryon: "The Attenuation of UHF Radio Signals by Houses," OT Rep. 76-98, August 1976.

The following figures cited on pp. 6-12 through 6-15, have been reproduced from Refs. 13 and 14.

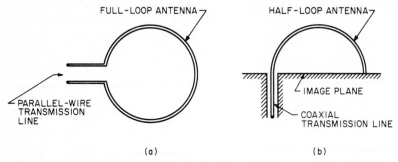

FIG. 5-10 Methods of driving the circular-loop antenna. (*a*) Full-loop antenna driven from parallel-wire transmission line. (*b*) Half-loop antenna driven from coaxial transmission line.

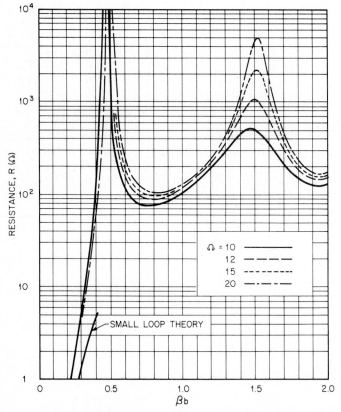

FIG. 5-11 Input resistance of circular-loop antenna versus electrical size (circumference / wavelength).

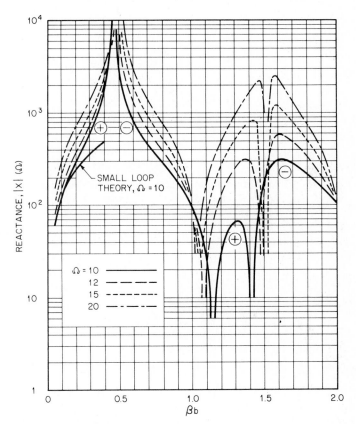

FIG. 5-12 Input reactance of circular-loop antenna versus electrical size (circumference/wavelength).

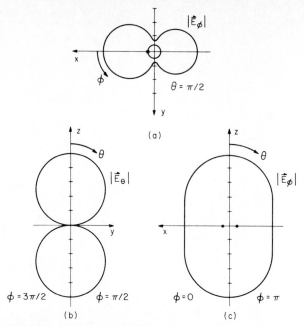

FIG. 5-14 Far-zone electric field for loop with $\beta b = 1.0$, $\Omega = 10$. (a) Horizontal-plane field pattern $|\mathbf{E}_\phi|$, $\theta = \pi/2$. (b) Vertical-plane field pattern $|\mathbf{E}_\theta|$, $\phi = \pi/2$, $3\pi/2$. (c) Vertical-plane field pattern $|\mathbf{E}_\phi|$, $\phi = 0$, π.

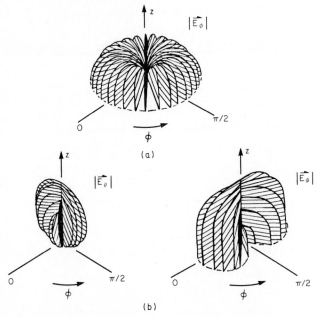

FIG. 5-15 Far-zone electric field patterns in upper hemisphere. (a) Electrically small loop, $\beta b \ll 1$. (b) Resonant loop, $\beta b = 1.0$.

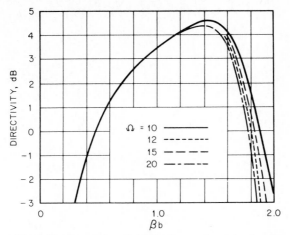

FIG. 5-16 Directivity of circular-loop antenna for $\theta = 0$, π versus electrical size (circumference/wavelength).

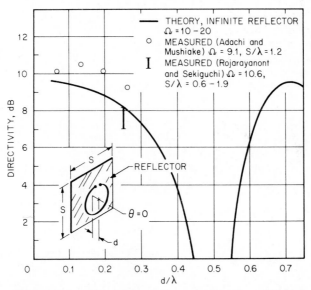

FIG. 5-17 Directivity of circular-loop antenna, $\beta b = 1.0$, for $\theta = 0$ versus distance from reflector d/λ. Theoretical curve is for infinite planar reflector; measured points are for square reflector.

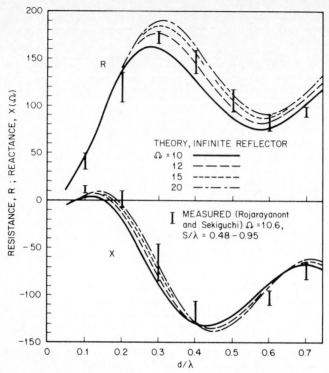

FIG. 5-18 Input impedance of circular-loop antenna, $\beta b = 1.0$ versus distance from reflector d/λ. Theoretical curves are for infinite planar reflector; measured points are for square reflector.

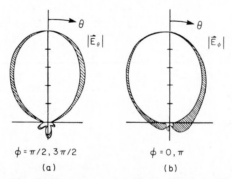

FIG. 5-19 Measured far-zone electric field patterns for loop with $\beta b = 1.0$ over reflector, $d/\lambda = 0.25$. Inner curve $s/\lambda = 0.95$; outer curve $s/\lambda = 0.64$. (*a*) Vertical-plane field pattern $|\mathbf{E}_\phi|$, $\phi = \pi/2, 3\pi/2$. (*b*) Vertical-plane field pattern $|\mathbf{E}_\phi|$, $\phi = 0, \pi$. (*Measured data from Rojarayanont and Sekiguchi.*)

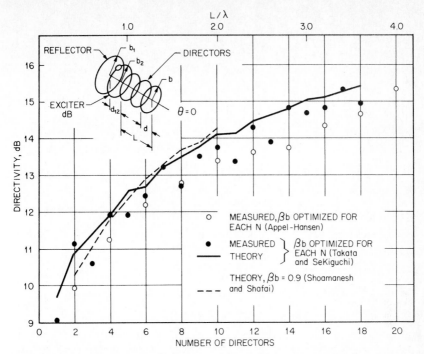

FIG. 5-21 Directivity of Yagi-Uda array of circular-loop antennas for $\theta = 0$ versus number of directors, director spacing $d/\lambda = 0.2$.

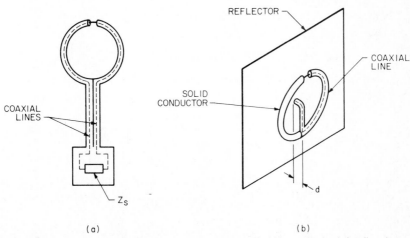

(a)   (b)

FIG. 5-23 (a) Balanced shielded-loop antenna and (b) method of feeding loop antenna in front of planar reflector.

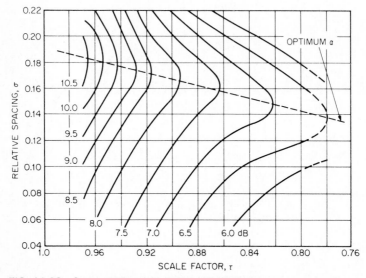

FIG. 14-26 Constant-directivity contours in decibels versus τ and σ; $Z_{0f} =$ 100 Ω, $h/a = 125$. Optimum σ indicates maximum directivity for a given value of τ. (*After Ref. 31.*)

Chapter 7

Microwave-Relay Antennas

Charles M. Knop

Andrew Corporation

Typical Microwave-Relay Path

The typical microwave-relay path consists of transmitting and receiving antennas situated on towers spaced about 30 mi (48 km) or less apart on a line-of-sight path. If we denote P_T, G_T, P_R, and G_R as the transmitted power, transmitting-antenna gain, received power, and receiving-antenna gain respectively, then these are related by the well-known free-space Friis transmission law (neglecting multipath effects):

$$P_R = P_T G_T G_R / (4\pi R/\lambda)^2 \qquad \textbf{(7-1)}$$

where R = distance between receiving and transmitting antennas

λ = operating wavelength in free space

A typical modern-day telecommunications link is designed to operate with $P_T =$ 5 W ($+37$ dBm) and $P_R = 5\ \mu$W (-23 dBm). (Note that -23 dBm is the operating level. Typical receivers still give adequate signal-to-noise ratios for fade margins 40 dB below this level, i.e., down to -63 dBm.) Thus, at a typical microwave frequency in the range of 2 to 12 GHz (say, 6 GHz) and $G_T = G_R$ (identical antennas at both sites), we see that antenna gains of about 10^4 (40 dBi) are required if no other losses exist. However, a typical installation (usually having several antennas on a given tower; see Fig. 7-1) has the antenna connected to a waveguide feeder running down the length of the tower (typically 200 ft, or 61 m, high) into the ground equipment. The total power received is then less than that given by Eq. (7-1) because of the losses in the waveguide feeder in the antenna (typically 0.25 dB) and in the tower feeder (typically 3.0 dB). This raises the above antenna gain requirement to about 43 to 44 dBi (all these values are at 6 GHz and change accordingly over the common-carrier bands, since the free-space loss and feeder loss increase with frequency).

FIG. 7-1 Typical tower with radio relay antennas. *(Courtesy Andrew Corp.)*

Specifications for Microwave-Relay Antennas

In addition to the above high gains, the radiation-pattern envelope (RPE; approximately the envelope obtained by connecting the peaks of the sidelobes) should be as narrow as possible, and the front-to-back ratio should be high to minimize interference with adjacent routes. Typical Federal Communications Commission (FCC) RPE specifications are shown in Fig. 7-2 (all patterns discussed here are horizontal-plane patterns; hence the *E*-plane pattern is horizontally polarized, and the *H*-plane is vertically polarized). Figure 7-2 also shows the RPE of a conical-cornucopia antenna, which will be discussed.

The requirement of simultaneous high gain and low sidelobes is, of course, inconsistent in antenna operation; hence some trade-off is necessary. Since the gain of an antenna is given by $G = \eta 4\pi A/\lambda^2$, where η is the total aperture efficiency (typically 0.50 to 0.70) and A is the physical area, we see that for circular apertures of diameter D, we require $D/\lambda \doteq 60$ for 43-dB gain and for $\eta = 0.5$; i.e., $D \doteq 10$ ft (3 m) at 6 GHz (with corresponding sizes at other bands). This means that the beamwidth between the first nulls, θ null, will be about θ null $= 120/(D/\lambda) \doteq 2°$. Thus, the supporting structure and tower must be very stable to ensure no significant movement of the beam (and consequent signal drop and cross-polarization degradation) even during strong winds (125 mi/h, or 201 km/h).

A further specification not found in radar or other communication systems is the extremely low voltage-standing-wave-ratio (VSWR; typically 1.06 maximum)

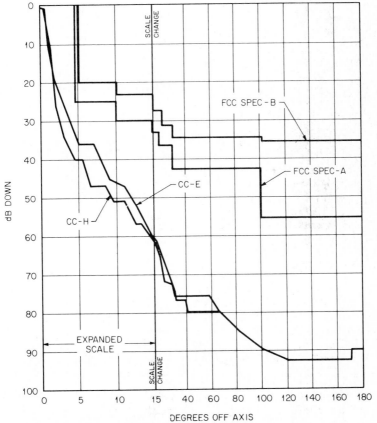

FIG. 7-2   FCC RPE specifications and conical-cornucopia RPE.

requirement at the antenna input. This is to reduce the magnitude of the round-trip echo (due to the mismatch at both the antenna and the equipment ends) in the feeder line. This echo produces undesirable effects: (1) cross talk in the baseband of a frequency-division-multiplexed-frequency modulation (FDMFM) analog system, especially in the upper telephone channels,[1] (2) ghosts on a TV link, and (3) an increase in the bit error probability in a digital system (which does, however, tolerate higher VSWR). Also, for dual-polarization operation at a given frequency, polarization purity expressed as cross-polar discrimination (XPD) between the peak of the main copolar beam and the maximum cross-polar signal is typically required to be -25 dB minimum across an angle twice the 3-dB beamwidth of the copolar pattern. Additionally, bandwidths of up to 500 MHz at any common-carrier band are required (these bands extend from 1.7 to 13.25 GHz, with FDMFM up to 2700 channels at 3.7 to 11.7 GHz and up to 960 channels of digital system above 10.7 GHz).

Added to the above are a low-cost requirement, a necessity to operate under virtually all weather conditions, and the low near-field coupling specifications between antennas mounted on the same tower (typically -120 dB back to back and -80 dB minimum side to side). Finally, no loose contacts or other nonlinear-contact phenomena (the so-called rusty-bolt effect) are allowed, because if signals of frequencies A and B arrive simultaneously, then a third signal of $2A$-B is generated. These $2A$-B products must be down to about -120 dBm or more.

Antennas in Use

The most common antennas in use today which meet all the above requirements are the symmetrical prime-fed circular paraboloidal dish (and its grid equivalent, though this is suitable for only single polarization and has much worse RPE), the offset paraboloid, the conical and pyramidal cornucopias, and the dual reflector (Cassegrain and gregorian). The principles of operation and designs of these antennas will now be reviewed, with major emphasis on the symmetrical prime-fed paraboloid, which is by far the most frequently used microwave-relay antenna.

7-2   THE PRIME-FED SYMMETRICAL PARABOLOIDAL DISH

Many excellent treatments describing the operation and design of this antenna are reviewed in the open literature[2-13] (see especially Silver,[2] Ruze,[3] and Rusch-Potter[4]). Almost all these references, however, were written prior to the ease of accessibility (for most antenna engineers) to the minicomputer (this, in turn, required analytical approximations to be made; for example, the feed-horn patterns were described by $\cos^n \psi$-type functions) and prior to the full evolvement of Keller's geometric theory of diffraction (GTD) to the uniform geometric theory of diffraction (UGTD) by Kouyoumijian[14] and the concept of equivalent rim currents of Ryan and Peters.[15] Alternatively, Rusch's asymptotic physical optics (APO)[16] in its corrected form (CAPO)[17,18] can be used instead of UGTD.[19-20] Because of space limitations, we present here only results based on the above[21] and other analyses.

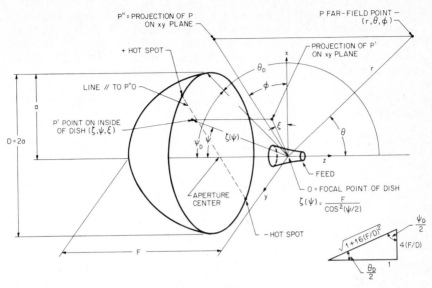

FIG. 7-3 Symmetrical-paraboloidal-dish geometry.

Basic Principle of Operation

Consider, then, the typical geometry of Fig. 7-3, depicting a prime feed illuminating a circular paraboloid. We start with the known horn E- and H-plane patterns in amplitude and phase in a computer file and assume that the horn produces a symmetrical pattern [see Eq. (1) of Ref. 17]. The surface current $\mathbf{K}$ produced on the dish surface is then found by using geometric optics (GO), i.e., by assuming that this surface is an infinite sheet and that the dish is in the far field of the horn, so that the E and H fields of the horn are related by the usual free-space impedance relationship. It is known that this GO approximation is questionable near the dish edge,[22] but this only has a consequence on the fields radiated in the 120° vicinity.[18]

The electric field $\mathbf{E}$ scattered by the dish is then obtained by integration of $\mathbf{K}$ over the *illuminated* surface of the dish up to and *including its edges*. The resulting expressions are considered in three regions: region 1, $0 \leq \theta \leq \theta_1 \doteq$ fifth sidelobe; region 2, $\theta_1 \leq \theta \leq \theta_2 \doteq 175°$; and region 3, $175° \leq \theta \leq 180°$. The region 1 expressions are double integrals [see Ref. 17, Eq. (2)], which can be reduced to a single integral via Bessel's orthogonality relation (Ref. 2, page 337). The region 2 UGTD (or CAPO) expressions are essentially asymptotic integrations (as given explicitly in Rusch[16] with the diffraction coefficients replaced by those of UGTD[19,20] or CAPO[18]) and come from the "hot spots" on the dish edge* (see Fig. 7-3). The region 3 fields come from the equivalent rim currents[15,19,20,22-24] [and are given by Eq. (6) of Ref. 23]. Also, in the range of $0 \leq \theta \leq \theta_D$ horn superposition is required.

---

*For the observation point in the horizontal plane, these hot spots are on the left and right edges of the dish.

The above expressions then give the patterns from the perfectly focused feed in a perfect paraboloid. To obtain the gain we compute the field on axis ($\theta = 0$), obtain the angular power density, and then divide by the total feed power over 4π sr. This also gives the aperture efficiency η. A program that computes both patterns and gain takes a few minutes to run on a minicomputer such as a DEC VAX 11/780.

The 100 Percent Feed Horn

Prior to considering actual feed horns, it is very instructive to create a hypothetical horn file having the pattern value* $1/\cos^2(\psi/2)$ for $0 \le \psi \le \psi_D$ and 0 for $\psi > \psi_D$ (as depicted in Fig. 7-4 for an $F/D = 0.30$ dish). This exactly negates the $(1/\rho)$ free-space loss [since $\rho = F/\cos^2(\psi/2)$ for a perfect paraboloid], so that the magnitude of the fields illuminating the dish surface is then constant. Hence, the aperture amplitude and phase are constant, giving $\eta = 100$ percent.

Using the expressions for the above three regions gives the patterns of Fig. 7-5a (near-in pattern) and 7-5b (wide-angle RPE†) for the E plane. The H-plane pattern is not shown because of lack of space, but it is similar except in the dish shadow $\theta > \theta_D$, where it is lower than the E-plane pattern since the H-plane diffraction coefficients here are smaller than those of the E plane. One notes that for small θ the patterns are close to $2J_1(U)/U$, where $U = C\sin\theta$, $C = \pi D/\lambda$.

Trade-Off between Gain and Radiation-Pattern Envelope

The above uniform case gives the highest possible gain, but its RPE is not low enough for most applications, since the dish edge illumination is 4.6 dB *above* that on the horn axis. Hence, a feed having lower edge illumination but still giving reasonable efficiency is required.

The opposite extreme of the 100 percent efficiency case is to illuminate the dish with a horn pattern having very low edge illumination. Now from Kelleher (see Ref. 49) we know that it is a universal characteristic of conventional single-mode horns (be they circular, rectangular, square, smooth-wall, corrugated, or whatever) that if their 10-dB beamwidths (i.e., let ψ_{10} be the angle off axis at which the power is 10 dB down from that on axis) are known, their power patterns from the 0- to about the 20-dB-down level can be described fairly well by Kelleher's formula:

$$\text{DBE}(\psi) = 10(\psi/\psi_{10E})^2$$

$$\text{DBH}(\psi) = 10(\psi/\psi_{10H})^2$$

$$(7\text{-}2)$$

where the subscripts E and H denote E and H planes respectively.

The above characteristic allows us to investigate quickly the trade-off between high gain and RPE. Four examples (all assumed to have back radiation of -40 dB and $\psi_{10E} = \psi_{10H} = \psi_{10}$ and to have zero phase error) are shown in Fig. 7-4. These

---

*E and H planes are assumed to be equal and with zero phase error.

†Note that for $10° \lesssim \theta \le \theta_D = 100°$, the RPE of Fig. 7-5$b$ is unrealistic, since this ideal feed has no spillover.

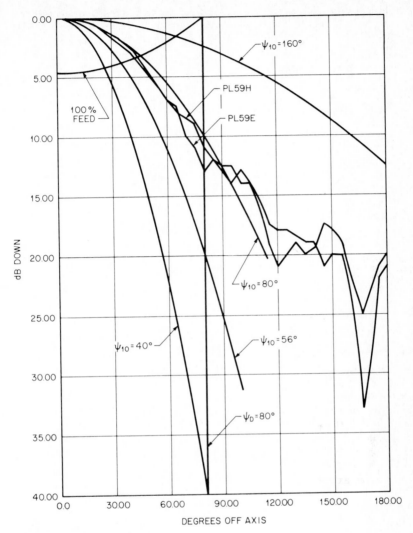

FIG. 7-4 Various feed patterns.

feeds produce the *E*-plane patterns also shown in Fig. 7-5*a* and *b*. A plot of η percent versus ψ_{10}/ψ_D (also in Fig. 7-5*b*) confirms the rule (e.g., see Silver,[2] page 427) of choosing about a 10-dB edge taper to achieve peak gain.* One notes from Fig. 7-5*b* that for $\theta > \theta_D$ (here $\theta_D \doteq 100°$) the 100 percent feed case has the highest RPE, whereas from Fig. 7-5*a* it has the narrowest main beam and the highest gain. Also, it

---

*Note that actual gains, because of feed defocusing, dish rms, feed-support and strut scattering, etc., are lower than that of Fig. 7-5*b* by typically 5 to 10 percent, as discussed below.

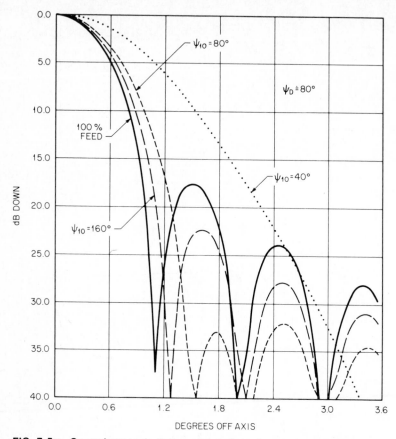

FIG. 7-5a   On- and near-axis *E*-plane dish patterns for various feeds.

is seen from Fig. 7-5*b* that an excellent (wide-angle, $\theta \gtrsim 10°$) RPE but (from Fig. 7-5*a*) a very low gain are provided by the extremely underilluminated feed ($\psi_{10} = 40°$).

Practical Feeds

Actual feeds[25] have main-beam patterns close (with 1 to 2 dB at the 20-dB-down level) to the Kelleher description; some examples are given in Table 7-1. The most common feed used for terrestrial antennas consists of the circular-pipe TE_{11}-mode feed, sometimes with a recessed circular ground plate having quarter-wave-type chokes. These chokes control the external *E*-plane wall current that normally would flow on the horn's exterior surface, and they tend to make the *E*-plane pattern equal to that of the *H* plane, with both being flatter over the dish and having a faster falloff near the

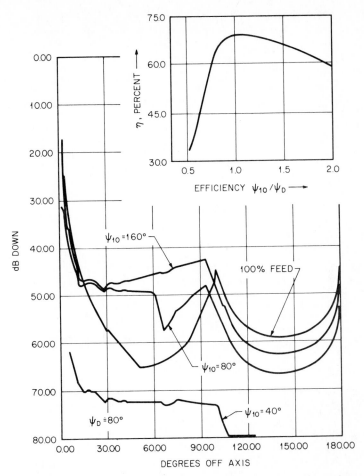

FIG. 7-5b *E*-plane RPE and aperture efficiency of dish with various feeds.

dish edge, causing improved efficiency. This horn has been patented by Yang-Hansen;[26] similar work by Wohlleben has been reported (see Love[25]). The rectangular horn is also commonly used since ℓ and w can be adjusted (at one band) to give almost equal *E*- and *H*-plane patterns.

Required *F/D*

For terrestrial radio work, the F/D ratios usually fall in the range of 0.25 (i.e., $\psi_D = 90°$, focal-plane dish) to about 0.38. For deeper dishes ($F/D < 0.25$), the requirement of building a feed having $\psi_{10} > 90°$ with low spillover is difficult. For shallower dishes

TABLE 7-1   Approximate 10-dB Angles for Practical Feed Horns*

Horn	Mode	ψ_{10E} °	ψ_{10H} °
Rectangular $\ell \times w$ (**E** perpendicular to ℓ)	TE_{10}	$42/(w/\lambda)$	$59/(\ell/\lambda)$
Circular,* $C \geq 1.84$	TE_{11}	$52/(DH/\lambda)$	$62/(DH/\lambda)$
Circular[25] (Potter), $C \geq 3.83$	$TE_{11} + TM_{11}$	$62/(DH/\lambda)$	$62/(DH/\lambda)$
Circular (corrugated, "$\lambda/4$ HE_{11} teeth") zero or small (less than 10°) phase error $C \geq 2.40$	HE_{11}	$62/(DH/\lambda)$	$62/(DH/\lambda)$
Circular (corrugated, "$\lambda/4$ HE_{11} teeth") large (180° or more) phase error, half angle α_0 $(0 \leq \alpha_0 \leq 85°)$	HE_{11}	$\alpha_0/\sqrt{2}$ (approximate)	$\alpha_0/\sqrt{2}$ (approximate)
Diagonal[25] (A. W. Love), $D/\lambda \geq 0.50$ (D = side length of aperture $D \times D$); (small phase error)	Two TE_{11}	$50.5/(D/\lambda)$	$50.5/(D/\lambda)$

*All circular horns have inner-diameter DH, $C = \pi DH/\lambda$. All zero- or small-phase-error horns have their phase center on or very close to the aperture; the large-phase-error horns have it close to the horn apex. Note that the E- and H-plane phase centers are usually not exactly coincident; their average position is made to coincide with the dish focal point. As seen, the zero- or small-phase-error horns are *diffraction-limited;* i.e., they have beamwidths which decrease with increasing frequency, whereas the large-phase-error corrugated horn has equal E- and H-plane patterns which are frequency-independent and hence is called a *scalar feed.*

$(F/D > 0.38, \psi_D < 67°)$, the feed is too far from the dish, which makes it impractical to use a radome; feed losses also are higher. Thus, F/D typically is about 0.3, and we therefore seek feeds having $\psi_{10} \doteq 80°$.

FIG. 7-6 A standard buttonhook feed in a paraboloidal dish. *(Courtesy Andrew Corp.)*

Typical Case

A typical feed is a rectangular horn (as described in Table 7-1) fed by a buttonhook feed (similar to that shown for a circular-horn feed in Fig. 7-6). The feed patterns (the phase patterns of which are virtually flat over the dish and hence are not shown) of this rectangular horn are shown in Fig. 7-4 (averaged over both sides in the absence of the buttonhook) and are designated by PL59 (we note they have about a 10-dB edge taper, so the horn has $w/\lambda \doteq 0.53$, $\ell/\lambda \doteq 0.74$). The resulting measured and predicted E-plane dish patterns are shown in Fig. 7-7. We see that the prediction and measurements are close in the main beam, in the first few sidelobes, and in the rear, but in the approximately -10-dBi (here the 0-

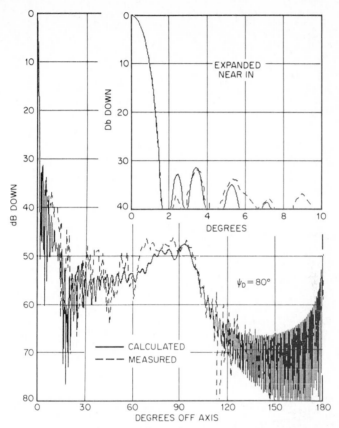

FIG. 7-7 Measured and predicted *E*-plane patterns of a paraboloidal dish with a buttonhook feed.

dBi level ≐ 43 dB down) level ($10° \lesssim \theta \lesssim 70°$) the measured fields are typically ± 5 dB in range about the predicted envelope. The major reasons for these discrepancies are twofold: dish imperfections and feed-support scattering, as discussed below.

Pattern and Gain Dependence on Distance

The above far-field formulations are (by definition) for an infinite-observation distance. We wish to determine the effects of a finite distance r in the range of $0.2 \, (D^2/\lambda) \leq r \leq \infty$, since such short paths are necessary in many applications. To do this we use the double-integral (exact-phase) expressions (the same result, off axis, can also be obtained by GTD[27]). The *E*-plane results are shown in Fig. 7-8*a* for the case of the typical buttonhook feed; the results for the 100 percent feed are shown in Fig. 7-8*b*. Figure 7-9 shows the on-axis gain drop relative to that at infinity for these cases

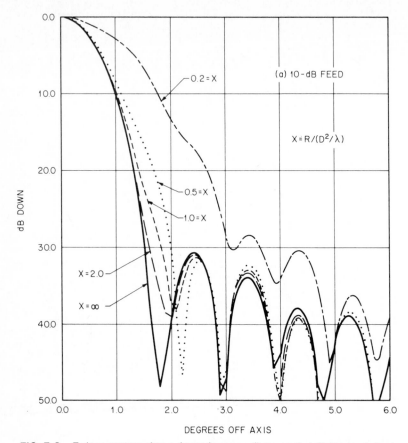

FIG. 7-8   *E*-plane-pattern-shape dependence on distance. (*a*) Buttonhook feed.

owing to this finite-observation distance. The gain drops because the main beam widens, the sidelobes rise, and the nulls fill in. In most practical installations, r is greater than D^2/λ; thus, the gain is degraded by at most 0.3 dB. These results are in agreement with those of Silver[2] (page 199, for the uniform case) and Yang[28] (for $\cos^n \psi$-type feeds). If two identical antennas are used, the total gain drop is then double the amount given by Fig. 7-9.

Effect of Feed Displacement

In general, the feed is inadvertently defocused (that is, the feed's phase center is not coincident with the focal point of the dish). This comes about because in most cases the feed is attached to a waveguide feeder, which in turn is attached to the hub of the dish, which may be distorted slightly (especially axially) from paraboloidal. It has been shown (e.g., by Imbriale et al.; see Love[25]) that this defocusing effect can be

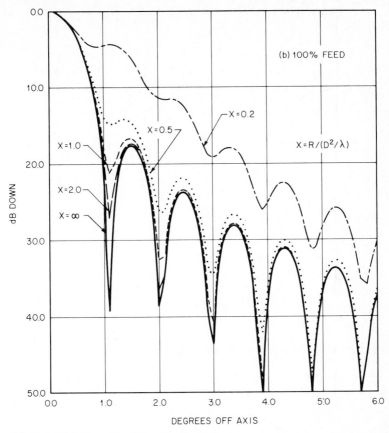

FIG. 7-8 (Continued) (*b*) 100 percent feed.

accounted for by inserting a phase-error term, exp $(j\Phi_D)$, multiplying the dish surface current, where (letting $k = 2\pi/\lambda$)

$$\Phi_D = k \cdot DT \sin \psi \cos (\xi - \xi_D) - k \, DZ \cos \psi \qquad \textbf{(7-3)}$$

where DT, ξ_D, and DZ define the feed displacement and are respectively the transverse ($DT > 0$), azimuthal angular, and axial displacement ($DZ > 0$ for feed movement away from the dish, and $DZ < 0$ for feed movement toward the dish).

The pattern degradation and gain drop resulting from axial shifting of the above typical buttonhook feed (in the above $F/D = 0.30$, $D/\lambda = 60$ dish) are shown in Figs. 7-10a and 7-11 respectively.[29] It is seen that the pattern distortion results in intolerable (≥ 0.2-dB) gain drops for $|DZ|/\lambda \gtrsim 0.15$. For transverse (lateral) displacements, the beam squints, but the pattern shape is not distorted as much for the same wavelength displacement as for the axial case (see Fig. 7-10b). The corresponding gain drops are also shown in Fig. 7-11, and it is seen that for $DT/\lambda < 0.15$ the gain

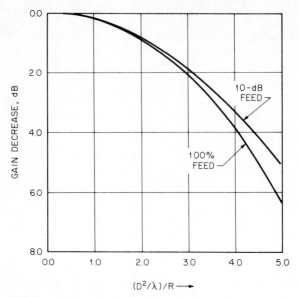

FIG. 7-9   Gain drop with distance.

drop is less than 0.04 dB. These results hold for any angular displacement ξ_D. The beam squint ($D\theta°$) for the subject antenna is also shown in Fig. 7-11. Although not shown here, repeating the above for a feed with the same edge taper in a deeper dish results in more distortion and gain drop than for a shallower dish.*

Effect of Surface Errors

By measuring at a sufficient number of points the normal difference Δn (ψ,ξ) between an actual dish and a fixture representing the paraboloid to be made, it is found that these surfaces are locally parallel. As such, a phase error of amount $360\Delta n$ (ψ,ξ)/λ cos ($\psi/2$), representing the additional path length to the reflector surface, must be included in **K**. This phase error degrades the gain, RPE, and XPD. Performing the necessary *double* integration reveals that the resulting gain drop is very close to that given by Ruze's formula[30,10] (see also Zucker[31]):

$$\Delta G_{rms} = -10 \log_{10} [\exp(-4\pi\epsilon_{rms}/\lambda)^2] \quad \text{dB} \quad \text{(7-4)}$$

where ϵ_{rms} is the root-mean-square deviation (in the normal direction) of the measured data points from the best-fit paraboloidal surface of these points [actually, Eq. (7-4) is based on a large F/D dish and hence is an upper bound on the gain drop]. Since we

---

*Similarly, for a given F/D dish and a given feed displacement, a feed with higher edge illumination produces more pattern distortion and gain drop than one with lower edge illumination (not shown here).

strive for $\Delta G_{rms} \leq 0.15$ dB, we must make $\epsilon_{rms} \leq \lambda/70$ [e.g., $\epsilon_{rms} < 0.030$ in (0.76 mm) at 6 GHz] for gain purposes. However, use of the Δn (ψ,ξ) file and double integration of many antennas reveals a rapid degradation of RPE (only in the $\theta < \theta_D$ region) with ϵ_{rms} (since this file contains many spatial harmonics[32]). This work and some analytical work by Dragone and Hogg[33] using this approach, as shown in Fig. 7-12 (for a 10-dB feed and 15ψ harmonics representing the actual surface), discloses that deviations from the parabolic should be held within about $\pm\lambda/100$ [i.e., $\epsilon_{rms} \lesssim \lambda/300$ or $\epsilon_{rms} \lesssim 0.007$ in (0.1778 mm) at 6 GHz] if approximately 3-dB or less increases from the perfect-case RPE are desired at about the 60-dB-down level (see also Fig. 7-7 for this increase in RPE in the 10° to 70° region due to rms). The XPD for realistic dishes is typically more than +50 dB down from the main beam.*

Front-to-Back Ratio

Using the above analytical results to evaluate the front- and back-field expressions explicitly for the general case, one obtains the expression for f/b:[23]

$$f/b = G + T + K - G_{HORN} \qquad dB \qquad (7\text{-}5)$$

where G = gain of dish = $10 \log_{10}(\eta C^2)$, T = average taper of feed at dish edge, K = $20 \log_{10}\{\sqrt{1 + 16(F/D)^2}/4(F/D)\}$, G_{HORN} = on-axis gain of horn, all in decibels. Since typical values are $T = 10$ dB, $F/D = 0.3$, and $G_{HORN} \doteq 5$ dB, we arrive at the simple result

$$f/b \doteq G + 7 \qquad (7\text{-}6)$$

which has been empirically known for many years[34] and is good, in practice, within 2 to 3 dB.[23]

In many installations, f/b ratios of 65 dB or larger are required; we then see from Eq. (7-6) that (for typical 40- to 45-dB-gain dishes) it is impossible to achieve this requirement with standard circular paraboloids. To overcome this fact, *edge-geometry schemes* are employed (here the circular edge is replaced with a polygonal one so as to destroy, in the back of the dish, the in-phase addition from each increment of the edge). Alternatively, an absorber-lined cylindrical shroud is used (see Fig. 7-1); this decreases the RPE significantly in the shadow of the shroud.

Effect of Vertex Plate

A plane circular plate (vertex plate[2,35] of diameter DV and thickness T having $DV = 2\sqrt{\lambda F/3}$ and $T = 0.042\lambda$; these dimensions are usually fine-tuned experimentally) is invariably used at the center of the dish (see Fig. 7-6) to minimize the VSWR contribution from the dish. The vertex-plate reflections essentially cancel the remaining dish reflections. This plate also affects the pattern and gain of the dish antenna since it introduces a leading phase error. The result is to raise the close-in radiation levels

---

*This is due to rms effects only; other effects (especially mechanical-feed asymmetry) increase this to 25 to 30 dB down.

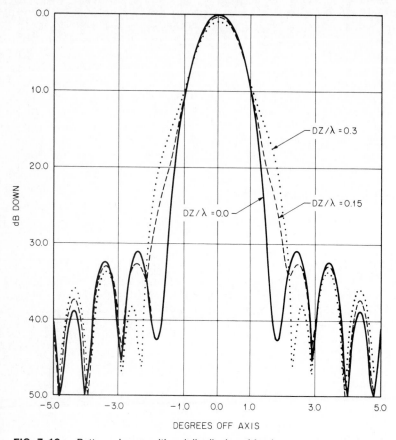

FIG. 7-10a Pattern change with axially displaced feed.

slightly and to cause a slight drop in gain. In some cases, edge diffraction from the plate degrades the RPE somewhat.

Effect of Feed-Support and "Strut" Scattering

The waveguide-bend feed support (see Fig. 7-6) is seen to interfere with the horn fields prior to their striking the dish; this results in amplitude, phase, and symmetry changes. Also, upon reflection, the dish fields pass by the same feed support and the vibration dampers (usually thin wires), where they are again diffracted or reflected. Analyses of these effects have been attempted[36] with some success, but in most cases they are usually best corrected by empirically determining the best positioning of absorber layers on the waveguide feed support and by making the vibration dampers from insulating material.

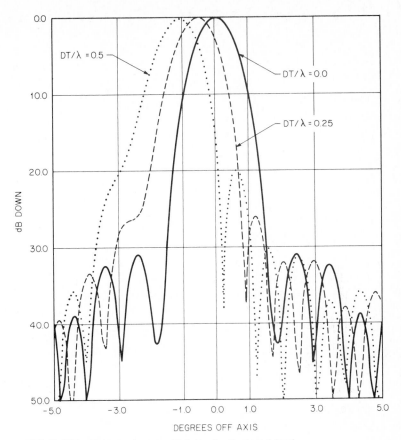

FIG. 7-10b Pattern change with laterally displaced feed.

Effect of Radome

The radome is used to protect the antenna against the accumulation of ice, snow, and dirt. In relay applications the fundamental radome problems are (1) the VSWR produced by the radome, (2) its total losses, and (3) its pattern influence. In the case of a planar radome (as seen on the antennas in Fig. 7-1), the VSWR increase is cured by tilting the radome a few degrees (moving the top further from dish) to prevent specular return to the feed horn; the loss and RPE influence are held low by using thin, low-loss material (e.g., Teglar, a fiberglass material coated on both sides with Teflon which sheds water readily[37]). Fiberglass conical or paraboloidal radomes are also used; these reduce wind loading, although since they are thicker than planar radomes, they have higher loss and degrade the RPE somewhat.

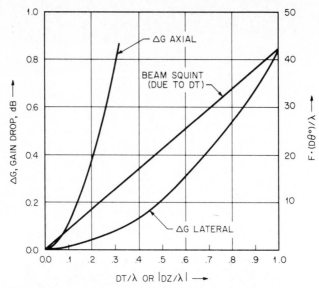

FIG. 7-11   Gain drop and beam squint due to defocusing.

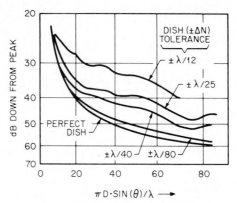

FIG. 7-12   Typical effect of surface errors on RPE. *(From Dragone and Hogg.[33])*

7-3   THE PARABOLOIDAL GRID REFLECTOR

Here metal (usually aluminum) tubes of outer radius r and center-to-center spacing s (where $s/\lambda \leq 0.3$ and $s/r \gtrsim 5$) and of paraboloidal contour replace the solid metal sheet (see Fig. 7-13) and act as a moderately good reflector. Typically, $s/\lambda = 0.3$ (maximum spacing possible so as to reduce cost) and $s/r = 5.5$ to 6 (to allow use of

FIG. 7-13 Paraboloidal grid antenna. *(Courtesy Andrew Corp.)*

commercial aluminum tubes). Moullin's theory[38] based on thin wires (for which s and r are adjusted to have the self-inductance and mutual inductance cancel) has been shown[39] to give good results; however, it is now known that it is not essential to meet his criterion. The efficiencies of these grid reflectors is typically 5 or so percent below that of a solid reflector (of the same F/D and D and fed with the same horn), and the front-to-back ratio is typically 3 dB lower.* The grid openings cause the wide-angle RPE (especially for $\theta > \theta_D$) to be quite inferior to that of a solid paraboloid (also the tubes have a small kr, where $k = 2\pi/\lambda$, and, as such, scatter readily to the rear). These antennas are used only in remote, uncongested areas not requiring narrow RPE and have the advantage of overall lower cost since their wind loading is much smaller than that of a corresponding solid reflector. This allows less expensive towers (the grid antenna is, however, actually more expensive than its solid counterpart). These antennas are not used above about 3 GHz because of achievable fabrication tolerances and are usable only for a single polarization.

---

*Fortunately, however, when the grid is mounted on a typical triangular cross-section tower, the back signal is "smeared," bringing the f/b ratio back up to about its solid-dish value.

7-4   THE OFFSET PARABOLOID

To eliminate feed blockage (and its consequent drop in efficiency of about 8 to 10 percent due primarily to the degradation of the prime-feed pattern by the waveguide feed support), one can use the prime-fed offset paraboloid, the geometry of which is shown in Chap. 13. This antenna has been known for some time, and analyses by Chu-Turrin and later by Rudge (see Chap. 13 for these references) are available for predicting the front-region fields. Using Rudge's work and trying different feeds, one again arrives[40] at the conclusion that a good compromise between high gain and narrow RPE is about a 10- to 12-dB edge taper. The copolar patterns for a typical case are similar to those of the prime-fed symmetrical paraboloid, but they have lower near-in sidelobes owing to the absence of blockage. The penalty paid for the removal of blockage is that the asymmetrical plane (i.e., the horizontal plane for terrestrial antennas) now has a rabbit-ear cross-polar pattern with the peaks of the rabbit ears being about 20 dB down from the copolar peak and being located at about the 6-dB-down angle of the copolar pattern (see Chap. 13). Dual-polarization operation is, however, still achievable with XPD $\doteq$ -30 dB at the boresight null (the depth of this null being quite sensitive to reflector-surface error). The offset also has the disadvantage of high cost and higher I^2R loss associated with the typically longer feed support.

7-5   CONICAL AND PYRAMIDAL CORNUCOPIAS

These antennas have been used for some time[10,41,42] and probably represent the ultimate in terms of the best wide-angle RPE attainable. In recent years, the conical cornucopia (CC) has found preference over the pyramidal (primarily for economic reasons; also, the pyramidal horn has undesirably high diffraction lobes at 90° which require edge blinders to be suppressed[43,44]). A modern-day CC is shown in Fig. 7-14. The CC geometry has a feed offset of 90° (but also has a conical shield) and hence can be handled by the same analysis as in the previous section. However, the classic earlier analysis of Hines et al. (see Ref. 10) is generally used. Briefly, this work shows that the E fields in the projected circular aperture are the conformal transformation of the E fields in the circular aperture of the conical horn. That is, circles (and their perpendiculars) are mapped into circles (and their perpendiculars), except that concentric circles in the conical aperture are no longer concentric in the projected aperture. Double integration of the projected-aperture fields then shows that this produces cross-polarization rabbit ears in the horizontal (asymmetric) plane (again, like the general offset, of typically 20 dB down at the 6-dB angle of the copolar beam).

In most real cases, a shield with absorber lining is used to capture the horn spillover and to reduce (almost eliminate) multiple-scattering effects at wide angles. This absorber introduces a small gain drop (at most, ½ dB). Typically, total efficiencies are about 65 percent; a representative RPE for 6 GHz (also a specification for these antennas) is shown in Fig. 7-2, which is seen to be vastly superior to the A or B FCC specifications. The cornucopia has the advantage of narrow RPE, relatively high efficiency, and simultaneous multiband operation at typically 4-, 6-, and 11-GHz bands (and possibly 2-GHz as well[44]), though it is very costly.

7-6 SYMMETRICAL DUAL REFLECTORS

This antenna type is covered also in Chap. 13 and hence will only be briefly mentioned here. An old (but very useful) basic "back-of-the-envelope design" Cassegrain configuration is depicted in Fig. 7-15.* Typically, the horn is chosen (see Table 7-1) to produce an 8- to 10-dB taper at the hyperboloidal subreflector edge α_E, where $DE/\lambda \gtrsim 5$. Then a Potter flange is used to capture some of the subreflector spillover energy and effectively use it to scatter back to the dish.[45,46] In this way, directive efficiencies in the mid-60s to low 70s are achievable. A typical design based on Fig. 7-15 for a focal-plane dish of $D/\lambda = 60$ is as follows: choose $DE/\lambda = 6.84$, hence $DE/D = 0.114$; choose M (magnification factor) $= 2.81$, which gives $a/\lambda = 0.996$, $c/\lambda = 2.097$, and $\alpha_E = 39.2°$; choose a horn from Table 7-1 (either a corrugated or a Potter horn having zero phase error) of $DH/\lambda = 1.58$ so that $\psi_{10} = \alpha_E = 39.2°$; get $\theta_E = 19.1°$; choose $LE/\lambda = 2$ (not critical); therefore $DO/\lambda = 10.61$ and $DO/D = 0.176$, which is large but acceptable blockage. At 6 GHz this design gives over 67 percent measured directivity, and its patterns meet FCC specification A. To date, gregorians have found little terrestrial use; they are best suited for antennas with larger F/D ratios. All reasonably efficient dual reflectors suffer from large subreflector blockage problems, since for sharp drop-off of the subreflector pattern near the dish edge one should make $DO/\lambda \gtrsim 5$ (preferably $DO/\lambda \gtrsim 8$, though one should limit blockage to 20 percent or less, $DO/D \lesssim 0.20$). This causes high first (typically 12 to 14 dB), third, etc., sidelobes. To achieve better results (higher efficiencies and better RPEs by minimizing spillover and VSWR) shaping of the subreflector and main reflector is mandatory[47] (see Chap. 13).

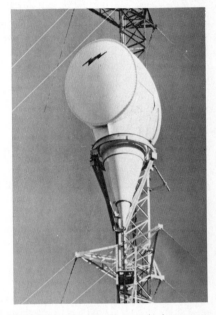

FIG. 7-14 Modern-day conical cornucopia. (*Courtesy Andrew Corp.;* for RPE see Fig. 7-2.)

---

*This figure does not show a small cone on the hyperboloid tip used to reduce the VSWR. The cone dimensions are usually determined empirically.

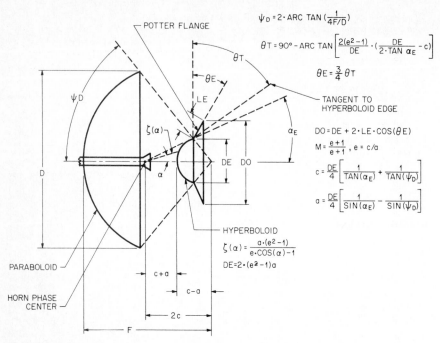

FIG. 7-15 "Back-of-the-envelope-design" Cassegrain configuration.

7-7 MISCELLANEOUS TOPICS

Most antennas usually have pressurized feeders (including the waveguide run from the antenna to the equipment building at ground location) in order to prevent moisture accumulation on the waveguide's inner walls and its ultimate increase in attenuation. Alternatively, for 2 GHz and lower, foam-filled coaxial-cable feeders are used. On paths which do not allow frequent repeater installation (e.g., when mountains or lakes intervene between two sites), gain is of the utmost importance. Here, oversized circular waveguide feeders are used (they, typically, have one-third of the attenuation of dominant-mode-size feeders). Being oversized, they can propagate higher-order modes (typically TM_{01} and TE_{21}), which are removed by mode filters. These paths invariably use Cassegrain antennas (with their typically 10 percent higher efficiency); this combination can give up to 3-dB extra gain per path.

7-8 COMPARISON OF ANTENNA TYPES

A concise summary roughly categorizing the above microwave-relay antennas regarding their efficiency, relative RPE, and cost is given in Table 7-2. Most antennas are designed to operate at their peak efficiency for a specified RPE so as to reduce their

TABLE 7-2   Summary of Microwave-Relay Antennas

Antenna	Total efficiency[a]	RPE[b]	Near-in[c] sidelobes	Cost[d]
Solid paraboloid (symmetric prime-fed)	55–60	3	2	1
Grid paraboloid (symmetric prime-fed)	50–55	4	2	2
Offset paraboloid (prime-fed)	60–70	2	1	4
Conical cornucopia	65[e]	1[e]	3	5
Cassegrain	65–78[f]	3	4	3

[a]Including all I^2R loss, feed loss, rms loss, etc.

[b]Excluding near-in sidelobe levels, 1 = lowest energy level.

[c]1 = lowest energy level.

[d]1 = lowest cost.

[e]With absorber in cylindrical shield.

[f]Shaped.

required dish diameter, since the total cost is approximately proportional to the diameter squared or more.[48]

REFERENCES

1   W. R. Bennett, H. E. Curtiss, and S. O. Rice, "Interchannel Interference in FM and PM Systems under Noise Loading Conditions," *Bell Syst. Tech. J.,* vol. 34, May 1955, pp. 601–636.

2   S. Silver (ed.), *Microwave Antenna Theory and Design,* MIT Rad. Lab. ser., vol. 12, Boston Technical Publishers, Inc., Lexington, Mass., 1964, chap. 12.

3   J. Ruze, "Antennas for Radar Astronomy," in J. Evans and T. Hagford (eds.), *Radar Astronomy,* part 2, McGraw-Hill Book Company, New York, 1970, chap. 8.

4   W. V. T. Rusch and P. Potter, *Analysis of Reflector Antennas,* Academic Press, Inc., New York, 1970.

5   D. L. Sengupta and R. E. Hiatt, "Reflectors and Lenses," in I. Skolnik (ed.), *Radar Handbook,* McGraw-Hill Book Company, New York, 1970, chap. 10.

6   C. J. Sletten, "Reflector Antennas," in R. E. Collin and F. J. Zucker (eds.), *Antenna Theory,* part 2, McGraw-Hill Book Company, New York, 1969, chap. 17.

7   L. Thourel, *The Antenna,* John Wiley & Sons, Inc., New York, 1960, chap. 12.

8   A. Z. Fradin, *Microwave Antennas,* Pergamon Press, New York, 1961, chap. VII.

9   N. Bui-Hai, *Antennes micro-ondes,* Masson, Paris, 1978. (In French.)

10   A. W. Love (ed.), *Reflector Antennas,* IEEE Press, New York, 1978.

11   R. C. Hansen, *Microwave Scanning Antennas,* vol. I, Academic Press, Inc., New York, 1966.

12   E. A. Wolff, *Antenna Analysis,* John Wiley & Sons, Inc., New York, 1966, pp. 312–335.

13   P. Clarricoats and G. Poulton, "High Efficiency Microwave Reflector Antennas—A Review," *IEEE Proc.,* vol. 65, no. 10, October 1977, pp. 1470–1504.

14 R. G. Kouyoumjian and P. H. Pathak, "A Uniform Geometrical Theory of Diffraction for an Edge in a Perfectly Conducting Screen," *IEEE Proc.*, vol. 62, November 1974, pp. 1448–1461.

15 C. E. Ryan and L. Peters, "Evaluation of Edge Diffracted Fields Including Equivalent Currents for the Caustic Regions," *IEEE Trans. Antennas Propagat.*, vol. AP-17, no. 3, May 1969, pp. 292–299; see also vol. AP-18, no. 2, March 1970, p. 275.

16 W. V. T. Rusch, "Physical Optic Diffraction Coefficients for a Paraboloid," *Electron. Lett., IEE (England)*, vol. 10, no. 17, Aug. 22, 1974, pp. 358–360.

17 C. M. Knop, "An Extension of Rusch's Asymptotic Physical Optics Diffraction Theory of a Paraboloid Antenna," *IEEE Trans. Antennas Propagat.*, vol. AP-23, no. 5, September 1975, pp. 741–743.

18 C. M. Knop and E. L. Ostertag, "A Note on the Asymptotic Physical Optic Solution to the Scattered Fields from a Paraboloidal Reflector," *IEEE Trans. Antennas Propagat.*, vol. AP-25, July 1977, pp. 531–534; see also correction, vol. AP-25, no. 6, November 1977, p. 912.

19 R. G. Kouyoumjian and P. A. J. Ratnasiri, "The Calculation of the Complete Pattern of a Reflector Antenna," Int. Electron. Conf., sess. 31, *Preconf. Dig.*, Pap. 69313, October 1969.

20 P. A. J. Ratnasiri, R. G. Kouyoumjian, and P. H. Pathak, "The Wide Angle Side Lobes of a Reflector Antenna," Tech. Rep. 2183-1, Ohio State University, Columbus, March 1970, pp. 17–25; also ASTIA Doc. AD707105, March 1970.

21 C. M. Knop and E. L. Ostertag, "The Complete Radiation Fields of a Parabolic Dish Antenna," Andrew Corp. Eng. Rep., July 1975.

22 G. L. James, *Geometrical Theory of Diffraction*, Peter Peregrinus Ltd., London, 1976, p. 148.

23 C. M. Knop, "On the Front to Back Ratio of a Parabolic Dish Antenna," *IEEE Trans. Antennas Propagat.*, vol. AP-24, no. 1, January 1976, pp. 109–111.

24 W. D. Burnside and L. Peters, Jr., "Edge Diffracted Caustic Fields," *IEEE Trans. Antennas Propagat.*, vol. AP-22, July 1974, pp. 620–623.

25 A. W. Love, *Electromagnetic Horn Antennas*, IEEE Press, New York, 1976.

26 "The Yang-Hansen Efficiency Plate," U.S. Patent 3,553,701, issued to Andrew Corporation, 1971.

27 M. S. Narisimhan and K. M. Prasad, "G.T.D. Analysis of Near-Field Patterns of a Prime-Focus Symmetric Paraboloidal Reflector Antenna," *IEEE Trans. Antennas Propagat.*, vol. AP-29, no. 6, November 1981, pp. 959–961.

28 R. F. Yang, "Quasi-Fraunhofer Gain of Parabolic Antennas," *IRE Proc.*, vol. 43, no. 4, April 1955, p. 486.

29 C. M. Knop and Y. B. Cheng, "The Radiation Fields Produced by a Defocused Prime-Fed Parabola," Andrew Corp. Eng. Rep., May 29, 1980.

30 J. Ruze, "The Effect of Aperture Errors on the Antenna Radiation Pattern," *Supplemento al Nuovo Cimento*, vol. 9, no. 3, 1952, pp. 364–380.

31 H. Zucker, "Gain of Antennas with Random Surface Deviations," *Bell Syst. Tech. J.*, October 1968, pp. 1637–1651.

32 N. I. Korman, E. B. Herman, and J. R. Ford, "Analysis of Microwave Antenna Sidelobes," *RCA Rev.*, September 1952, pp. 323–334.

33 C. Dragone and D. C. Hogg, "Wide-Angle Radiation Due to Rough Phase Fronts," *Bell Syst. Tech. J.*, September 1963, pp. 2285–2296.

34 A. Wojnowski and L. Hansen, Andrew Corporation, private communication.

35 H. Cory and Y. Leviatan, "Reflection Coefficient Optimization at Feed of Parabolic Antenna Fitted with Vertex Plate," *Electron. Lett.*, vol. 16, no. 25, Dec. 4, 1980, pp. 945–947.

36 P. Brachat and P. F. Combes, "Effects of Secondary Diffractions in the Radiation Pattern of the Paraboloid," *IEEE Trans. Antennas Propagat.*, vol. AP-28, no. 5, September 1980, pp. 718–721.

37 C. A. Sillar, Jr., "Preliminary Testing of Teflon as a Hydrophobic Coating for Microwave Radomes," *IEEE Trans. Antennas Propagat.,* vol. AP-27, no. 4, July 1979, pp. 555–557.
38 E. B. Moullin, *Radio Aerials,* Oxford University Press, New York, 1949, pp. 203–207.
39 E. F. Harris, "Designing Open Grid Parabolic Antennas," *Tele-Tech & Electronic Ind.,* November 1956.
40 C. M. Knop and Y. B. Cheng, "Analysis and Design of Corrugated Fog-Horn Feed Antennas," Final Rep. RP-883, Andrew Corp. Eng. Rep., Dec. 12, 1978.
41 H. T. Friis, "Microwave Repeater Research," *Bell Syst. Tech. J.,* April 1948.
42 R. W. Friis and A. S. May, "A New Broad-Band Microwave Antenna System," *AIEE Proc.,* March 1958, pp. 97–100.
43 D. T. Thomas, "Design of Multiple-Edge Blinders for Large Horn Reflector Antennas," *IEEE Trans. Antennas Propagat.,* vol. AP-21, March 1973, pp. 153–158.
44 J. E. Richards, "Horn Reflector Antenna Performance at 2 GHz with Simultaneous Operation in the 4, 6, 11 GHz," *IEEE Antennas Propagat. Symp. Dig.,* 1979.
45 P. D. Potter, "Application of Spherical Wave Theory to Cassegrain-Fed Paraboloids," *IEEE Trans. Antennas Propagat.,* vol. AP-15, no. 6, November 1967, pp. 727–736.
46 P. D. Potter, "Unique Feed System Improves Space Antennas," *Electronics,* vol. 35, June 22, 1962, pp. 36–40.
47 C. M. Knop, E. L. Ostertag, and H. J. Wiesenfarth, "The Analysis and Design of Advanced-Symmetrical-Dual-Shaped Reflector Antennas," Andrew Corp. Eng. Rep., October 1979.
48 T. Charlton, E. L. Brooker, and E. Book, Andrew Corporation, private communication.
49 K. S. Kelleher and G. Hyde, "Reflector Antennas," in R. C. Johnson and H. Jasik (eds.), *Antenna Engineering Handbook* 2/e, McGraw-Hill Book Company, New York, 1984, Chap. 17.

BIBLIOGRAPHY

Beckmann, P., and A. Spizzichino: *The Scattering of Electromagnetic Waves from Rough Surfaces,* Pergamon Press, New York, 1963.
Croswell, W. F., and R. C. Kirby: "Antennas and Wave Propagation," in D. G. Fink (ed.), *Electronic Engineers' Handbook,* McGraw-Hill Book Company, New York, 1975, sec. 18, par. 51–53.
Lee. S. H., and R. C Ruddock: "Numerical Electromagnetic Code (NEC)—Reflector Antenna Code," *Code Manual,* part II, Electroscience Laboratory, Ohio State University, Tech. Rep. 784508-16, September 1979.
Semplak, R. S.: "A 30 GHz Scale Model Pyramidal Horn-Reflector Antenna," *Bell Syst. Tech. J.,* vol. 58, no. 6, July–August 1979, pp. 1551–1556.
Swift, C. T.: "A Note on Scattering from a Slightly Rough Surface," *IEEE Trans. Antennas Propagat.,* vol. AP-19, no.4 , July 1971, pp. 561–562.

Chapter 8

Radiometer Antennas

William F. Croswell

Harris Corporation

M.C. Bailey

NASA Langley Research Center

8-1 INTRODUCTION

Radiometer antennas are widely used as part of remote-sensing systems to infer physical properties of planetary atmospheres and the surface.[1] Such antennas are similar in type to those used in radio astronomy except for several important differences. In the case of radio astronomy, many of the sources to be measured are stable in time for minutes or longer. In addition, the so-called radio stars or other natural sources are point sources or are of limited angular extent and immersed in a cold sky of several degrees kelvin.[2,3] For most remote-sensing applications, the radiometric system observes a distributed target of large angular extent and warm in temperature. For example, the ocean brightness temperatures in the microwave bands are in the order of 120 K, while land surfaces can be 200 K or more.[1] Hence, antenna systems for downward-looking radiometric systems must have small, close-in sidelobes out to the angles where the surface brightness diminishes. Such techniques as interferometry used in radio astronomy to improve resolution at the expense of high sidelobes are of little value in the remote sensing of atmospheres and surfaces in a downward-looking observation.

Radiometer-antenna design requires much more precise consideration of different parameters than most antenna engineers are accustomed to. For example, beam efficiency, antenna losses, and antenna physical temperature are extremely important as well as directional gain, low sidelobes, mismatch, and polarization. Therefore, this chapter will include an extended section on basic principles and a description of basic system types in present or proposed use on spacecraft or aircraft, along with basic radiometer-antenna types commonly in use.

8-2 BASIC PRINCIPLES

For microwave remote-sensing applications, the radiometer antenna is used with a very sensitive receiver to detect and provide a measurement of the electromagnetic radiation emitted by downwelling radiation and the earth's surface. The downwelling radiation is from the cosmic-background and the sky-background radiation due to atmospheric properties including moisture.[4,5,6] Depending upon knowledge of the roughness and dielectric properties of the surface and the relationship of the dielectric properties to the physical properties of the surface, an indirect or remote measurement of the physical property is feasible.

The received power detected by the radiometric system is by Swift:[6]

$$P = KT_A f \qquad (8\text{-}1)$$

where K is Boltzmann's constant, f is the radio-frequency bandwidth of the radiometric receiver, and T_A is the antenna temperature.

Antenna Temperature

The antenna temperature, stated in a manner suppressing polarization, is given by an expression of the form

$$T_A = \frac{\int_0^{2\pi} \int_0^{\pi} f(\theta, \phi) T_B(\theta, \phi) \sin\theta d\theta d\phi}{\int_0^{2\pi} \int_0^{\pi} f(\theta, \phi) \sin\theta d\theta d\phi}$$ **(8-2)**

where $f(\theta, \phi)$ is the normalized radiation pattern of a perfect plane-polarized antenna and $T_B(\theta, \phi)$ is the brightness temperature of the scene observed by the radiometer antenna with the same polarization. A typical scene in microwave remote sensing in which the antenna is pointed toward the surface is given in Fig. 8-1. T_{sky} is the downwelling sky radiation in the specular direction and includes thermal radiation from the atmosphere and the cosmic background. The brightness temperature observed by the radiometer antenna, therefore, is the sum of the brightness temperature T_{Bs} emitted by the surface and the amount of energy from the sky radiation scattered by the surface in the specular direction. This brightness temperature $T_B(\theta, \phi)$ is given by

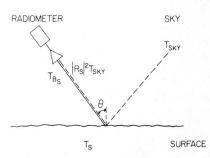

FIG. 8-1 Typical radiometer scene in microwave remote sensing.

$$T_B(\theta, \phi) = T_{Bs} + /R_s/^2 T_{sky}$$ **(8-3)**

where $/R_s/^2$ is the plane-wave reflection coefficient for the air-surface interface.

The emissivity of the surface is given by

$$\varepsilon_s = 1 - /R_s/^2$$ **(8-4)**

and the brightness temperature T_{Bs} is related to the physical temperature T_s by the relation

$$T_{Bs} = \varepsilon_s T_s$$ **(8-5)**

From Eqs. (8-3), (8-4), and (8-5),

$$T_B(\theta, \phi) = (1 - /R_s/^2) T_s + /R_s/^2 T_{sky}$$ **(8-6)**

It is emphasized that these equations were written in a conceptual form and do not include the entire vector phasor property of the antenna and the observed surface. The polarization property of the antenna and its relationship to the surface-polarization property are discussed in the next subsection.

Polarization

To include polarization properly in the antenna temperature expressions the polarization properties of both the antenna and the surface must be included. The polarization is defined relative to the plane of the aperture in the case of the antenna and to the incident plane in the case of the surface. Consider the following coordinate

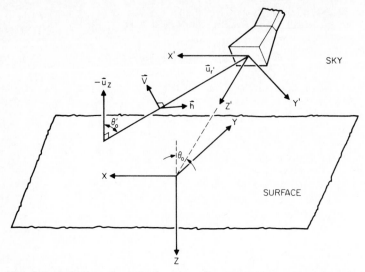

FIG. 8-2 Definition of polarization of a radiometer antenna with the surface and sky scene.

system given in Fig. 8-2, where the observed surface is in the xy plane and the antenna aperture is in the $x'y'$ plane. In this instance,

$$\mathbf{h} = \frac{\mathbf{u}_z \times \mathbf{u}_r'}{\sin \theta'} \qquad \mathbf{v} = \mathbf{h} \times \mathbf{u}_r$$

and ψ = the angle between $\mathbf{h}$ and $\mathbf{u}_\phi$. Also note that

$$\mathbf{h} \cdot \mathbf{u}_\phi' = \cos \psi = \frac{\sin \theta' \cos \theta_0 - \cos \theta' \sin \theta' \sin \theta_0}{\sin \theta_0'}$$

and

$$\mathbf{v} \cdot \mathbf{u}_\psi = \sin \psi = - \frac{\sin \theta_0 \cos \phi}{\sin \theta_0'}$$

so that

$$\begin{aligned} \mathbf{h} &= \cos \psi\, \mathbf{u}_\phi' - \sin \psi\, \mathbf{u}_\phi' \\ \mathbf{v} &= \sin \psi\, \mathbf{u}_\phi' - \cos \psi\, \mathbf{u}_\phi' \end{aligned} \qquad (8\text{-}7)$$

It should also be noted that $\mathbf{h}$ and $\mathbf{v}$ are the vector directions commonly referred to as horizontal and vertical polarization. This should allow the conversion of available reflection-coefficient data on natural surfaces to emissivity and produce brightness temperatures through the use of Eq. (8-6). If we assume that the far field from an aperture in the $x'y'$ plane can be broken into vertical and horizontal components, then

$$\mathbf{E} = \mathbf{h}[-\sin \psi\, E_\theta + \cos \psi\, E_\phi] + \mathbf{v}[\cos \psi\, E_\theta + \sin \psi\, E_\phi]$$

Using these results, the antenna temperature is defined as

T_A

$$
= \frac{\int_0^{2\pi} \int_0^{\pi} \{[-\sin\psi\, E_\theta + \cos\psi\, E_\phi]^2 T_{BH} + [\cos\psi\, E_\theta + \sin\psi\, E_\phi]^2 T_{BV}\} \sin\theta\, d\theta\, d\phi}{\int_0^{2\pi} \int_0^{\pi} \{[-\sin\psi\, E_\theta + \cos\psi\, E_\phi]^2 + [\cos\psi\, E_\theta + \sin\psi\, E_\phi]^2\} \sin\theta\, d\theta\, d\phi}
$$

$$\textbf{(8-8)}$$

If we assume, as given in Peake's derivation,[7] that $E_\theta = C\sqrt{f_\theta}\, e^{j\alpha}$ and that $E_\theta = C\sqrt{f_\phi}\, e^{j(\alpha+\delta)}$, where f_ϕ and f_θ are the normalized radiation patterns as a function of θ and ϕ and polarized in the θ and ϕ directions respectively, the $\mathbf{h}$ and $\mathbf{v}$ terms in Eq. (8-6) are given by

$$
\begin{aligned}
E_h &= C e^{j\alpha}[-\sin\psi\,\sqrt{f_\theta} = \cos\psi\,\sqrt{f_\phi}\, e^{j\gamma}] \\
E_v &= C e^{j\alpha}[\cos\psi\,\sqrt{f_\theta} + \sin\psi\,\sqrt{f_\phi}\, e^{j\gamma}]
\end{aligned}
$$

$$\textbf{(8-9)}$$

Using these expressions, the antenna temperature is given by[8]

$$
T_A = \frac{\begin{aligned}\int_0^{2\pi}\int_0^{\pi}\{[f_\theta\sin^2\psi + f_\phi\cos^2\psi]\,T_{BH} + [f_\theta\cos^2\psi + f_\phi\sin^2\psi]\,T_{BV} \\ + (T_{BV} - T_{BH})f_\theta f_\phi\sin2\psi\,\cos\gamma\}\sin\theta\,d\theta\,d\phi\end{aligned}}{\int_0^{2\pi}\int_0^{\pi}(f_\theta + f_\phi)\sin\theta\,d\theta\,d\phi}
$$

$$\textbf{(8-10)}$$

The form of Eq. (8-10) is a little different for other choices of coordinate systems, but the basic properties are the same. Notice in Eq. (8-10) that, in addition to the contributions from the clearly horizontal and vertical surface-polarization terms, there is a third term which contributes to antenna temperature that is related to the difference in horizontal and vertical brightness temperatures at given angles, the relative amplitudes of the radiation pattern, and the *phase difference between the polarization terms in the antenna pattern. If the cross-polarization level is small, this term can be neglected.* Depending upon the application, cross-polarization levels of -30 dB or more are usually satisfactory. Further discussion of this subject is given in Refs. 9 and 10 and in later sections of this chapter. It is interesting to note that for some surfaces[4,5,9] T_{BH} and T_{BV} have properties such that $T_{BH} - T_{BV} \approx 0$ out to 20 or 30°. Hence, if a narrow-beam low-sidelobe antenna is used for near-nadir observations, the antenna temperature as given in Eq. (8-10) will be independent of the polarization of the radiometer antenna. An additional reference useful in interpreting the importance of polarization is Classen and Fung.[11] For the antenna designer, the message at this point is rather clear. A radiometer antenna must be designed so that sidelobes, cross-polarization lobes, and backlobes do not pick up unwanted contributions. Common design parameters for comparing radiometer antennas are the beam efficiency and the cross-polarization index.

Beam Efficiency

The beam efficiency of an antenna is defined as

$$BE = \frac{\text{power radiated in cone angle } \theta_1}{\text{power radiated in } 4\pi \text{ sr}} \qquad \textbf{(8-11)}$$

or

$$BE = \frac{\displaystyle\int_0^{2\pi} \int_0^{\theta_1} f(\theta, \phi) \sin \theta \, d\theta \, d\phi}{\displaystyle\int_0^{2\pi} \int_0^{\pi} f(\theta, \phi) \sin \theta \, d\theta \, d\phi} \qquad \textbf{(8-12)}$$

It should be noted that polarization is not included in Eq. (8-12) for simplicity, but it will be included later. One can observe from Eqs. (8-11) and (8-12) that to relate one antenna to another in terms of beam efficiency a choice of the angle θ_1 must be made in a standard manner. A common choice for θ_1 is the angle between the beam axis and the first null. Another choice for comparing antennas, such as the wide-angle corrugated horn or the multimode horn in which the first null is poorly defined, is to use the criterion for θ_1 as 2½ times the half angle of the 3-dB beamwidth.

The beam efficiency of aperture antennas with variable amplitude distributions has received much attention in the past, curves to compare one aperture distribution to another being readily available.[12,13,14]

Examples of the effects that the shape of the amplitude distribution has upon beam efficiency are given in Fig. 8-3 for rectangular and circular apertures as a function of U ($U = Ka \sin \theta$). For uniform circular and rectangular amplitude distributions, the first several sidelobes cause pronounced ripples in the beam-efficiency curves, and values of $BE = 95$ percent are not achieved for large values of U. As more tapered distributions are assumed, the beam efficiency rises to large values rapidly and independently of the aperture shape.

This characteristic of ideal aperture distributions to approach large beam-efficiency values can be misleading in practical applications since the wide-angle sidelobes for many tapered aperture distributions are very small.[12,14] Indeed, some antennas such as the multimode, exponential, and corrugated horns have very low sidelobes and backlobes and hence excellent beam efficiencies. On the other hand, reflector antennas are sometimes used as radiometer antennas, and in this case the overall beam efficiency is influenced by spillover, cross-polarization, and blockage in addition to aperture illumination and reflector-surface roughness. These properties will be discussed later on.

The effects of polarization on beam efficiency can be stated as BE_{dp} (direct polarization):

$$BE_{dp} = \frac{\text{power at angle } \theta_1, \text{ direct polarization}}{\text{total received power, both polarizations}} \qquad \textbf{(8-13)}$$

Stated in equation form,

$$BE_{dp} = \frac{P_{\theta_1}, dp}{P_{dp} + P_{op}} \qquad \textbf{(8-14)}$$

where P_{dp} is the total power received in the direct polarization and P_{op} is the total power received in the orthogonal polarization. The effect of the orthogonal-polariza-

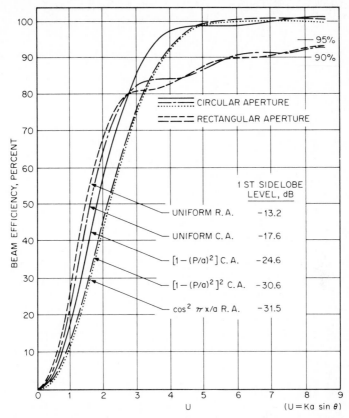

FIG. 8-3 Comparison of beam efficiencies of rectangular and circular apertures with various aperture distributions.

tion energy therefore is to decrease the direct-polarization beam efficiency of the antenna. A cross-polarization index (CPI) can be defined as[10]

$$CPI = BE_{dp} \frac{P_t}{P_{op}} \qquad \text{(8-15)}$$

For example, for $BE_{dp} = 85$ percent and on the assumption that the antenna has the property $P_{op}/P_t = 26$ dB, the cross-polarization isolation index is 25 dB.

The significance of cross-polarization levels in a radiometer antenna can be vividly demonstrated by reviewing the earlier discussion on polarization resulting in Eq. (8-10). The brightness temperature of a particular scene is polarization- and angle-dependent. In general, except at near-nadir angles the brightness temperature of a given surface will be different for, say, vertical and horizontal polarization at the same observation angle. To demonstrate the effect of cross-polarization properties of an antenna upon the radiometric measurement of a scene the curves in Fig. 8-4 are presented. The bias error ΔT_A in this figure is the error in brightness temperature produced by the integrated cross-polarization lobes observing the scene. This approxi-

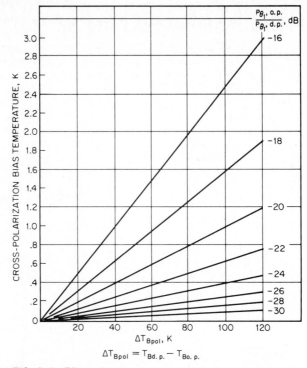

FIG. 8-4 Effect of antenna cross-polarization power on radiometer bias as a function of scene polarization difference temperature.

mation is valid for only narrow-beam antennas in which emission from the surface at a given polarization is constant over 2½ times the 3-dB beamwidths of the antenna. For precision measurements of ocean temperature in which accuracies of 0.3 K are desired, values of cross-polarization energy less than 28 dB are required.

Stray Radiation

For narrow-beam antennas, $T_B (\theta, \phi)$ is relatively constant over the main beam, and Eq. (8-2) can be written as

$$T_A = T_B(\theta, \phi)BE_{\theta_1} + \frac{\displaystyle\int_0^{2\pi} \int_{\theta_1}^{\pi} f(\theta, \phi) T_B(\theta, \phi) \sin \theta d\theta d\phi}{\displaystyle\int_0^{2\pi} \int_0^{\pi} f(\theta, \phi) \sin \theta d\theta d\phi} \qquad \textbf{(8-16)}$$

If the nonphysical assumption that $T_B(\theta, \phi)$ is constant in the second term of Eq. (8-16) is made, this equation may be written as

$$T_A = T_B(\theta, \phi)BE_{\theta_1} + T_B(\theta, \phi)[1 - BE_{\theta_1}] \qquad \textbf{(8-17)}$$

The second term in Eq. (8-17) is the so-called stray-radiation contribution sometimes used in radio-astronomy applications. For such applications, the near-in and relatively far-out sidelobes observe cold sky surrounding the radio source, thus justifying the assumptions made in Eq. (8-17). For radio-astronomy applications, therefore, much design emphasis must be placed upon backlobes which point toward the hot earth. For remote-sensing applications, the close-in and wide-angle sidelobes observe brightness temperatures in the same range as the main beam (100–270 K), while in many configurations the backlobes point to the cold sky. Hence, for remote-sensing applications the stray-radiation term has added significance. In general, such concepts as stray radiation are useful in that they give some indication of the ultimate accuracy of the T_A measurement. Simplistic forms to compute stray radiation such as Eq. (8-17) should be avoided unless great care has been exercised to understand the amplitude and angular distribution of brightness temperatures in the scene.

Ohmic and Reflection Losses

The ohmic losses in the antenna will modify the apparent temperature observed by the radiometric system. These ohmic losses will modify the observed temperature by the relation[5]

$$T_a = (1 - \ell)T_A + \ell T_0 \qquad \textbf{(8-18)}$$

where T_a is the apparent antenna temperature of a source whose lossless observed antenna temperature of the scene is T_A, T_0 is the physical temperature of the antenna, and ℓ represents the fractional power loss in the antenna.

The significance of physical losses in the antenna upon the absolute accuracy of remotely sensed surface properties may be obtained from the following example. Assume that antenna physical temperature is that of room temperature, i.e., $T_0 \approx$ 300 K. Typical ocean and land scenes exhibit lossless antenna temperatures between 100 and 300 K. Using this information, the effect of antenna losses upon measurement accuracy is given in Fig. 8-5. For some ocean-temperature applications ($T_A \approx 120$ K) absolute accuracies of 0.3 K are of importance. Hence, very small losses (≤ 0.005 dB) are of interest. For antennas in which losses are significant, antenna loss is generally treated as a fixed bias. This method is acceptable, but in such cases the physical temperature must be known and maintained to great precision. The Potter horn built for a precision S-band radiometer[5] exhibits a loss of ≤ 0.1 dB. Studies of corrugated and multimode horns[15] indicate small but significant losses when these are used as radiometric antennas. Hence, for calibration purposes both the loss and the antenna physical temperature must be known. To ensure stability in calibration, antennas may be enclosed in a thermally stable box with a very-low-loss radome.

The effect of mismatch in the input to the radiometer upon the observed antenna temperature can be expressed as[5]

$$T_a = (1 - \ell - \rho)T_A + \ell T_0 + \rho T_R \qquad \textbf{(8-19)}$$

where ρ is the reflection coefficient of the antenna and T_R is the microwave temperature seen looking into the receiver. The effects of small mismatches are very important for precision measurements. Stability of this mismatch will allow one to treat this error as a bias. Again, as in the case of physical loss, the thermal stability of the antenna impedance is important.

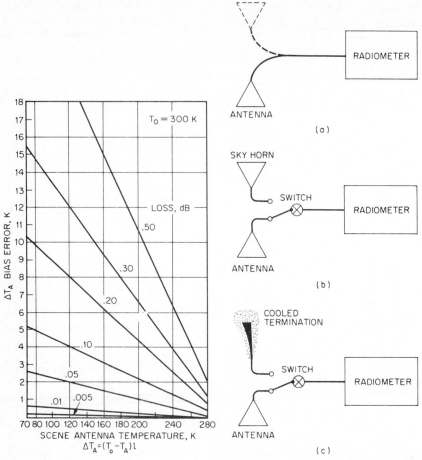

FIG. 8-5 Effects of losses upon scene antenna temperature.

FIG. 8-6 Radiometer calibration methods. (*a*) Steerable antenna. (*b*) Reference antenna and switch. (*c*) Cooled termination and switch.

Calibration

To provide a precision method of calibration, the normal procedure is to calibrate the radiometer antenna and radiometer as a single unit. This allows the user to obtain a relatively simple calibration factor for the entire system. Three simple methods of calibration, all of which have certain advantages and disadvantages, are given in Fig. 8-6. The steerable-antenna method (see Fig. 8-6*a*) requires that the radiometer antenna be pointed at the galactic pole periodically for a stable, cold reference temperature. This method has the additional advantage that the antenna can be pointed at a warm, stable surface feature such as a deep, thermally stable lake for an additional calibration point. The disadvantage of this method for spacecraft applications is that the spacecraft must be rotated. For ground applications the atmospheric attenuation is too large to use this method for upper-microwave or millimeter-wave frequencies. The switchable two-antenna method (see Fig. 8-6b) normally uses a so-called sky horn pointing to the galactic center and therefore eliminates the spacecraft-control problem. The primary disadvantage of this method is that the extra switch has losses that may change as a function of time, depending upon the switch technology and thermal control. A secondary disadvantage is that the two antennas are usually dissimilar with different losses. The third method (see Fig. 8-6c), the so-called cooled-termination method, has the advantage of not requiring a sky horn and may be very useful for millimeter- and higher-frequency ground or airborne systems in which atmospheric effects are very significant. Of course, the switch losses are common to the methods in Fig. 8-6b and c. A disadvantage of the system is the required provisions for a cooled transmission-line load.

A fourth calibration method, which is a modification of the first method, is to point the radiometer antenna toward a special free-space load[5] constructed from a porous microwave absorber located in a container filled with liquid nitrogen. This type of load has been successfully used in airborne radiometers when the radiometers have been designed to be very stable over long periods of time. Such a method also has advantages for millimeter and submillimeter applications.

8-3   SYSTEM PRINCIPLES

Radiometer Types

A variety of radiometer types have been used for radio astronomy and remote sensing. A very thorough tutorial discussion of radiometer designs is given by Hidy et al.[5] The three basic types commonly in use for remote sensing are described in simplistic block-diagram form in Fig. 8-7, which is adapted from Fig. 9.1 of Ref. 5.

The simplest radiometer, the absolute-power type, is shown in Fig. 8-7a. The output voltage of this radiometer, which is not electronically modulated, can be expressed as

$$V = G(T_A + T_N) \tag{8-20}$$

where T_A is the antenna temperature, T_N is the system noise temperature referred to the antenna input terminals, and G is the gain of the receiver system. The field calibration of this system is dependent upon maintaining stability after the calibration procedure by using such sources as a hot and cold load. If the latest technology in radio-frequency components is employed, this system is good but usually is not adequate to achieve accuracies of 1 K over a period of time. For some spacecraft systems the calibration of this system is improved by spinning the spacecraft antenna so that the cold sky is observed during each revolution as in the calibration method indicated in Fig. 8-6a. This spinning-spacecraft-antenna system is equivalent to the signal-modulated Dicke radiometer discussed next, except that the switching circulator is eliminated and the modulation frequency is reduced to much lower frequencies.[16,17]

The next radiometer type commonly used is the Dicke radiometer,[18] as shown simplified in Fig. 8-7b. The basic improvement of the signal-modulated Dicke radiometer is that the stability of the noise from the receiver T_N is eliminated so that the output voltage is given by

$$V = G[T_A(1 - \ell_s) - T_{\text{ref}}(1 - \ell_R) + (\ell_s - \ell_R)T_0] \tag{8-21}$$

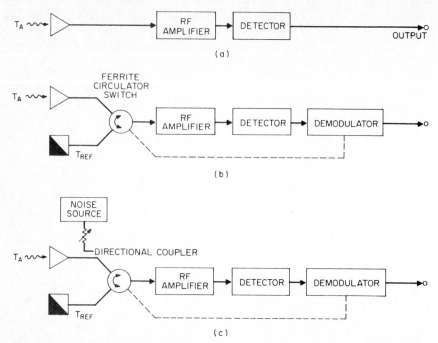

FIG. 8-7 Simplified diagram of basic radiometer types. (*a*) Absolute-power radiometer. (*b*) Dicke signal-modulated radiometer. (*c*) Noise-injection signal-modulated radiometer.

where ℓ_s and ℓ_r are the ohmic losses in the signal and reference arms and T_0 is the physical temperature of the radiometer components.

The next and latest improvement in radiometers is the noise-injection signal-modulated type shown in simplified form in Fig. 8-7*c*. The noise-injection method eliminates gain instability by allowing known amounts of noise to be put into the input of the radiometer. Various methods of performing this variable noise injection, including variable calibrated attenuators as shown in Fig. 8-7*c*, can be used. Variable-height pulse modulation of a noise diode feedback circuit can also be employed, as discussed in Ref. 5. A digital version of the noise-injection modulated radiometer has been devised and analyzed[19] and implemented.[20]

Radiometer-System Types

The most common radiometer system used in remote sensing is the scanning-beam-antenna type depicted in Fig. 8-8. This method of scanning on a spacecraft or an aircraft is usually achieved by bidirectional scanning of a reflector-antenna-feed-radiometer system relative to a stable platform. The advantage of this system is that wide-swath coverage of the surface scene can be obtained. The spatial resolution of this bidirectional scanning system is limited by an antenna size that can fit within an aircraft or launch-vehicle fairings. When finer spatial resolutions are required, deployable antennas which are limited in scan rate and can require the spacecraft to spin with the radiometer antenna as a single unit may be employed. Owing to the combination

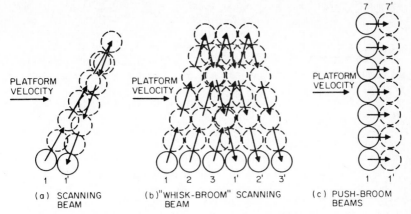

FIG. 8-8 Ground spot patterns of radiometer-antenna systems.

of scan rate and spacecraft velocity, a single-beam radiometer antenna of the scanning type may not produce contiguous resolution cells on the surface at the beginning and end of each rotation. To produce this contiguous coverage, the so-called whisk-broom radiometer-antenna system can be employed, as shown in Fig. 8-8b. Here each beam in the whisk broom requires an independent antenna port and radiometer.[21] The beam patterns can be produced by multiple feeds in a reflector system.

As even larger radiometer antennas are required to obtain better surface resolution, physical movement to obtain swath coverage may not be possible. To provide swath-width coverage for nonscanning systems, the so-called push-broom radiometer system may be used, as depicted in Fig. 8-8c. In this design, an independent radiometer is connected to each antenna port. Movement of this push-broom beam along the surface can be time-gated to produce a surface-radiometric-brightness-temperature map.

Because of requirements to separate surface parameters such as salinity and temperature for the ocean, multiple frequencies are commonly required in radiometric systems that must be integrated into the scanning,[16] whisk-broom, or push-broom radiometric systems. Such multifrequency systems are usually designed as multiple feeds in a main reflector. Combining frequencies in a single broadband corrugated horn is feasible over nearly 2:1 bands with low-loss properties. A very broadband horn with over 5:1 bandwidth has been designed and used in space with a scanning system.[17] This antenna exhibits large losses, which must be compensated for as very large bias errors.

8-4   ANTENNA TYPES

Horns

A straightforward design for a moderate-beamwidth (10 to 30°) radiometer antenna is the electromagnetic horn; however, because of the requirement for high beam efficiency (and, therefore, for low sidelobes) specialized horn designs are necessary for radiometer applications. Radiometer horn antennas are designed so that the normally high E-plane sidelobes are reduced to an acceptable level. The H-plane sidelobes (typically -23 dB) are already low enough to achieve a sufficiently high beam efficiency for moderate beamwidths. The E-plane-sidelobe reduction can be accomplished by several techniques. The simplest technique is to utilize multimodes[22] in the radiating-horn aperture to provide sidelobe cancellation in the far-field pattern. The higher-order modes are excited by a step or sudden change in the cross section of the waveguide or horn taper. The bandwidth of a multimode or dual-mode horn is limited to a few percent because of the difference between the modal phase velocities in the tapered waveguide horn section. However, a shortened version[23] of the Potter horn can exhibit good pattern characteristics over a bandwidth approaching 10 percent. The performances of the square multimode horn and the conical dual-mode horn are quite similar.

For radiometer applications requiring a larger bandwidth, the class of corrugated horns[24] is attractive. As with the multimode-type horn, the purpose of the corrugated-horn design is to reduce the normally high E-plane sidelobes. This is accomplished in the corrugated horn by designing the corrugations so as to decrease the current along the corrugated wall, thus producing a tapered (approximately cosine) E-plane aperture field distribution. This tapered distribution can be maintained over a bandwidth approaching 2:1 but is usually limited to the operating bandwidth of the feed waveguide.

Another type of horn which shows promise as a wideband radiometer antenna is the exponential horn with a specially flared aperture.[25] A similar wideband horn is one which is flared like a trumpet.[26] The design approach of these wideband horns is to eliminate sharp discontinuities and provide a smooth transition between the horn modes and free space.

Beam-efficiency calculations are given in Fig. 8-9 for a circular aperture with a radial aperture distribution equivalent to that of the TE_{11}-mode H-plane distribution and with a quadratic phase taper, as being representative of the beam efficiencies obtained from radiometer horns. The figure also illustrates that, in order to achieve high beam efficiencies, the horn must be designed for small phase taper, either by decreasing the horn flare angle or by using a phase-correcting aperture lens.

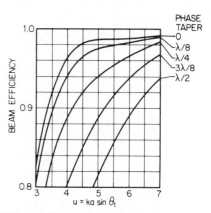

FIG. 8-9 Effect of aperture phase taper on beam efficiency for a conical corrugated or dual-mode horn.

Horn-Reflector Systems

For radiometer applications requiring narrow beamwidths, reflector antennas are more appropriate since highly tapered reflector illuminations can yield the much lower sidelobes necessary for high beam efficiency. The feed elements should be designed for spillover minimization, which represents additional beam-efficiency degradation.

The reflector surface must be constructed very accurately (usually a machined and possibly polished surface is required) to minimize loss in beam efficiency due to statistical roughness.

The beam efficiency for a reflector antenna can be expressed as

$$BE = BE_s BE_f BE_\delta \qquad (8\text{-}22)$$

where BE_s is the beam efficiency obtained by integration of the reflector secondary pattern, neglecting the back radiation, BE_f is the beam efficiency of the feed evaluated at the angle of the edge illumination, and BE_δ is the reduction factor due to reflector-surface roughness. Calculations of beam efficiencies BE_s for a circular aperture with a parabola-on-a-pedestal and with a parabola-on-a-pedestal-squared distribution are presented in Figs. 8-10 and 8-11 as representative of reflector illumination with an edge taper of -20, -10.5, -8, and -6 dB. The dashed curve for uniform illumination is included for reference. Figure 8-12 shows the effect of edge taper on the feed or spillover efficiency BE_f. The calculations in the figure are for a conical dual-mode horn with no phase error as being typical of feed patterns used in horn-reflector radiometer antennas. The reduction in beam efficiency BE_δ due to reflector-surface roughness can be obtained from the analysis of Ruze[27] as

$$BE_\delta = BE_s \exp(-\delta^2) + \Delta BE \qquad (8\text{-}23)$$

where $\delta = 4\pi\epsilon/\lambda$, ϵ is the rms surface roughness, and ΔBE is a correction term which accounts for the nonzero correlation length c of the surface error; i.e.,

$$\Delta BE = \exp(-\delta^2) \sum_{n=1}^{\infty} (\delta^{2n}/n!)[1 - \exp(-(uc/D)^2/n)] \qquad (8\text{-}24)$$

where D is the reflector diameter. Figure 8-13 shows a plot of the reduction factor versus rms surface roughness and correlation length. It should be noted from Eq. (8-

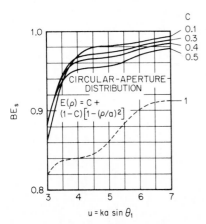

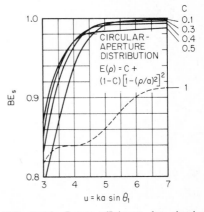

FIG. 8-10 Beam efficiency for circular aperture with parabola-on-a-pedestal distribution.

FIG. 8-11 Beam efficiency for circular aperture with a parabola-on-a-pedestal-squared distribution.

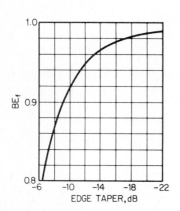

FIG. 8-12 Feed-beam-efficiency factor for a reflector antenna.

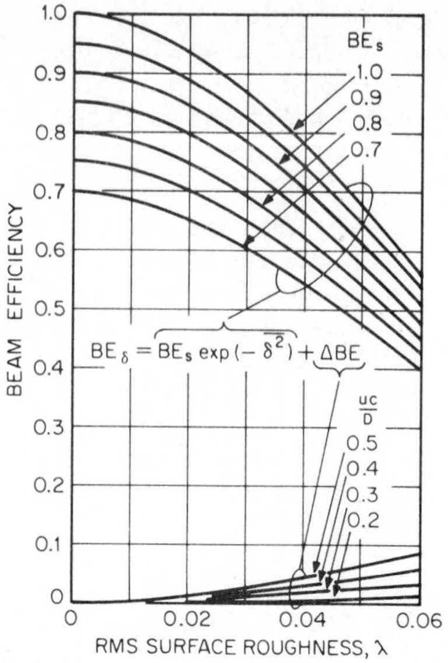

FIG. 8-13 Beam-efficiency factor of a random rough-surface reflector.

23) that for a beam efficiency greater than 90 percent the reflector surface should be $\lambda/50$ or smoother.

An additional concern for offset reflectors is the cross-polarization level, which could complicate the interpretation of antenna temperature data, as discussed earlier. The cross-polarization for reflectors decreases for larger focal-length-to-diameter ratios and smaller offset;[28] however, these parameters must be optimized to minimize feed- and spar-blockage effects. The BE_s curves in Figs. 8-10 and 8-11 neglect the effects of feed and spar blockage. Indeed, these effects cannot be treated simply as an aperture blockage as in gain calculations, and the secondary-pattern beam efficiency BE_s therefore should be recomputed with the wide-angle scattering from feed and spars included.

Phased Arrays

Phased arrays have been successfully employed as radiometer antennas[29,30] when beam scanning is required or volume constraints indicate that an array is the appropriate antenna type. High beam efficiency is obtained through amplitude tapering within the feed-distribution network. In the design of feed networks for radiometer array antennas, internal line losses should be minimized. If line losses are not excessive, stabilization of the losses through temperature control may be used in combination with calibration techniques to correct the radiometric temperature data. Resistors and terminations within the feed network should be avoided or used with discretion, since noise emitted by such components into the receiver could negate the radiometric measurement.

REFERENCES

1   D. H. Staelin and P. W. Rosenkranz, "High Resolution Passive Microwave Satellites," MIT Res. Lab. Electronics, Apr. 14, 1977.

2   P. W. Bounton, R. A. Stokes, and D. T. Wilkinson, "Primeval Fireball at $\lambda = 3$ mm," *Phys. Rev. Lett.,* vol. 21, 1968, p. 462.

3   J. R. Shakeshaft and A. S. Webster, "Microwave Background in Steady State Universe," *Nature,* vol. 217, 1968, p. 339.

4   W. H. Peak, "The Microwave Radiometer as a Remote Sensing Instrument," Electrosci. Lab. Rep. 1907-8, Ohio State University, Columbus, Jan. 17, 1969.

5   G. M. Hidy, W. F. Hall, W. N. Hardy, W. W. Ho, A. C. Jones, A. W. Love, J. Van Melle, H. H. Wang, and A. E. Wheeler, "Development of a Satellite Microwave Radiometer to Sense the Surface Temperature of the World Oceans," NASA CR-1060, National Aeronautics and Space Administration, Washington, February 1972.

6   C. T. Swift, "Passive Microwave Remote Sensing of the Ocean—A Review," *Boundary-Layer Meteorology,* vol. 18, 1980, pp. 25–54.

7   W. H. Peake, "Radar Return and Radiometric Emission from the Sea, Electrosci. Lab. Rep. 3266-1, Ohio State University, Columbus, October 1972.

8   C. T. Swift, private notes, 1981.

9   F. B. Beck, "Antenna Pattern Corrections to Microwave Radiometer Temperature Calculations," *Radio Sci.,* vol. 10, no. 10, October 1975, pp. 839–845.

10   W. H. Kummer, A. T. Villeneuve, and A. F. Seaton, "Advanced Microwave Radiometer Antenna System Study," NASA Cont. NAS 5-20738, Hughes Aircraft Company, Antenna Department, Culver City, Calif., August 1976.

11   J. P. Classen and A. K. Fung, "An Efficient Technique for Determining Apparent Temperature Distributions from Antenna Temperature Measurements," NASA CR-2310, National Aeronautics and Space Administration, Washington, September 1973.

12   A. F. Sciambi, "The Effect of the Aperture Illumination on the Circular Aperture Antenna Pattern Characteristics," *Microwave J.,* August 1965, pp. 79–31.

13   R. T. Nash, "Beam Efficiency Limitations of Large Antennas," *IEEE Trans. Antennas Propagat.,* vol. AP-12, November 1964, pp. 691–694.

14   J. Ruze, "Circular Aperture Synthesis," *IEEE Trans. Antennas Propagat.,* vol. AP-12, November 1964, pp. 691–694.

15   R. Caldecott, C. A. Mentzer, L. Peters, and J. Toth, "High Performance S-Band Horn Antennas for Radiometer Use," NASA CR-2133, National Aeronautics and Space Administration, Washington, January 1973.

16   T. Walton and T. Wilheit, "LAMMR, a New Generation Satellite Microwave Radiometer—Its Concepts and Capabilities," IEEE International Geoscience and Remote Sensing Society, Washington, June 8–10, 1981.

17   E. G. Njoku, J. M. Stacey, and F. T. Banath, "The Sea Sat Scanning Multichannel Microwave Radiometer (SMMR): Instrument Description and Performance," *IEEE J. Oceanic Eng.,* vol. OE-5, no. 2, April 1980, pp. 100–115.

18   R. H. Dicke, "The Measurement of Thermal Radiation at Microwave Frequencies," *Rev. Sci. Instrumen.,* vol. 17, 1946, pp. 268–275.

19   W. D. Stanley, "Digital Simulation of Dynamic Processes in Radiometer Systems," Final Rep., NASA Cont. NASI-14193 No. 46, Old Dominion University, Norfolk, Va., May 1980.

20   R. W. Lawrence, "An Investigation of Radiometer Design Using Digital Processing Techniques," master's thesis, Old Dominion University, Norfolk, Va., June 1981.

21   "A Mechanically Scanned Deployable Antenna Using a Whiskbroom Feed System," Cont. NAS 5-26494, Harris Corporation, Melbourne, Fla., February 1982.

22   P. D. Potter, "A New Horn Antenna with Suppressed Sidelobes and Equal Beamwidths," *Microwave J.,* June 1963, pp. 71–78.

23 M. C. Bailey, "The Development of an L-Band Radiometer Dual-Mode Horn," *IEEE Trans. Antennas Propagat.*, vol. AP-23, May 1975, pp. 439–441.

24 R. E. Lawrie and L. Peters, Jr., "Modification of Horn Antennas for Low Sidelobe Levels," *IEEE Trans. Antennas Propagat.*, vol. AP-14, September 1966, pp. 605–610.

25 W. D. Burnside and C. W. Chuang, "An Aperture-Matched Horn Design," *IEEE Antennas Propagat. Symp.*, Quebec, June 2–6, 1980, pp. 231–234.

26 J. C. Mather, "Broad-Band Flared Horn with Low Sidelobes," *IEEE Trans. Antennas Propagat.*, vol. AP-29, November 1981, pp. 967–969.

27 J. Ruze, "Antenna Tolerance Theory—A Review," *IEEE Proc.*, vol. 54, April 1966, pp. 633–640.

28 T.-S. Chu and R. H. Turrin, "Depolarization Properties of Offset Reflector Antennas," *IEEE Trans. Antennas Propagat.*, vol. AP-21, May 1973, pp. 339–345.

29 T. Wilheit, *The Electronically Scanning Microwave Radiometer (ESMR) Experiment: The Nimbus 5 User's Guide*, NASA Goddard Space Flight Center, Greenbelt, Md., November 1972.

30 B. M. Kendall, "Passive Microwave Sensing of Coastal Area Waters," AIAA Conf. Sensor Syst. for 80s, Colorado Springs, Colo., Dec. 2–4, 1980.

Chapter 9

Radar Antennas

Paul E. Rawlinson
Harold R. Ward

Raytheon Company

Radar (radio detection and ranging) is a technique for detecting and measuring the location of objects that reflect electromagnetic energy. This technique was first demonstrated in practice by Christian Hülsmeyer in 1903 by detecting radio reflections from ships.[1] With the development of the magnetron by the British in 1940 and the incentives of World War II, the radar principle was used effectively in many applications. These early radars generally operated at microwave frequencies, using a pulsed transmitter and a single antenna that was shared for transmission and reception.

The radar antenna is an important element of every radar system and is intimately related to two fundamental parameters: coverage and resolution. A radar's volume of coverage, in particular its maximum range, is one of its most basic parameters. The radar-range equation shows that the product of transmitted power and antenna gain squared is proportional to the fourth power of range.[2] Because of this dependency, economic considerations in the design of a radar usually conclude that the most cost-effective system will have 20 to 40 percent of the system cost budgeted for the antenna.

Resolution, the ability to recognize closely spaced targets, is another important radar property. The better the radar's resolution, the better it is able to separate desired returns from the returns of other objects. The size of the radar antenna measured in wavelengths is inversely proportional to its beamwidth and hence determines the radar's angular resolution. While radar applications vary, antenna beamwidths typically fall between 1 and 10°.

The radars developed to date span a wide range of size and importance. Sizes range from proximity fuzes used in artillery shells to phased-array radars housed in multistory buildings for detecting and tracking objects in space. In any one application the size of the radar and also its cost may be limited either by the physical space available or by the importance of the radar information in relation to other competing techniques or operational alternatives.

The radar applications listed in Table 9-1 illustrate the wide variety of problems that radar has been called upon to solve. Few of these applications have required large production quantities, so the radar industry is characterized by a great diversity of models and small production quantities of each model. A result is an industry that is development-intensive and requires considerably more engineering personnel than most other electronic industries. The many applications combined with a large community of development engineers has resulted in many antenna developments originating for radar applications. The phased arrays developed for satellite surveillance, weapon control, and precision approach control for aircraft landing are but a few examples.

Most of the radars designed for the applications listed in Table 9-1 operate in the microwave-frequency band. While certain radars, such as long-wave over-the-horizon radar and millimeter-wave radar for short-range applications, operate outside this region, the majority of the systems operate between 1 and 10 GHz. The considerations that bound the frequency choice for a given radar application are antenna size and the angular resolution at the low end and atmospheric attenuation and the availability of radio-frequency (RF) power at the high end. The particular frequency bands allocated for radar use by international agreement are listed in Table 9-2. The letter designations, originating during World War II for security reasons, have been

TABLE 9-1   Radar Applications

Acquisition	Navigation
Air defense	Over-the-horizon applications
Air search	Personnel detection
Air traffic control	Precision approach
Airborne early warning	Remote sensing
Airborne intercept	Satellite surveillance
Altimeter	Sea-state measurement
Astronomy	Surface search
Ballistic-missile defense	Surveillance
Civil marine applications	Surveying
Doppler navigation	Terrain avoidance
Ground-controlled interception	Terrain following
Ground mapping	Weather avoidance
Height finding	Weather mapping
Hostile-weapon location	Weapon control
Instrumentation	

used in the years since then as familiar designators of the particular radar-band segments.

Because of the great variety of radar applications, radar antennas are required to operate in many different environments. Each of these environments, listed in Table 9-3, has a special impact on a radar's antenna design and the parameters listed in Table 9-4. Land-based systems are classed as fixed-site, transportable, or mobile. At

TABLE 9-2   Standard Radar-Frequency Letter-Band Nomenclature*

Band designation	Nominal frequency range	Specific radio-location (radar) bands based on International Telecommunications Union assignments for Region 2
HF	3–30 MHz	
VHF	30– 300 MHz	138–144 MHz
		216–225 MHz
UHF	300–1000 MHz	420–450 MHz
		890–942 MHz
L	1000–2000 MHz	1215–1400 MHz
S	2000–4000 MHz	2300–2500 MHz
		2700–3700 MHz
C	4000–8000 MHz	5250–5925 MHz
X	8000–12,000 MHz	8500–10,680 MHz
Ku	12–18 GHz	13.4–14.0 GHz
		15.7–17.7 GHz
K	18–27 GHz	24.05–24.25 GHz
Ka	27–40 GHz	33.4–36.0 GHz
mm	40–300 GHz	

*From Ref. 1.

TABLE 9-3   Radar Environments

Location	Climate
Surface-based	Arctic
Airborne	Desert
Space-borne	Marine
Mobility	Electromagnetic environment
Fixed	Electromagnetic interference (EMI)
Transportable	Electromagnetic pulse (EMP)
Mobile	Electronic countermeasure (ECM)
Portable	

TABLE 9-4   Radar-Antenna Parameters

Peak power	Cross-polarization rejection
Average power	Scan volume
Gain	Scan time
Beamwidths	Pointing accuracy
Sidelobe levels	Size
Bandwidth	Weight
Loss	Environment
Mismatch	Cost
Polarization	

fixed sites the larger radar antennas are often protected by a radome, especially in arctic regions that experience heavy winds, ice, and snow. Transportable systems generally require that the antenna be disassembled for transport. Mobile systems are required to move rapidly from place to place and usually do not allow time for antenna disassembly.

Marine radar systems carried by surface craft also have special requirements. The lack of space above deck often forces the radar to a smaller antenna and larger transmitter than would be used in a comparable land-based application. The rolling platform leads to requirements for broad elevation beams or stabilization in the elevation coordinate in addition to the demands of a salt and smoke environment. Airborne and space-borne radar antennas also have special requirements peculiar to their operating environments.

In addition to the physical and climatological requirements, the electromagnetic environment has an important impact on radar-antenna design. There is a distinct contrast between radar systems designed for civil applications and those designed for military applications. While the electronic environment for civil systems consists of unintentional interference with radars operating at assigned frequencies, in military applications the threat of electronic countermeasures increases the need for an antenna to have wide bandwidth and low sidelobes.

Antennas designed for radar applications have many special requirements, some of which are not encountered in other applications. Most radar applications employ pulsed transmitters that allow the radar to resolve distant targets from nearby clutter. Peak powers from a few tens of kilowatts to a few megawatts and operating at duty ratios from 0.1 to 10 percent are typical. These high peak powers require special attention to the RF path through the antenna to the point at which power is distributed over the antenna aperture. Pressurized waveguide and feed horns are often used to prevent RF breakdown. In active array antennas the high peak power is not concentrated in a single microwave path, thus lessening the severity of this requirement.

Most radar applications require an antenna capable of scanning its beam, either mechanically or electronically or both, to search for targets in a volume of space. Scanning techniques for single or multiple beams may be classified as either mechanical or electronic. The early radar antennas developed during and after World War II relied primarily on mechanical scanning. More recently the development of high-power RF phase shifters and frequency-scanning techniques have allowed the beams to be scanned more rapidly by avoiding the inertia associated with moving mechanical components.

Accurate pointing is a requirement inherent in all radar applications that measure target location. Radar requirements are generally more demanding than those of communications antennas, especially when precision measurements of target positions are required.

Radar applications tend to be divided between two functions, search and track. The search function requires that the radar examine a volume of space at regular intervals to seek out targets of interest. In this case the volume must be probed at intervals ranging typically from 1 to 10 s and every possible target location examined.

The radar tracking function operates in a manner quite different from search. Here one or more targets are kept under continuous surveillance so that more accurate and higher-data-rate measurements may be made of the target's location. Often the interesting targets detected by a search radar will be assigned to and acquired by a tracking radar. Certain radar systems combine search and track functions by time-sharing the agile beam of a phased-array antenna.

The remainder of this chapter describes a few examples of the more common radar applications which are indicative of today's state of the art. Too few are included to give a proper perspective of the great variety of radar applications. However, those described do illustrate the differences between some of the basic types. For greater depth on the subject of radar antennas the reader is directed to Refs. 1 through 5.

9-2 SEARCH-RADAR ANTENNAS

Search radars scan one or more beams through a volume in space at regular intervals to locate targets of interest. The search function is sometimes referred to as *acquisition,* as in a missile-defense system in which detected targets are eventually tracked and intercepted. It is also called *surveillance* when targets are being monitored, as in the control of air traffic. But regardless of the term used to describe the search function, the common requirement is to scan a specified region of space at regular intervals.

The most important parameters determining the size and cost of the search radar are the coverage volume, scan time, and target size. For a given application with fixed target size and scan time, it has been shown that the requirement may be met by specifying the product of average transmitter power and antenna-aperture area.[2] It is a fundamental property of search radars that their coverage capability is proportional to their power-aperture product and does not depend on the radar's operating frequency. The benefits of operating at a lower frequency are that (1) higher-power RF transmitters are available, (2) antenna tolerances are less severe for a given antenna size, and (3) the peak power-handling capacity of waveguide and components in the high-power RF path is higher. Radars designed to perform only the search function therefore tend to operate at the lower microwave frequencies, and just how low is often determined by a trade-off between antenna size and desired angular resolution.

This section describes four examples of air search radars in operation today. These examples of current technology fall into two categories: two-dimensional (2D) and three-dimensional (3D) air search radars. The 2D systems use a fan beam that is broad in elevation and narrow in azimuth because only the target range and azimuth coordinates are required. In the 3D systems elevation is also measured to determine target altitude, and in this case the antenna's beams must be relatively narrow in both angular coordinates.

2D Systems

The two examples chosen to illustrate 2D search radars are both designed for air-traffic-control applications. In these applications, the radar must scan the air space every few seconds. Aircraft targets are detected and their range and azimuth measured. Modern systems use computers to track the aircraft from scan to scan to provide continuity in the radar data provided to the air traffic controllers.

Radar antennas used in air traffic control usually have an elevation beam with a pattern similar to that shown in Fig. 9-1. The particular pattern shape is matched to the desired radar coverage. Since aircraft have a maximum altitude, the pattern is shaped to provide a constant altitude cutoff; however, this is usually modified at shorter ranges to compensate for the use of sensitivity time control in the receiver.

The beamwidth in the azimuth coordinate is typically between 1 and 2° to provide sufficient resolution with today's air traffic densities. Since the beam has a much larger beamwidth in the elevation coordinate, it is clear that the width of the antenna aperture must be larger than its height.

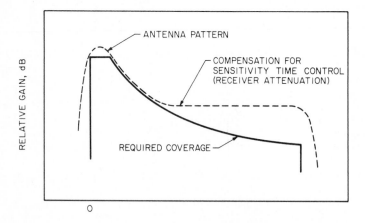

FIG. 9-1 Typical elevation pattern of a 2D surveillance radar antenna.

FIG. 9-2   ARSR-3 2D antenna. *(Courtesy of Westinghouse Electric Corp.)*

Other antenna requirements in air traffic control usually include selectable polarization and a second, high-elevation, receive-only beam. Circular polarization, if used during rainy weather, can provide between 10- and 20-dB rejection of rain return. However, circular polarization also has an associated target loss (about 3 dB), so a switch is often provided to allow the system to be operated with maximum sensitivity when rain rejection is not required.

A high-elevation receive-only beam is in common use in air-traffic-control radar antennas to give improved rejection of ground return at short range. An electronic switch connects the receiver to the high beam in the ground-clutter regions and then switches to the lower-elevation transmit beam pattern for the remainder of active range. Both polarization switching and beam switching add complexity to the feed of the feed-reflector antennas typically used in this application. Two examples which illustrate current designs follow.

ARSR-3

The ARSR-3 radar system was developed by Westinghouse for use in the en route air-traffic-control system of the Federal Aviation Administration (FAA).[6,7] The L-band radars (1.25 to 1.35 GHz) are used to maintain surveillance over aircraft in the high-altitude air routes between terminals. They provide 2D search data to a range of 366 km with a 12-s data interval. For this application, an antenna must have a shaped elevation coverage, a second high-elevation receive beam, and selectable linear or circular polarization.

Figure 9-2 shows the horn-reflector antenna used in the ARSR-3. The 6.9- by 12.8-m reflector forms a fan beam 1.25° wide in the azimuth coordinate. The antenna and pedestal are supported by a steel open-structure tower and are protected from the external environment by a space-frame radome.

The shaped elevation coverage, shown in Fig. 9-3, is formed by a doubly curved reflector. The upper portion of the reflector is nearly paraboloidal to direct energy to

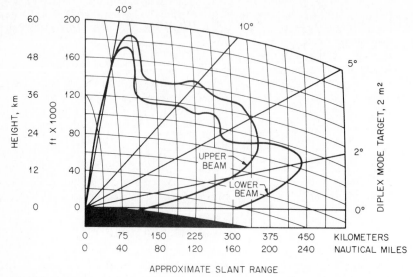

FIG. 9-3   ARSR-3 vertical coverage. *(Courtesy of Westinghouse Electric Corp.)*

the peak of the low-beam pattern, while the lower portion of the reflector departs from the parabola to direct more energy to elevation angles between 5 and 40°. The antenna features a rapid beam cutoff of the lower vertical pattern to minimize the lobing caused by surface reflections.

The dual-feed horns shown in Fig. 9-2 are used to provide two receive beams. One is a low-elevation beam used during transmission and reception, while the second is a higher-elevation beam which is switched in only during the time of arrival of the short-range returns in order to suppress land-clutter returns. For ranges beyond the clutter, the receiver is connected to the low-beam port to give better low-elevation coverage at far range. An important requirement of a two-beam design is the allowable gain difference in the direction of the horizon. If the gain difference is too large, the horizon range of the high-elevation beam will be shorter than the range of the farthest clutter, and hence the high beam cannot be fully utilized. This gain ratio is held to 16 dB in the ARSR-3 design by using the outputs of both horns shown in Fig. 9-2 to form the high-elevation beam.

Circular polarization (CP), needed to cancel rain return, complicates antenna and feed design. A switch is required to select either vertical or circular polarization; but, more important, symmetry is required between horizontal and vertical polarizations in order to achieve a high integrated cancellation ratio from the CP mode. An integrated cancellation ratio of better than 18 dB is realized by this design.

AN/TPN-24

The AN/TPN-24 is an airport surveillance radar developed by Raytheon for the U.S. Air Force as part of the AN/TPN-19 ground-controlled approach system.[8] This S-band radar (2.7 to 2.9 GHz) provides 2D surveillance of the air space in the vicinity of an airfield. Air traffic controllers seeing this 2D information displayed on a plan position indicator (PPI) vector landing aircraft onto the runway approach while communicating with the pilots via radio. For this application, a shorter-range and higher-data-rate radar is required than in the en route surveillance application described above, so the AN/TPN-24 gives coverage to 111 km with a 4-s data interval.

Figure 9-4 shows the AN/TPN-24 radar. The shelter contains all the radar electronics as well as a display, a radio, and microwave-relay equipment. The antenna mounted atop the shelter is disassembled and stowed inside the shelter for transport. When erected, the antenna reflector is 4.27 m wide and 2.4 m high, which at S band gives an azimuth beamwidth of 1.6°. The elevation beam pattern is shaped to match the desired elevation coverage, as shown in Fig. 9-5.

The requirement for transportability provides an incentive to use the smallest possible vertical aperture that will give the desired elevation pattern. In this antenna the entire reflector has a paraboloidal shape, and the elevation-pattern shaping is achieved by controlling the amplitude and phase of the power distributed to each of the 12 feed horns.

The particular feed illumination is determined by the microwave power divider mounted on the back of the reflector. Incorporated into this divider is a pin-diode switch that allows the amount of power coupled to the first feed horn (the horn contributing to the pattern near the horizon) to be varied during the receive mode. The

FIG. 9-4 AN/TPN-24 2D antenna. *(Courtesy of Raytheon Company.)*

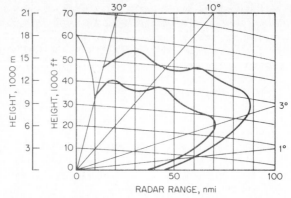

FIG. 9-5 AN/TPN-24 vertical coverage for single-channel and two-channel diversity. *(Courtesy of Raytheon Company.)*

switch allows the beam to be raised to reduce clutter at short range. The underside of the elevation beam pattern is tailored to produce a rapid cutoff and low sidelobes, thus minimizing lobing effects.

Linear or circular polarizations are selected by polarizers in series with each of the 12 feed horns. The polarizers are ganged and rotated mechanically to change polarization. To achieve the 20-dB integrated cancellation ratio provided by the antenna, mutual coupling between horns is minimized and compensated so as to equalize the feed pattern in both vertical and horizontal polarizations.

3D Systems

The need for 3D search radars originates in the requirements of military air defense systems. These systems protect an air space by detecting intruders and vectoring defending aircraft to intercept these intruders. This process requires that the interceptor know the altitude of an intruder, and therefore an elevation measurement by the radar is needed. Early systems used a 2D radar in combination with a separate height-finder radar that was designated to selected targets to measure altitude.[1] More recent systems combine the 2D search and height finding into a single 3D radar system with the capability of measuring a target's position in all three coordinates.

The antenna for a 3D radar must have a narrow receive beamwidth in the elevation as well as in the azimuth coordinate. While the required vertical coverage is about the same as that shown in Fig. 9-3 for the ARSR-3, the antenna must be capable of forming a narrow elevation beam within that coverage. Since the operational requirement is for a fixed-altitude accuracy, it is possible to allow the elevation error (and hence the elevation beamwidth) to increase at higher elevation angles.

Since these 3D radars are for military applications, the antennas must have low-azimuth sidelobes and wide bandwidth to operate effectively in an electronic-countermeasure (ECM) environment. The low sidelobes reduce jamming received from directions other than that of the target, and the wide bandwidth allows the use of frequency agility, which forces the jammer to dilute its power by spreading it over a wider band.

Various antenna techniques have been used to satisfy the 3D requirement. Two common types are stacked-beam and pencil-beam antennas. Stacked-beam systems transmit through a fan beam to illuminate the entire vertical coverage region and then receive returns through a stack of simultaneous receive beams. The pencil-beam system has a single beam, narrow in azimuth and elevation, which is used for both transmit and receive to scan the required elevations sequentially. The examples described below illustrate these 3D techniques.

AN/TPS-43

The AN/TPS-43 is a stacked-beam 3D surveillance radar developed by Westinghouse for the U.S. Air Force. It is one of the most successful 3D systems in use today, with over 120 radars manufactured and operating in 19 countries throughout the world.[9]

The antenna, shown in Fig. 9-6, is an important element of this S-band system. It has a 6.2-m-wide by 4.3-m-high paraboloidal reflector that forms a beam 1.1° wide in azimuth and 1.5° high in elevation. The reflector is illuminated by a stack of 15 vertically polarized horns combined to form a stack of six separate elevation receive beams covering from the horizon to 20° elevation. In the transmit mode, power is divided among the horns so as to illuminate the entire vertical coverage. On receive, the six beam outputs are received and processed in parallel to detect targets and measure their elevation, from which altitude is computed. The antenna completes a 360° azimuth scan every 10 s.[10,11]

This stacked-beam antenna requires considerably more microwave hardware above the rotary joint than is needed in a 2D antenna. The feed network, located at the base of the antenna structure, contains 13 duplexers and 6 RF receivers along with a transmit power divider and a receive power-combiner matrix. The received signals

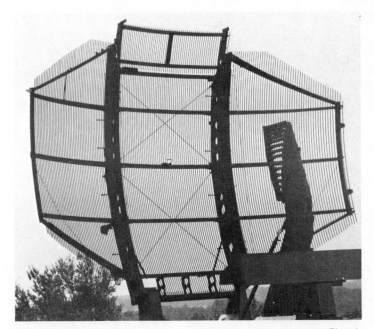

FIG. 9-6 AN/TPS-43 3D antenna. *(Courtesy of Westinghouse Electric Corp.)*

are down-converted to intermediate frequency (IF) so that they may be brought by slip rings to the stationary portion of the antenna pedestal.

This AN/TPS-43 antenna also contains a separate secondary radar antenna mounted below the S-band feed.

AN/TPS-59

The AN/TPS-59 is a long-range 3D surveillance radar developed by General Electric for the U.S. Marine Corps. This L-band radar uses a single pencil beam formed by an active phased array to scan elevation sequentially as the antenna rotates mechanically in azimuth. The system's most distinctive feature is its solid-state transmitter, which is distributed throughout the active array antenna.[12,13]

The antenna shown in Fig. 9-7 is 9.1 m high and 4.6 m wide. It is made up of 54 rows of horizontally polarized elements and their associated feed networks and row electronics. Each row contains an azimuth power divider, a solid-state row transmitter, two RF receivers, and two low-power phase shifters. Column feed networks combine the row outputs to form the pencil-shaped transmit beam and a three-beam monopulse cluster for receive. A receiver and an RF exciter are also included on the rotating portion of the antenna so that only power, IF, and control signals need be transferred to the antenna pedestal.

FIG. 9-7   AN/TPS-59 3D antenna. (*Courtesy of General Electric Co.*)

9-3   TRACKING RADAR

In radar applications for which target data with greater accuracy or a higher data rate than can be provided by a search radar are required, a tracking radar is used. The tracking-radar antenna usually has a pencil beam that follows a target in track to provide continuous measurement of the target's position in range, azimuth, and elevation. In some applications target signature data in the form of amplitude and phase variations of the target return are also measured. Targets are usually first detected by a search radar and then designated to the tracker for measurement. To acquire a target, the tracker must search a small volume about the designated point. When the target is detected, range and angle tracking loops close, and the range gate and beam are held centered on the target until target data are no longer required and the track is broken off.

Tracking radars tend to operate at higher microwave frequencies than search radars. It has been shown that tracking-radar sensitivity is proportional to average transmitter power times the antenna aperture area squared times the operating frequencies squared.[14] The increased importance of aperture and frequency compared with the search-radar case provides a greater incentive to use larger apertures and to operate at higher frequencies. An upper limit on the choice in these parameters is often set by atmospheric attenuation or antenna cost in relation to transmitter cost. The exceptions are radar applications for which target signature or propagation data are needed at a specific frequency, independently of cost considerations.

Fundamental to all tracking-radar antennas is the ability to measure a target's position within the beam so that the beam may be centered on the target. The two techniques used for this purpose are conical scan and monopulse. Conical scan consists of rotating the pencil beam in a circular pattern around the target in track.[1] If the target lies off the center of the circular scan pattern, the target return is amplitude-modulated by the scanning beam. The phase and amplitude of this modulation are detected and used to measure the target's position within the beam. This process requires that at least one revolution of the beam be completed before the measurement can be made. Monopulse is another technique for measuring the target's position in the beam that is capable of measuring azimuth and elevation on a single pulse return, but it requires two additional receiver channels.[1] Monopulse uses a three-beam cluster consisting of a pencil beam plus two superimposed difference beams to measure the target off-axis error in both angular coordinates.

For examples of current tracking-radar antenna designs the reader is referred to Chap. 11, "Tracking Antennas." The tracking function as incorporated into multi-function phased arrays is discussed in the next section.

9-4   MULTIFUNCTION ARRAYS

Radar search and track functions are normally realized with separate antennas, optimized in frequency and aperture size. However, with the development of high-power RF phase shifters and advanced digital-processing techniques, phased arrays have been employed to furnish nonmechanical means of antenna beam scanning and, because of the highly agile characteristics of the beam positioning, to provide an

TABLE 9-5   Features of Multifunction Arrays

Versatility: search and track
Multiple independent beams
Inertialess beam agility
Computer control
Planar or conformal type
Blast resistance
Independent control of transmit and receive illumination
Potential use of distributed power amplifiers
Graceful degradation

opportunity to combine search and track functions in a single radar. The digital-processing and beam-steering aspects of the multifunction phased array are beyond the scope of this text but are significant contributors to both the successful operation of the radar and its cost.[15,16] For the most severe target environments, the use of the multifunction array can be justified, but only after careful consideration of alternatives such as mechanically scanning antennas and hybrid approaches. Table 9-5 lists some of the features available from the multifunction array. The primary disadvantages are high cost, complexity, and compromise in the choice of frequency between the optima for the search and track functions.

The high cost of these arrays has been significant in restricting the systems to which this technology has been applied. Nevertheless, there has been gradual, though sometimes unsteady, growth in the number of phased arrays in operation. Sperry Corp.'s AN/FPS-85 at ultrahigh frequency (UHF) used a separate transmit and thinned-receive aperture[17,18] to obtain a high target resolution at reduced costs. Aperture thinning was also used by Raytheon for COBRA DANE at L band and the all-solid-state PAVE PAWS at UHF. Raytheon's mobile Patriot radar at C band uses an optical RF power-distribution system (space feed) to reduce both cost and weight.

For applications in which beam scanning in one or both planes is less than 120°, several radiating elements can share one phase shifter, as in the AN/TPQ-37 by Hughes, or hybrid scanning techniques such as a combination of a reflector and an array can be utilized, as in the AN/GPN-22 by Raytheon. The Sperry dome antenna[17] incorporates a planar array space-feeding a dome and achieves hemispheric coverage with a significant reduction in radiating elements and phase shifters.

From an antenna engineer's viewpoint, costs are controlled by optimization of the aperture, using the minimum number of radiating elements and phase shifters to accomplish the radar mission. After the array has been optimized, the next task is to design a cost-effective radiating-element and feed-distribution network (including the phase shifters, power- and signal-distribution network, and beam-steering computer). For examples of current phased-array design practice the reader is referred to Ref. 30, "Phased Arrays." Examples of phased-array radar applications are discussed in the following subsections. For additional information the reader is directed to the Refs. 15 through 19.

COBRA DANE

The COBRA DANE system, designated the AN/FPS-108, is a large phased-array radar installed on Shemya Island, Alaska, near the western end of the Aleutian

Islands chain. With its single radiating antenna face directed northwest toward the Bering Sea, it has as its prime mission collecting data on missile systems launched toward the Kamchatka Peninsula and the north Pacific Ocean. It is also capable of providing early warning of ballistic-missile attack on the continental United States as well as satellite detection and tracking. It is capable of detecting and tracking targets as small as 1 m^2 at ranges exceeding 8000 km. The system replaces the AN/FPS-17 and AN/FPS-80, individual search and track radars. Development was undertaken in 1973 by Raytheon with system testing completed in 1976.[11,20]

The building housing the radar, located on the northwest corner of Shemya (as shown in Fig. 9-8), is 34 m high, 33 m wide, and 26 m deep. The array face is 29 m in diameter and contains 34,769 elements, of which 15,360 are active radiators. A near-field horn radiator and screening fence, which are used to monitor the phase and gain of the active elements, can be seen on the left side of Fig. 9-8.

The radar-system characteristics are shown in Table 9-6, and a simplified radar-system block diagram is presented in Fig. 9-9.[21] The active array elements are density-tapered (thinned) across the aperture to provide a 35-dB Taylor weighting on transmit. This thinning technique[18] provides lower near-in sidelobes and a narrower beamwidth than would have been possible from a full array with the same number of uniformly illuminated active radiators. The inactive (dummy) elements provide a constant mutual-coupling environment for the active elements. Work published prior and subsequent to the development of COBRA DANE indicates that a small reduction in gain and an increase in the sidelobe level results if these dummy elements are eliminated.[19,22]

The active elements are arranged into 96 subarrays, each containing 160 active elements. Because of density tapering, the number of dummy elements per subarray varies from a minimum near the center of the array to a maximum for the peripheral subarrays. Each of the 96 traveling-wave transmitter tubes is fed through a coaxial line to a 1:160 microwave equal-power splitter which drives the elements of a subarray.

The received signals are combined by the same 160:1 splitter-combiner. The

FIG. 9-8   COBRA DANE. *(Courtesy of Raytheon Company.)*

TABLE 9-6 COBRA DANE System Parameters[21]

Frequency	
Narrowband	1215–1250 MHz
Wideband	1175–1375 MHz
Antenna	
Total number of elements	34,769
Number of active elements	15,360
Type of feed	Corporate
Aperture diameter	29 m
Scan coverage	120° cone
Transmitter	
Peak power	15.4 MW
Average power	0.92 MW
Power amplifier	96 traveling-wave tubes

transmit-receive duplexing function is achieved by changing the phase by 180° between transmit and receive for half of the elements within any subarray. The received signal thus appears on the difference port of the final 2:1 waveguide magic T, where it is amplified by a low-noise amplifier. The receiver outputs are combined in quadrantally symmetric subarrays. Each group of four symmetric subarrays is combined in monopulse comparators to form a sum- and two difference-channel signals. The transmit-and-receive sum beam-illumination taper is achieved by the density taper of the active elements. The difference-channel monopulse beams are further tapered through the use of fixed attenuators at the outputs of the 24 monopulse comparators.

Phase steering of the array in both planes is accomplished through the use of 3-bit diode phase shifters at each active element and time-delay steering, for wideband operation, at each subarray. Since the COBRA DANE array diameter is considerably greater than the desired range resolution, some form of time delay is required to com-

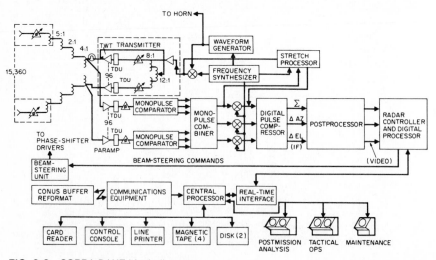

FIG. 9-9 COBRA DANE block diagram.

pensate for the time dispersion across the array for off-axis targets. Different time-delay units are used in the transmitting and receiving microwave feeds. Each time delay has two parts, 5 bits of time delay totaling one wavelength used as a phase shifter for narrowband subarray steering and 6 bits of time delay with a lowest significant bit of one wavelength.

Patriot

The Patriot radar, AN/MPQ-53, designed and built by Raytheon is a mobile C-band phased-array radar which performs target search and track, missile search and track, communications during midcourse guidance, and target-via-missile (TVM) terminal guidance. This tactical air defense system is designed to operate in a hostile ECM environment as a replacement for the improved Hawk and Nike-Hercules systems, which typically require up to nine radars.[11] Studies for this system were initiated in 1965, with an engineering development phase beginning in 1972. Initial production was funded in 1979, with a planned additional 12 units in 1982 and 18 per year through fiscal 1986.

The unique features of the radar antenna can be seen in Fig. 9-10. The antenna consists of a space-fed main array, a smaller TVM array, several ECM arrays, and an identification, friend or foe (IFF), antenna.[23,24] The main array, approximately 2.5 m in diameter, is filled with 5161 elements, which are contained in an array lens structure that is stowed in a horizontal position during transport and is erected hydraulically for system operation. Each element consists of a circular-waveguide dielectrically loaded front radiator, a flux-driven latching ferrite phase shifter, and a rectangular-waveguide dielectrically loaded rear radiator.

FIG. 9-10 Patriot-array lens. *(Courtesy of Raytheon Company.)*

FIG. 9-11 Patriot feed assembly. *(Courtesy of Raytheon Company.)*

The phase shifter utilizes a dielectrically loaded nonreciprocal garnet formed into a toroid and located in the center of the rectangular waveguide. Phase commands are in the form of row and column start-and-stop pulses whose time intervals are equivalent to 4-bit phase commands. Prior to any phase command, the garnet is driven into saturation to establish a reference phase from which all phase increments are set. The amount of phase shift is obtained by applying a voltage pulse of constant amplitude and variable width.

The feed assembly shown in Fig. 9-11, which is located approximately 2.5 m from the array, consists of transmitter horns and a multimode, multilayer receive feed. Through this arrangement, the duplexing function is implemented by the space separation of the feeds rather than through the use of conventional duplexers. The receive feed is located on the array axis. Phase-shift commands for transmit and receive differ to account for the different placements of the transmit-receive feeds. The receive feed is a five-layer (*E*-plane), multimode (*H*-plane) monopulse horn with independent control of the sum-and-difference patterns. Dielectric lenses in the feed aperture provide phase and amplitude correction of the *H*-plane excitation.

A TVM array (shown below and to the right of the main array in Fig. 9-10) produces a receive sum output from the TVM downlink signals and provides main-array sidelobe blanking and cancellation. There are 253 elements arranged in an aperture approximately half a meter in diameter. The array utilizes a stripline corporate feed. Uniform illumination is provided through a primary-distribution-network 12:1 power combiner, which in turn is connected to an aperture illumination network. This is followed by a transition to the waveguide phase-shift element, which is identical to that used in the main array.

Up to five ECM arrays are provided, three along the bottom of the antenna and two, one on each side, below the main array. These arrays each contain 51 elements identical to those used in the main and TVM arrays. The corporate feed is a single stripline layer which provides uniform amplitude and in-phase signals at each element.

The IFF antenna is located just below the main array.

Limited-Scan Arrays

Many radar applications for which antenna beam scanning in the order of $\pm 10°$ is required, such as ground-controlled approach and weapon locating, can use multifunction array antennas that are considerably less complex than those previously described. Limited-scan or limited-field-of-view (LFOV) antennas have been designed to reduce the number of phase shifters in the array.[25,26] The most common approach is either to provide one phase shifter for a subarray of elements or an oversized element or to use a small array in combination with either a reflector or a lens. It is interesting to note that this latter hybrid approach of using a lens to reduce the number of radiating elements and phase shifters also promises to reduce the cost of wide-angle scanning (hemispheric), as demonstrated in the Sperry dome antenna.[27]

Two examples of the implementation of multifunction arrays to the limited-scan application are given. In the first, the AN/GPN-22 utilizes a hybrid approach, and in the second, the AN/TPQ-37, one phase shifter per several elements is used in one plane of scan.

AN/GPN-22[28]

The antenna used by Raytheon for the ground-controlled approach AN/GPN-22 system is similar in its concept to the AN/TPN-25 antenna. The antenna and its radar shelter are shown in Fig. 9-12. After the development of the mobile version, AN/TPN-25, in 1969 and the subsequent production of 11 systems, the antenna system

FIG. 9-12   AN/GPN-22 precision approach radar. (*Courtesy of Raytheon Company.*)

TABLE 9-7 AN/GPN-22 Antenna Parameters[28]

Type	Limited-scan phased array
Gain	42 dB
Beamwidth	
Azimuth	1.4°
Elevation	0.75°
Scan volume	
Azimuth	±10°
Elevation	8°
Polarization	Circular
Array elements	443
Phase-shifter bits	3
Reflector size	4 m by 4.7 m

was redesigned electrically and mechanically in 1975 to reduce costs further for fixed-site applications and was redesignated the AN/GPN-22. Approximately 50 of these systems have been produced. The radar searches a volume of 20° in azimuth and 8° in elevation out to a range of 36.5 km while simultaneously tracking up to six targets.

For this antenna, the parameters of which are listed in Table 9-7, a small space-fed array illuminates a large reflector. The antenna gain and beamwidths are determined by the reflector size, while the number of beam positions is determined by the number of array elements. The small phased array illuminates the reflector and, by scanning on the reflector surface with properly adjusted phase shifts, can cause the antenna's far-field beam to scan over a limited sector. Since the AN/GPN-22 system is to be used at fixed sites, the antenna reflector was made larger than in its mobile version. This, along with a reduced scan volume, reduced the number of array elements from 824 to 443. At the same time, the array was relocated above the reflector.

The antenna is on a pedestal mounted on a concrete base separate from the shelter. The pedestal base is capable of rotating the antenna through 280° of azimuth to permit its use for multiple-runway coverage. The feed array consists of 443 three-bit ferrite phase shifters space-fed from a monopulse multimode horn. The array RF monopulse receivers and phase-shifter power supplies are mounted in the base of the antenna.

AN/TPQ-37

The AN/TPQ-37 radar system is a tactical S-band phased-array system capable of being transported by surface vehicles or by helicopter lift. The combination of the AN/TPQ-37 and the AN/TPQ-36, which has a similar architecture but uses different antenna techniques at a higher frequency, is referred to as the Firefinder. The combined system provides automatic first-round location of hostile artillery positions and is designed to locate simultaneous fire from numerous weapons on the battlefield. The AN/TPQ-37 will normally be sited behind the battle area to locate opposing long-range artillery fire. Emplacement time is estimated at 30 min, with a displacement time of 15 min. Initial development was undertaken by the Hughes Aircraft Company in 1973, with limited production authorized in 1976 and an expected production of 72 systems.[11]

FIG. 9-13 AN/TPQ-37 radar set. *(Courtesy of Hughes Aircraft Company.)*

The antenna, shown in its erected position above its trailer in Fig. 9-13, remains stationary in normal operation but has phase scanning in both planes to provide a 90° azimuth sector scan and an elevation scan of a few degrees. Since the weapon-locating mission can be accomplished with a limited elevation scan, the number of phase shifters in the elevation plane is greatly reduced by having one phase shifter feed a vertical subarray of six elements.

Figure 9-14 is a block diagram of the antenna feed network.[29] The number of elements in the phased array is 2154, consisting of 359 vertical subarrays of 6 elements each. A vertical subarray module is shown in Fig. 9-15. Each module contains three separate microstrip diode phase shifters with the following phase states:

Phase shifter	Phase states
A	22.5°, 45°, 90°, 180°
B	25°, 50°
C	25°

The phase shifter is fabricated by using thick film techniques on an alumina substrate. The rest of the module is a 1:6 power divider using ring hybrids and trans-

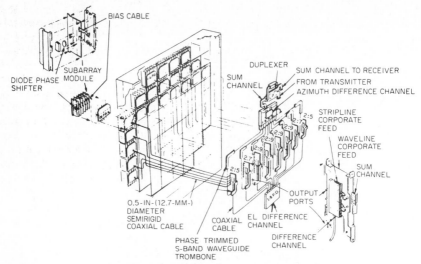

FIG. 9-14 AN/TPQ-37 antenna-feed block diagram. *(Courtesy of Hughes Aircraft Company.)*

mission lines to the six dipole feeds. These transmission-line lengths are designed to provide a progressive phase shift between dipoles, causing the radar beam to be tilted in the elevation plane.

Subarrays are arranged in groups of six fed by 1:6 air stripline power dividers. These groups are in turn stacked vertically into eight columns. The number of groups

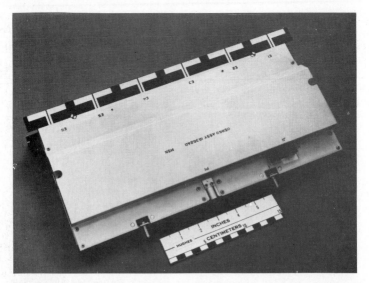

FIG. 9-15 AN/TPQ-37 antenna subarray module. *(Courtesy of Hughes Aircraft Company.)*

in a column increases from five at the edges to nine in the center. The 1:6 power dividers for subarrays in the last two columns on each side have unequal power split. Signals from the groups in any one column are connected through semirigid coaxial cable to a column-waveguide corporate feed with two outputs per column. One output, the sum arm, possesses even symmetry, and the other output, the difference, possesses odd symmetry in the elevation plane. The column difference signal is amplitude-weighted in a stripline feed assembly to control the elevation difference sidelobe level.

The difference signals for all eight columns are connected by a coaxial line to an 8:1 corporate feed with amplitude weighting for azimuth sidelobe control to form an elevation difference beam. The column sum signals which have been amplitude-weighted in their column networks to control the elevation sum sidelobe level are connected by waveguide to an 8:1 waveguide corporate feed to form a sum beam. The azimuth difference beam is formed by taking the difference of the eight column signals and combining them in the same manner as the elevation difference beam.

REFERENCES

1   M. I. Skolnik, *Introduction to Radar Systems*, 2d ed., McGraw-Hill Book Company, New York, 1980.

2   D. K. Barton, *Radar System Analysis*, Prentice-Hall, Inc., Englewood Cliffs, N.J., 1964.

3   M. I. Skolnik (ed.), *Radar Handbook*, McGraw-Hill Book Company, New York, 1970.

4   C. J. Richards (ed.), *Mechanical Engineering in Radar and Communications*, Van Nostrand Reinhold Company, London, 1969.

5   S. Silver, *Microwave Antenna Theory and Design*, MIT Rad. Lab. Ser., vol. 12, Boston Technical Publishers, Inc., Lexington, Mass., 1964.

6   P. C. Ratliff and L. F. Meren, "Advanced Enroute Air Traffic Control Radar System (ARSR-3)," *Eascon Conv. Rec.*, Institute of Electrical and Electronics Engineers, New York, 1973.

7   *The ARSR-3 Story*, Westinghouse Electric Corp., Baltimore, Md., n.d.

8   H. R. Ward, C. A. Fowler, and H. I. Lipson, "GCA Radars: Their History and State of Development," *IEEE Proc.*, vol. 62, no. 6, 1974.

9   P. L. Klass, "TPS-43 Radar Improvements Tested," *Aviation W.*, Aug. 18, 1980.

10   "Military Electronics," *Electronics*, Oct. 16, 1967.

11   R. T. Pretty (ed.), *Jane's Weapon Systems, 1980–81*, Jane's Publishing Company Ltd., London, 1980.

12   C. M. Lain and E. J. Gersten, "AN/TPS-59 Overview," *IEEE Int. Radar Conf. Rec.*, Institute of Electrical and Electronics Engineers, New York, 1975.

13   L. E. Bertz and L. J. Hayes, "AN/TPS-59 — A Unique Tactical Radar," *IEEE Mechanical Eng. Conf. Radar*, Institute of Electrical and Electronics Engineers, New York, 1977.

14   D. K. Barton, "Radar Equations for Jamming and Clutter," *Eascon Conv. Rec.*, Institute of Electrical and Electronics Engineers, New York, 1967.

15   E. Brookner (ed.), "Practical Phased-Array Systems," *Microwave Journal Intensive Course*, Dedham, Mass., 1975.

16   P. J. Kahrilas, *Electronic Scanning Radar Systems (ESRS) Design Handbook*, Artech House, Inc., Dedham, Mass., 1976.

17   E. Brookner, *Radar Technology*, Artech House, Inc., Dedham, Mass., 1977.

18   R. E. Willey, "Space Tapering of Linear and Planar Arrays," *IRE Trans. Antennas Propagat.*, vol. AP-10, July 1962.

19   A. A. Oliner and G. H. Knittel (eds.), *Phased Array Antennas*, Artech House, Inc., Dedham, Mass., 1972.

20   R. W. Coraine, "COBRA DANE (AN/FPS-108) Radar System," *Signal*, May-June 1977.

21 E. Filer and J. Hartt, "COBRA DANE Wideband Pulse Compression System," *Eascon Conv. Rec.,* Institute of Electrical and Electronics Engineers, New York, 1976.

22 F. Beltran and F. King, "Elimination of the Dummy Elements in Thinned Phased Arrays," *IEEE Antennas Propagat. Int. Symp. Rec.,* New York, 1981.

23 E. J. Daly and F. Steudel, "Modern Electronically Scanned Array Antennas," *Electronic Prog. (Raytheon Co.),* winter, 1974.

24 D. R. Carey and W. Evans, "The PATRIOT Radar in Tactical Air Defense," *Eascon Conv. Rec.,* Institute of Electrical and Electronics Engineers, New York, 1981.

25 J. M. Howell, "Limited Scan Antennas," *IEEE Antennas Propagat. Int. Symp. Rec.,* Institute of Electrical and Electronics Engineers, New York, 1974.

26 R. J. Mailloux (chairman), "Antenna Techniques for Limited Sector Coverage," *IEEE Antennas Propagat. Int. Symp. Rec.,* Institute of Electrical and Electronics Engineers, New York, 1976.

27 P. M. Liebman, L. Schwartzman, and A. E. Hylas, "Dome Radar—A New Phased Array System," *IEEE Int. Radar Conf. Rec.,* Institute of Electrical and Electronics Engineers, New York, 1975.

28 H. R. Ward, "AN/TPN-25 & AN/GPN-22 Precision Approach Radars," *IEEE Int. Radar Conf. Rec.,* Institute of Electrical and Electrical Engineers, New York, 1980.

29 D. A. Ethington, "The AN/TPQ-36 and AN/TPQ-37 Firefinder Radar Systems," *Eascon Conv. Rec.,* Institute of Electrical and Electronics Engineers, New York, 1977.

30 R. Tang and R. W. Burns, "Phased Arrays," in R. C. Johnson and H. Jasik (eds.), *Antenna Engineering Handbook* 2/e, McGraw-Hill Book Company, New York, 1984, Chap. 20

Chapter 10

Microwave Beacon Antennas

Jean-Claude Sureau

Radant Systems, Inc.

10-1 INTRODUCTION

Microwave beacon systems are used whenever there is a need to enhance the target return signal with regard to strength and/or information content. As such, these systems are highly reliable and accurate surveillance systems and, in most cases, provide some data-link capability.

Beacon systems typically consist of transponders and interrogators. *Transponders* are the active devices associated with the targets and provide the enhanced echo. On the basis of certain criteria which are system-dependent, they will selectively reply to interrogations by using a recognizable encoded message format with different degrees of information content. Transponders are used either on moving targets to assist in their surveillance or at surveyed points for self-location of the interrogator. *Interrogators* are the devices which elicit and process the replies for surveillance and message decoding; they represent the users of the beacon system, and they tend to be functionally sophisticated.

By far the most widely deployed beacon-system complex is the military identification, friend or foe (IFF), Mark X and Mark XII system and its civilian surveillance-system derivative, the air traffic control radar beacon system (ATCRBS), also known as secondary surveillance radar (SSR). More recently, these have been evolving in the United States into the discrete-address beacon system (DABS) and, in the United Kingdom, into a similar system called address-selective (ADSEL) SSR. All these systems share a common frequency allocation, 1030 MHz for interrogation and 1090 MHz for reply, as well as other features in waveform format.[1,2,3] The antennas associated with transponders are typically omnidirectional on receive and transmit. From a design viewpoint, they offer no unique issues other than those related to their installation. On aircraft, for example, they are frame-mounted blades or annular slots which are located so as to minimize shadowing.[4,5] In contrast, interrogator antennas are generally more complex in their operation and, as a result, offer unique design features. These constitute the central focus of this chapter.

10-2 INTERROGATOR ANTENNAS: DESIGN PRINCIPLES

The principal system requirements imposed on interrogator antennas fall into three basic categories:

1 Support each one-way link from a power-budget as well as a time-on-target viewpoint.

2 Elicit replies only from main-beam interrogations, and process replies received only in the main beam.

3 Provide target-bearing estimates from the replies.

For ground-based interrogators, minimization of ground multipath as it relates to these three categories has been a central design consideration. Constraints resulting from colocation with a primary radar also are often factors. Such constraints become the principal issue in airborne interrogators because they impose limitations on possible performance.

Vertical Pattern Design

In ground-based (and shipborne) interrogators, link reliability can be radically affected by multipath-induced lobing in the vertical plane. An example of this lobing is shown in Fig. 10-1 for a linear array with a vertical aperture similar to that widely used for civilian air traffic control (ATC) and contrasted with that for a typical large vertical aperture providing a sharper horizon cutoff.

While the structure of the lobes is dependent on the height of the antenna above ground, the envelope of the amplitude at the minima is not but depends on the shape of the elevation pattern near the horizon. A simplistic but useful to first-order measure of this shape is the slope of the pattern, measured in decibels per degree at the horizon. The common practice has been to point antennas so that the horizon is at the −6-dB point whenever possible. Figure 10-2 shows the resulting envelope of the lobing minima as a function of this horizon cutoff rate. Since higher cutoff rates require larger vertical apertures, reduction of lobing is a major trade-off issue.

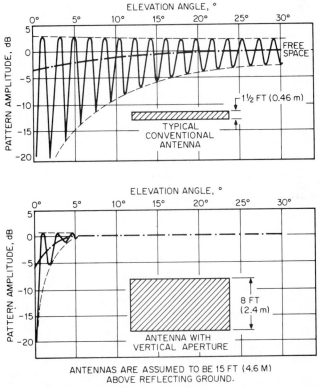

FIG. 10-1 Influence of vertical aperture on elevation lobing patterns.

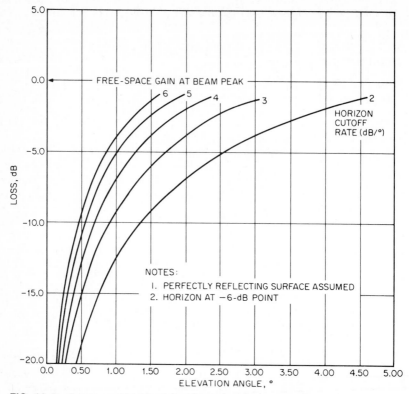

FIG. 10-2 Envelope of lobing minima for various horizon cutoff rates.

In some installations which require high cutoff rates, interrogator antennas with a large vertical aperture are directly implemented, while in others this feature is obtained by sharing the usually large primary radar reflector with the beacon system. These are called *integral beacon feed systems;* examples of both types will be discussed later in this chapter.

The most significant contribution of the development of interrogator antennas to the theory of antenna design has been the refinement of techniques for synthesizing sharp cutoff beam patterns, i.e., patterns which have constant gain over a prescribed angular sector (this tends to be the preferred shape for beacon operation). Specifically, these refinements have examined the relationship between the aperture size (or number of elements) and the rate of pattern roll-off at the beam edges as a function of sidelobe levels and gain ripple within the main beam. Evans[6] has adapted a procedure originally developed for the synthesis of digital filters. The procedure he describes allows independent specification of sidelobe level and gain ripple and results in a maximum roll-off slope for a given number of elements. For engineering estimates, Lopez[7] has provided a normalized curve (independent of aperture size), shown in Fig. 10-3, that exhibits the impact of sidelobe level and gain ripple on the roll-off rate. This rate is standardized as being measured at the -6-dB point and is given in units of decibels per degree per (D/λ), where D is the aperture size and λ is the wavelength. For exam-

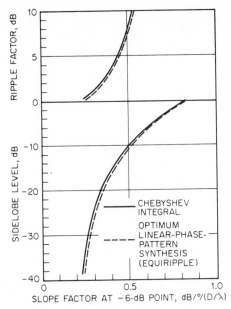

FIG. 10-3 Normalized horizon cutoff rate.
(From Ref. 7, © 1979 IEEE.)

ple, if a -24-dB sidelobe level and a 3-dB ripple-factor design are desired, the maximum slope factor is estimated by averaging the slope factors for the -24-dB sidelobe level (0.30) and the 3-dB ripple factor (0.38); the result is a slope factor of 0.34 dB/ $°/(D/\lambda)$.

Sidelobe Suppression

The procedure by which replies are elicited only from main-beam interrogations is called *interrogation sidelobe suppression* (SLS). It is generally accomplished through the sequential transmission of a pulse over the directional beam, followed by another pulse over a "control" pattern.[1] The transponder replies only when the relative amplitudes meet predetermined criteria which indicate that it is being interrogated in the main beam. A variation called *improved sidelobe suppression* (ISLS) attempts to go one step further and force suppression of transponders in sidelobes by transmitting the first pulse jointly over the sum (Σ) and control (Ω) beams. For DABS interrogations, suppression is obtained by masking the synchronization phase reversal through simultaneous transmission over the Σ and Ω beams. Some beacon systems also include receive sidelobe suppression (RSLS), in which the interrogator multichannel receiver compares, for each pulse, the amplitudes corresponding to the directional and control beams and determines whether or not the pulse should be accepted as being received through the main beam.

When the directional and control beams are derived from the same aperture *(integral suppression pattern)*, they have common phase centers. Consequently, the multipath-induced vertical lobing will be the same in both beams and, to first order,

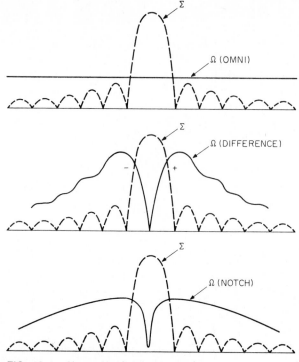

FIG. 10-4 Generic control-pattern types.

no false SLS operation results. However, when the phase centers are vertically displaced, the vertical lobing patterns do not track and faulty SLS operation results. This can take the form of false suppression and/or false interrogation when the directional beam is in a fade. This problem is mitigated when both beams have elevation patterns with sharp horizon cutoff even if they have displaced phase centers. The various types of control patterns which are typically used are illustrated in Fig. 10-4.

Omnidirectional Pattern This pattern is implemented with a separate antenna (with a few exceptions). Its primary advantage is that it does not need to rotate and therefore does not require an additional channel in the rotary joint. The vertical pattern is often designed to match that of the directional antenna. Unfortunately, such installations tend to have vertically displaced phase centers and hence exhibit differential lobing. In addition, nearness of the directional antenna causes some blockage, which is modulated by scanning. Such systems often exhibit marginal SLS operation.[8,9]

Difference Pattern This pattern is used primarily in airborne military interrogators because it is convenient, given installation constraints, and also because the interrogation window can be controlled, giving the effect of beam sharpening. It provides the desired narrow interrogation resolution that is somewhat better matched to the resolution of the associated radar than its inherent broad beam. This pattern is also

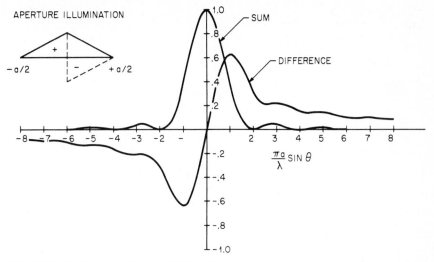

FIG. 10-5 Typical sum-difference SLS-pattern pair.

used in some integral-feed-type ground-based interrogators. Because of the central null and because such systems have no differential lobing (common phase centers), false main-beam suppression is virtually eliminated. Sidelobe breakthrough, resulting in false interrogations, can still occasionally occur. This type of suppression beam is usually implemented by taking the difference between the two halves of the aperture while the directional beam is taken as the sum. Whereas this arrangement generally results in suboptimal conventional monopulse performance, it is ideal for SLS purposes, as illustrated in Fig. 10-5.

Notch Pattern   This type offers the best available combination of common phase center, central null, and low probability of sidelobe punch-through. It can be implemented only by array-type interrogators because of the way in which such patterns are generated: the central element of the array provides the pattern's broad coverage and is combined, in antiphase, with an appropriately attenuated array directional pattern to create the central notch. The notch-and-difference-type control patterns often need to be complemented by a back-fill auxiliary radiator to cover the directional beam's backlobes.

Bearing Estimation

Estimation of the bearing to the transponder is determined by the interrogator antenna and the associated receiver system by one of two techniques, beam-split and monopulse. The beam-split technique requires that while the antenna is scanning through the transponder, several valid replies be received. Various standard algorithms for such angle estimation are used.

The monopulse technique used is inherently of the off-boresight type, and an estimate can be made on each reply pulse. Since the system typically operates at a high signal-to-noise ratio, estimation errors are due predominantly to inaccurate monopulse calibration. Experience indicates this error to be between one-fiftieth and one-hundredth of the beamwidth.[10] In particular, the monopulse slope tends to be dependent on elevation angle (approximately proportional to its cosine). This phenomenon is a direct geometric consequence of the definition of *bearing angle* as the projected direction into a horizontal plane of the true direction vector. The estimation errors are therefore elevation-dependent and also proportional to the off-boresight angle. It is therefore an advantage for the estimation algorithm to favor replies near boresight if more than one are available from a given transponder. From the viewpoint of antenna design, optimization of monopulse performance is standard with regard to slope maximization while low sidelobes are maintained for interference minimization.

For electronically phase-scanned fan-beam interrogators, the previously mentioned geometric effect takes on more serious proportions since the beam boresight exhibits a bearing angle ϕ which is elevation-angle-(α)-dependent (this phenomenon is then called *coning*) and is related to the nominal zero-elevation boresight angle Φ_0 by

$$\sin \Phi(\alpha) = \frac{\sin \Phi_0}{\cos \alpha} \qquad \textbf{(10-1)}$$

If an elevation correction is not included, bearing errors will be made for both split and monopulse systems.

10-3 INTERROGATOR ANTENNAS: PRACTICE

Ground-Based Systems

For many years the standard configuration for interrogators (ATC-IFF) consisted of a linear array referred to as a *hog trough* because of its shape and a stationary omnidirectional antenna. The array consists of a vertical-dipole-excited sectoral horn fed by a corporate power divider with an azimuth beamwidth of 2.3°; the omnidirectional antenna is typically a biconical horn. Figure 10-6 shows such an installation at a Federal Communications System (FAA) site. A different version, developed for the

FIG. 10-6 Typical hog-trough antenna. *(Courtesy of Department of Transportation, Federal Aviation Administration.)*

FIG. 10-7 Typical ADSEL-compatible SSR antenna. *(Courtesy Cossor Electronics, Ltd.)*

ADSEL system[11] and shown in Fig. 10-7, features an integral notch-type SLS pattern and also a difference pattern for monopulse azimuth estimation and receive sidelobe suppression. The military versions of the hog trough, as exemplified by that shown in Fig. 10-8, are often much shorter but provide azimuth beam sharpening and sidelobe suppression by a difference pattern. All these linear arrays have little elevation beam shaping and a horizon cutoff rate near zero, and they are subject to different degrees, depending on the SLS pattern, to the problems of multipath.

To reduce multipath, recent designs have emphasized increased vertical aperture. The *open-array concept,*[12] an implementation of which is shown in Fig. 10-9, simultaneously satisfies the need for near-horizon shaping and physical compatibility with the existing primary radar (weight, wind loading, etc.). The use of a resonant design for the ground plane permits a wider-than-usual separation between ground-plane reflector elements and minimizes wind loading. This antenna features a notch-type SLS pattern with a back-fill radiator and a low sidelobe monopulse difference pattern. All three beams (Σ, Δ, and Ω) have the same vertical pattern, with a sharp horizon cutoff and a common phase center. Typical radiation patterns for this antenna are shown in Figs. 10-10 and 10-11. This open array is being substituted for the hog trough at many FAA terminal surveillance sites.[13]

Another approach to obtaining a sharp horizon cutoff has been to use the associated primary radar reflector (integral-feed implementation). The two principal design problems associated with this approach are the constraints on beacon patterns

FIG. 10-8 Typical military IFF antenna. *(Courtesy Hazeltine Corporation.)*

FIG. 10-9   Open-array SSR antenna. *(Courtesy Texas Instruments, Inc.)*

imposed by the radar reflector and the physical integration of beacon feed with radar feed. The latter problem is more acute with L-band radars than with S-band radars. At S band this technique has been implemented primarily for military installations (ATC-IFF), in which the 4° beacon beamwidth (inherent because of the standard S-band radar-antenna size) is effectively reduced through beam sharpening by using a difference-type SLS pattern. An example of such an implementation is shown in Fig.

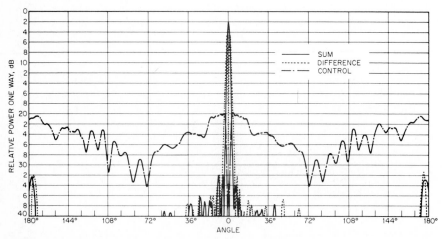

FIG. 10-10   Open-array azimuth patterns. *(Courtesy Texas Instruments, Inc.)*

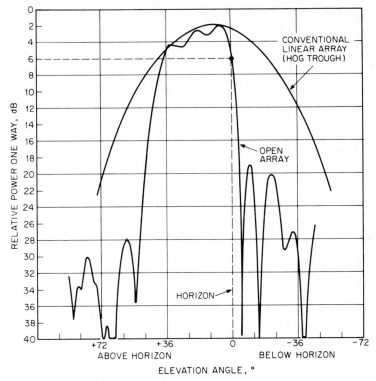

FIG. 10-11 Open-array elevation pattern. *(Courtesy Texas Instruments, Inc.)*

10-12. The horizon cutoff rate obtained is about 2 dB/°. The reflector, shaped for the radar, tends to produce an elevation pattern whose cosecant-squared upper-angle behavior is sometimes undesirable. By vertically stacking several beacon feeds (see Fig. 10-13), thereby obtaining some form of multiple-beam synthesis, the effective elevation pattern can be modified to resemble more closely a sector-type shape.

The integration of a beacon feed with an L-band radar is more difficult because of the closeness of the two frequency bands. The approach taken for the ARSR-3 radar (see Figs. 10-14 and 10-15) has been to offset the beacon feed sideways. This results in an azimuth beam shift between the radar and the beacon which must be compensated for in the information processing, since both skin and transponder returns are usually combined into one surveillance report. The feed elements are slots, several of which are vertically stacked for elevation-pattern control. The SLS function is provided by a separate omnidirectional antenna mounted on top of the reflector (it rotates with the rest of the system) and with an elevation pattern shaped to provide optimal sidelobe coverage at all elevation angles. Since the spoiled reflector has worse azimuth sidelobes at high elevation angles (see Fig. 10-16), the omnidirectional pattern cannot be allowed to drop off as much as the directional-beam principal-plane pattern.

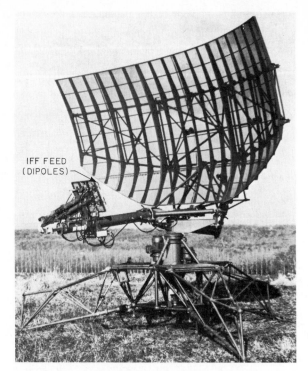

FIG. 10-12 Tactical S-band radar antenna with integral IFF feed. *(Courtesy Thomson-CSF.)*

FIG. 10-13 Integral beacon feed for elevation-pattern shaping. *(Courtesy AEG-Telefunken.)*

For the TPS-43, a tactical air defense radar, a feed-mounted linear array (see Fig. 10-17) has been adopted. The use of a difference-type suppression pattern together with a back-fill radiator provides an effective 4° interrogation beamwidth as desired (shown in Fig. 10-18).

A limited number of beacon antennas featuring electronic scanning have taken the form of cylindrical arrays. The version shown in Fig. 10-19, implemented by Hazeltine Corporation, Wheeler Laboratory,[14] is designed to be compatible with an S-band radar tower installation. It features an integral omnidirectional SLS pattern obtained by a uniform excitation of all columns and a monopulse dif-

FIG. 10-14 ARSR-3 long-range surveillance radar antenna. *(Courtesy Westinghouse Electric Corp.)*

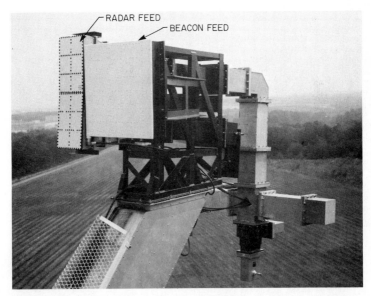

FIG. 10-15 Integral beacon feed for the ARSR-3. *(Courtesy Westinghouse Electric Corp.)*

10-13

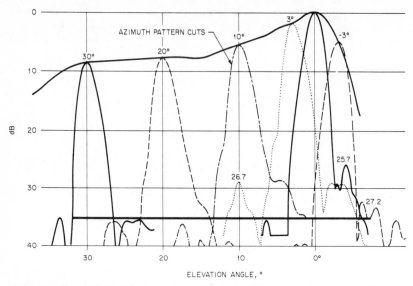

FIG. 10-16 Typical radiation patterns for the ARSR-3 beacon antenna. *(Courtesy Westinghouse Electric Corp.)*

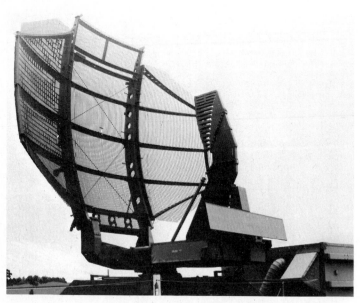

FIG. 10-17 IFF array for the TPS-43 antenna. *(Courtesy Westinghouse Electric Corp.)*

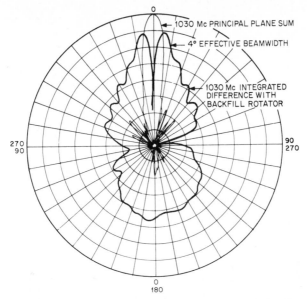

FIG. 10-18 Sum-difference suppression patterns for the TPS-43 IFF array. *(Courtesy Westinghouse Electric Corp.)*

ference pattern. The vertical pattern is of the sector beam type with a sharp horizon cutoff. In addition, there is an electronic hop-over feature in which the elevation beam can be lifted in specified azimuth directions to avoid multipath-producing obstacles. The azimuth beamwidth is maintained nearly constant as a function of elevation angle (see Fig. 10-20) by a multiple-angle collimation scheme. Scanning, performed by a

FIG. 10-19 E-SCAN cylindrical array. *(Courtesy Hazeltine Corporation.)*

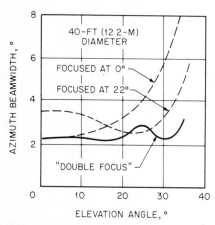

FIG. **10-20** Beamwidth variation of E-SCAN array versus elevation angle. (*Courtesy Hazeltine Corporation.*)

FIG. **10-21** Tactical IFF-ATM cylindrical array. (*Courtesy Bendix Communication Division.*)

combination of switches (sector steps) and phase shifters (fine incremental steps), can be either of the uniform continuous type to emulate a mechanically scanned system (for compatibility with existing processing) or of the agile type to support more advanced interrogation-management schemes.

The array shown in Fig. 10-21 has been developed for tactical applications. Scanning in increments of 5.6° of the 65° azimuth beam is done by a modal Butler-matrix type of feed in which all eight dipole columns are excited. The SLS and azimuth monopulse functions share the same difference-type pattern synthesized to provide backlobe coverage. All beams have a common elevation pattern with a high horizon cutoff rate.

The cylindrical array shown in Fig. 10-22 has been developed for shipborne ATC-IFF application. The directional beam, complemented by a stationary omnidirectional SLS beam, is positioned by a Lockheed-proprietary Trimode Scanner System.[16]

Airborne Systems

Traditionally, airborne beacon interrogation systems have been military IFF systems. These are typically used in conjunction with a radar which is often at X band. As mentioned earlier, the design problem is primarily one of real estate. Figure 10-23 shows a typical implementation in which L-band dipoles for the beacon interrogation antenna are mounted on the surface of the radar reflector. The dipoles use resonant techniques to make them invisible at the radar frequency. Beam sharpening and SLS are obtained by a difference pattern.

FIG. 10-22 Shipborne ATC-IFF cylindrical array AN/OE 120/UPX.
(Courtesy Lockheed Electronics Co.)

Recently some airborne interrogators have been implemented for the civilian traffic-alert and collision-avoidance system (TCAS-II), which is based on air-derived surveillance of an aircraft's near airspace. A TCAS-II antenna is shown in Fig. 10-24. This antenna provides a capability to form either a directional beam or a notched beam in any one of eight azimuth positions. During the beacon interrogation the normal sidelobe-suppression pulse is transmitted over the notch pattern so that only transponders in a 45°-wide sector reply. The TCAS-II design can also transmit omnidirectional interrogations. Transponder replies are received omnidirectionally with a 360° instantaneous direction-finding capability to an 8° rms accuracy. Top-loaded

FIG. 10-23 IFF array for airborne fire-control radar. *(Courtesy Hazeltine Corporation.)*

FIG. 10-24 Fuselage-mounted interrogator array for collision-avoidance system. *(Courtesy Dalmo Victor Operations, Textron Inc.)*

monopoles are utilized in this design to provide a low-drag antenna. Four monopoles spaced approximately one-quarter wave apart are used. Antenna-mode control is provided by using stripline circuitry consisting of hybrid couplers, pin-diode switches, and phase shifters.

REFERENCES

1 M. I. Skolnik (ed.), "Beacons," *Radar Handbook,* McGraw-Hill Book Company, New York, 1970, chap. 38.
2 P. R. Drouilhet, "The Development of the ATC Radar Beacon System: Past, Present and Future," *IEEE Trans. Com.,* vol. Com-21, 1973, pp. 408–421.
3 D. Boyle, "DABS and ADSEL—The Next Generation of Secondary Radar," *Interavia,* March 1977, pp. 221–222.
4 K. J. Keeping and J.-C. Sureau, "Scale Model Pattern Measurements of Aircraft L-Band Beacon Antennas," MIT Lincoln Lab. Proj. Rep. ATC-47 (FAA-RD-75-23), Apr. 4, 1975.
5 G. J. Schlieckert, "An Analysis of Aircraft L-Band Beacon Antenna Patterns," MIT Lincoln Lab. Proj. Rep. ATC-37 (FAA-RD-74-144), Jan. 15, 1975.
6 J. E. Evans, "Synthesis of Equiripple Sector Antenna Patterns," *IEEE Trans. Antennas Propagat.,* vol. AP-24, May 1976, pp. 347–353.
7 A. R. Lopez, "Sharp Cut-Off Radiation Patterns," *IEEE Trans. Antennas Propagat.,* vol. AP-27, November 1979, pp. 820–824.
8 D. L. Sengupta and J. Zatkalik, "A Theoretical Study of the SLS and ISLS Mode Performance of Air Traffic Control Radar Beacon System," *Int. Radar Conf. Proc.,* April 1975, pp. 132–137.
9 N. Marchand, "Evaluation of Lateral Displacement of SLS Antennas," *IEEE Trans. Antennas Propagat.,* vol. AP-22, July 1974, pp. 546–550.
10 D. Karp and M. L. Wood, "DABS Monopulse Summary," MIT Lincoln Lab. Proj. Rep. ATC-72 (FAA-RD-76-219), Feb. 4, 1977.
11 M. C. Stevens, "Cossor Precision Secondary Radar," *Electron. Prog.,* vol. 14, 1972, pp. 8–13.
12 P. Richardson, "Open Array Antenna for Air Traffic Control," *Texas Instruments Equip. Group Eng. J.,* September–October 1979, pp. 31–39.

13 C. A. Miller and W. G. Collins, "Operational Evaluation of an Air-Traffic Control Radar Beacon System Open Array Antenna," *Eascon Conv. Rec.*, Institute of Electrical and Electronics Engineers, New York, 1978, pp. 176–185.

14 R. J. Giannini, J. Gutman, and P. Hannan, "A Cylindrical Phased Array Antenna for ATC Interrogation," *Microwave J.*, October 1973, pp. 46–49.

15 J. H. Acoraci, "Small Lightweight Electronically Steerable Antenna Successfully Utilized in an Air-Traffic Management System," *IEEE Nat. Aerosp. Electron. Conf.*, vol. 1, 1979, pp. 43–49.

16 U.S. Patent 3,728,648, Apr. 17, 1973.

Chapter 11

Tracking Antennas

Josh T. Nessmith

Georgia Institute of Technology

Willard T. Patton

RCA Corporation

11-1 INTRODUCTION

Tracking antennas are used in communications, direction finding, radio telescopes, and radar. The first three applications are covered in Chaps. 13, 16, and 18 respectively. This chapter is concerned with radar tracking antennas whose function is to provide accurate estimates of target location in two orthogonal angular coordinates. The reflector tracking antenna and the planar phased-array tracking antenna will be used for illustration.

Angle Estimation

Accuracy in estimating target direction is a fundamental requirement for the tracking antenna. It depends on systematic and random errors in the design of the antenna and mount, correction of these errors by calibration, and stability of the antenna and mount with respect to time. In addition, receiver linearity, the signal-to-noise ratio developed out of the signal processor, and the leads or lags in the difference-signal estimate, introduced by filters in the data processor, also affect the overall radar-system angular accuracy. Only the errors associated with the antenna and its mount are considered here.

Tracking a target requires a real-time feedback system that will direct the antenna beam to the target's angular direction when there is a difference* between the electrical reference direction, or boresight axis, and the actual target direction, as shown in Figs. 11-1 and 11-2.

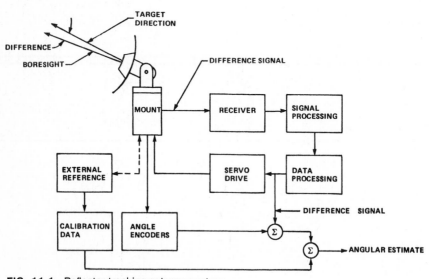

FIG. 11-1 Reflector-tracking-antenna system.

---

*Although the term *error* has been used in much of the literature to describe this quantity, the term *difference* will be used in this chapter to avoid confusion when discussing errors in tracking-antenna output data.

11-2

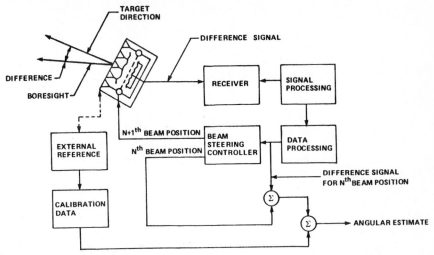

FIG. 11-2   Phased-array-tracking-antenna system.

The difference signal derived from the tracking antenna provides the information to drive the antenna boresight axis toward the target direction and also provides a correction factor[1] in the angular estimate when boresight direction leads or lags target direction. A reflector-tracking-antenna system has an output that is the sum of its mechanical orientation, as measured by angle encoders, and the electrical-angle difference, derived from the location of the target with respect to the boresight axis. Both the mechanical position and the electrical-angle-difference data are corrected as necessary from calibration data. The phased-array tracking system provides an angle estimate that is the calibration-corrected sum of the beam-steering order, the electrical-angle difference developed between the boresight axis and the target direction, and the mechanical orientation of the array face.

Because of inertia, the reflector tracking antenna usually is limited to tracking a single target, although angle estimates may be made on several targets that are in the same beam as the target being tracked. The phased-array antenna can provide accurate angle estimates on a large number of targets at widely different angular spacing. The limitation on the number of targets that may be tracked by the phased-array tracking antenna is due not to the antenna itself but to the radio-frequency (RF) energy available, the amount of energy required to provide the accuracy of estimation for each target, and the dwell time required for each target.

Tracking Techniques

Two basic classes of tracking techniques are used to obtain the angular estimate: sequential-lobe comparison and simultaneous-lobe comparison. Both provide the magnitude and direction of the difference between target and beam position necessary to perform the tracking function. Although tracking involves the entire radar system, the type of tracking is defined by the method of sensing angular-difference magnitude and direction by the antenna.

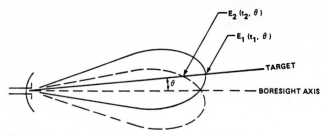

FIG. 11-3 Sequential-lobe comparison of squinted beams.

Sequential-Lobe Comparison Sequential lobe operates by displacing the antenna beam on successive transmissions and measuring the difference in relative signal strength, expressed by $\Delta = E(t_1,\theta) - E(t_2,\theta)$. The signal strength E is a function of both time t and angular position θ. The amplitude of the difference Δ indicates the magnitude of the angular difference, and the sign, positive or negative, indicates angular direction in the lobing plane.

Sequential lobing can be performed, as shown in Fig. 11-3, by using a single feed with symmetrical mechanical or electrical displacements from the boresight axis to give the two beam positions or by electrically or mechanically switching between two offset feeds.

Conical scanning is another sequential technique that can be used. One method is to use a feed that is electrically or mechanically displaced from the boresight axis. The feed maintains this displacement as it is nutated about the boresight axis.[2] A constant return occurs for a target on the boresight or mechanical axis, and a sinusoidal return occurs for a target that is off axis. The amplitude of the sinusoid indicates the magnitude of the difference between boresight axis and target position, and the feed position at peak response provides the angular direction of the difference. Conical scanning also can be accomplished by fixing the feed and rotating a tilted reflector, as is sometimes done in small tracking systems, or by electronically combining the outputs from a monopulse feed.

Because the sequential technique is time-dependent, changes in target-signal amplitude with time can modulate the difference signal. The influence of such amplitude fluctuations can be minimized by speeding up the beam switching or the conical-scan rate. Both the switching and the scan rates, however, are limited by the pulse-repetition period of the radar.

Simultaneous-Lobe Comparison Simultaneous-lobe comparison utilizes two different beams at the same time. Two identical beam shapes can be generated by using a reflector antenna and two feed horns that are symmetrically displaced from the boresight axis, as shown in Fig. 11-4. These "squinted" beams are identical to those received sequentially, but reception is simultaneous. Only a single signal channel following the antenna is required to make the difference estimate for sequential lobing, but two or more signal channels are required for simultaneous lobing.

If a microwave hybrid comparator is added to the feed network, a sum response, $\Sigma = E_1(t_1,\theta) + E_2(t_1,\theta)$, and a difference response, $\Delta = E_1(t_1,\theta) - E_2(t_1,\theta)$, are generated simultaneously as shown in Fig. 34-5a and b. Depending upon the comparator used, the Σ and Δ signals may occur in phase or in quadrature. If they occur in quadrature, a 90° phase shift is added to one of the signals to bring the signals in

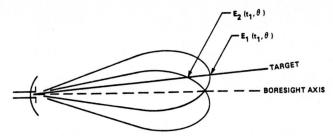

FIG. 11-4   Simultaneous-lobe comparison of squinted beams.

phase before comparison in a product detector. In the tracking technique referred to as *amplitude monopulse*,[3] the product detector forms the vector dot product of the two normalized signals, $\Sigma/|\Sigma| \bullet \Delta/|\Sigma|$, as shown in Fig. 11-5$c$. Normalization often is obtained by using the sum signal for gain control of both the sum and the difference channels. Estimates out of the product detector are dependent only on the angle off axis and are independent of amplitude fluctuations of the target.

Certain objections exist to describing monopulse in terms of beams $E_1(t_1,\theta)$ and $E_2(t_1,\theta)$ because only the Σ and Δ beams are generated at the antenna ports. These expressions, however, serve as a convenient model for basic understanding. The term *difference pattern* resulted from taking the difference between the patterns of E_1 and E_2.

The separation of the two feed horns in order to obtain an amplitude comparison introduces a small phase difference when the target is off the boresight axis. If the two feed horns are separated far enough on the same aperture or if two separate antennas on a common mount are used, then an estimate of the angle between boresight direction and target direction may be derived by comparing the relative phase of the two signals received simultaneously.[3] In the dual-aperture implementation of this technique, known as phase-comparison monopulse, the two beams are pointed in the same direction rather than having an angular displacement, as in the amplitude-comparison

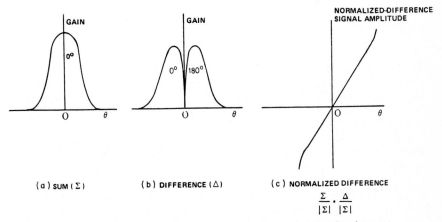

FIG. 11-5   Amplitude monopulse pattern and normalized difference-signal curve.

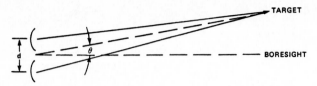

FIG. 11-6   Phase comparison.

method. For a target that is off axis, a phase difference occurs between the two signals, as shown in Fig. 11-6. This phase difference can be expressed as

$$\Delta\phi = \frac{2\pi}{\lambda} d \sin \theta \qquad (11\text{-}1)$$

where d is the distance between antennas, λ is the wavelength, and θ is the angle between boresight and target direction. For small angles, $\sin \theta$ is equivalent to θ, and the phase difference may be used as a linear estimate of the difference between the boresight and target direction.

Choice of Tracking Technique

The investment required to design and fabricate a phased-array antenna and the limited time available in most systems point to the choice of monopulse as the tracking technique but does not preclude any other choice of technique. The monopulse technique provides Σ and Δ outputs that may be used either as an amplitude or as a phase monopulse system or, with specific orders to the beam-steering controller, for sequential lobing or conical scan.

For reflector systems, the choice is most often a trade-off between costs and accuracy. Monopulse was developed as a means of eliminating the scintillation or amplitude noise effects on tracking accuracy. Although targets with scintillation may be tracked with sequential-lobing systems, accuracy is considerably degraded, and very high scintillation may cause loss of track. This phenomenon has been the basis for radar-jamming techniques because the modulation that is introduced enables an enemy intercept receiver to readily interpret the type of radar and employ the jamming techniques to which the radar would be most vulnerable.

Consideration must be given to the gain and sidelobe levels achievable because both are influenced by the choice of tracking technique. There is an improvement in signal-to-noise ratio of 2 to 3 dB with monopulse for targets on or near the boresight axis compared with lobe switching or conical scan. Monopulse also can achieve lower levels of sidelobes than can conical scanning or other sequential-lobing methods because of the designer's ability to optimize separately the sum and difference patterns of a monopulse system.

In estimating the angular difference between target direction and the boresight axis, the standard deviation of the error in the estimate is inversely proportional[4] to the slope of the normalized angular-difference signal:

$$\sigma_{A,E} = \frac{\theta_B}{K_m \sqrt{2 \cdot S/N}} \cdot \sqrt{1 + \left(K_m \frac{\theta}{\theta_B}\right)^2} \qquad (11\text{-}2)$$

where $\sigma_{A,E}$ = standard deviation in estimate of azimuth or elevation angular difference

K_m = normalized monopulse difference slope

S/N = integrated signal-to-noise ratio

θ_B = half-power, one-way sum-pattern beamwidth (scan-angle-dependent for phased arrays)

θ = angle between boresight and target direction

Data in Skolnik[3] show that monopulse has a difference slope potential greater than conical scan, and a more precise estimate of the angular difference may be made for the same beamwidth and signal-to-noise ratio.

Although the tracking technique of choice appears to be monopulse, sequential techniques can be used when cost is the major factor and if the targets are reasonably well behaved. The same accuracy can be achieved with sequential techniques as for monopulse for targets with high signal-to-noise ratios, low scintillation, and low dynamic performance. If the targets have considerable angle noise or glint, all techniques suffer in accuracy. The attributes of monopulse in such cases may not override the economical advantages of sequential scan.

11-2   REQUIREMENTS

In specifying tracking-antenna requirements, such parameters as gain, sidelobes, and losses are often secondary to accuracy and are specified at values that support the accuracy requirements. Accuracy is expressed in terms of the allowable angular error in each axis between the true target direction, with respect to tracking-antenna location, and the direction estimate derived from antenna pattern and mount data. The specified error is generally subdivided into a budget that identifies major contributors to the error and the size for each contribution.

Errors

The term *precision,* defined as "the quality of being exactly or sharply defined or stated,"[5] often is incorrectly substituted for the term *accuracy,* which is "the quality of correctness or freedom from error." Thus, a tracking antenna may be precise but not accurate, whereas an accurate tracking antenna must be precise.

In specifying error in each axis, the root-sum-square (rss) value of both residual systematic errors and random errors is used, although in error budgets individual errors are expressed in terms of root-mean-square (rms) values. In much of the literature, the term *bias* is substituted for *systematic,* and *noise* is substituted for *random.* In this chapter, the terms *systematic* and *random* will be used, and electrical noises due to thermal effects will be discussed as a subset of random errors.[6]

Systematic Errors   A *systematic error* is a constant or a variable error that can be predicted as a function of antenna beam position, frequency of operation, or environmental factors. Such errors may be measured by calibration procedures and can be expressed as algorithms or in tables to correct antenna output data. Because only a finite number of measurements can be made, algorithms or tables developed for correction are approximations. Consequently, systematic residual errors remaining after data correction include error due to approximation or interpolation, error inherent in instrumentation, and error on the part of the observer. The rms value of each residual error for each contributor in each angular coordinate is the value given in the tracking-antenna error budget under systematic residual errors. The rms value of each error contributor and the rss value of the total are usually expressed in milliradians (mrad). Sometimes these values are expressed in millisines for errors that are a function of the scan angles in phased arrays. Care must be taken in summing error budgets so that the dimensions are not intermixed. In this chapter, values will be expressed in milliradians.

Random Errors The angular error due to thermal noise can be predicted on the basis of the known electromagnetic characteristics of the system. Maximum amplitude of backlash in data gearing or coupling to an encoder can be measured, but since it cannot be related to an angular position, it must be treated as a random error. Errors in a phased array due to phase-shifter quantization or rounding off in steering orders are also treated as random errors.

Random errors cannot be calibrated out of the system. They are grouped, and their rms values are squared and summed. The square root of this sum is the rss value for random error given in the antenna specification for each coordinate. The permitted error in each coordinate is the square root of the sum of the squares of total systematic residual and random errors. Though the process may be open to question from a strict mathematical interpretation, such procedures appear to be valid since measurements of targets of known trajectories and the use of regression analysis lead to similar estimates for the total error in each of the coordinates.

Thermal-Noise Errors Although the tracking error due to thermal noise is a component of the random errors, the fact that it may be the dominant error in the tracking-antenna system makes it of special importance. In Eq. (11-2), the standard deviation or error in the single-pulse angular estimate due to target noise at the boresight is inversely proportional to the square root of the signal-to-noise ratio. The signal-to-noise ratio from the radar-range equation is

$$S/N = \frac{P_t \cdot G_t \cdot A_r \cdot \sigma}{(4\pi)^2 \cdot R^4 \cdot k \cdot T_0 \cdot \Delta f \cdot \overline{NF} \cdot L_r \cdot L_p} \qquad \textbf{(11-3)}$$

where P_t = peak power of transmitter
$\quad G_t$ = gain of transmitted pattern (scan-angle-dependent in phased arrays)
$\quad A_r$ = effective receiving-aperture area (scan-angle-dependent in phased arrays)
$\quad \sigma$ = target cross section
$\quad R$ = range of target
$\quad k$ = Boltzmann's constant ($1.38 \cdot 10^{-23}$ J/K)
$\quad T_0$ = temperature in degrees Kelvin (290 K reference)
$\quad \Delta f$ = bandwidth of receiving channel
$\quad \overline{NF}$ = noise figure of system
$\quad L_r$ = ohmic and nonohmic receiving losses
$\quad L_p$ = propagation losses (two-way)

The standard deviation of the target-angle estimate due to the signal-to-noise ratio is often used as a figure of merit for tracking radars and is also used to specify antenna performance. It is often used incorrectly in the literature to describe the accuracy of a radar system. This is correct only when thermal noise is the dominant error, for instance, when the radar tracks small targets at long ranges and available power and directivity limit the signal-to-noise ratio.

Error Budget for Reflector Tracking Antennas

Such parameters as gain, frequency, bandwidth, and losses enter into the error budget by means of Eqs. (11-2) and (11-3). Requirements on sidelobes are generally specified to minimize multipath for instrumentation tracking systems and also to minimize sidelobe-jammer impact for tactical tracking systems. Barton[7] points out that the difference-pattern sidelobes must be down by at least 27 dB when there is a ground-reflection coefficient of 0.3 so that the multipath return will be within a reasonable value to meet an overall accuracy requirement of 0.1 mrad.

Elements of the error budget that are electrical in nature are discussed in Sec. 11-3. Mechanical errors in the budget for an elevation-over-azimuth mount are discussed in Sec. 11-4.

Servodrive requirements for reflector-tracking-antenna systems usually are such that the torque available and the mechanical bandwidth of the antenna and mount are sufficient to keep a target of maximum dynamics well within the 3-dB sum-pattern beamwidth or within the linear region of the normalized difference signal. Because of the $\sqrt{1 + (K_m\theta/\theta_B)^2}$ factor of Eq. (11-2) and because of cross-polarization errors to be discussed later, it is desirable to track the target as near the boresight as possible.

A typical error budget for azimuth and elevation for a highly accurate reflector tracking antenna is given in Tables 11-1 and 11-2. The first column in each table gives the systematic residual errors after calibration. The second column indicates random errors that cannot be reduced by calibration. Both columns represent figures obtainable under favorable conditions, including a firm foundation to which the antenna mount is attached. The systematic residual errors are primarily a function of the calibration procedure, the capability of the observer, and the stability of the system. If the environment changes, the systematic residual errors will change, and recalibration should be initiated.

Error Budget for Phased-Array Tracking Antennas

Many of the mechanical errors associated with the determination of the tracking axis of a reflector antenna are also present in determining the direction of the normal to the array face. In addition, the phased array has errors associated with the angular displacement of the boresight axis from the direction of the array normal by means of electronic scanning. These latter errors are both dependent and independent of the amount of the displacement commanded by the beam-steering controller. However, the variation in the systematic dependent errors with scan can be accommodated in the calibration process. The measurement process by which these and the independent errors are determined is not dependent upon the scan angle, and therefore the systematic residual error remaining after the scan-dependent systematic error is corrected is independent of scan angle.

The reference system for a phased-array tracking antenna is shown in Fig. 11-7.

TABLE 11-1 Reflector-Tracking-Antenna Azimuth Error Budget, 0–85° Elevation

Contributor*	Systematic residuals, mrad rms	Random, mrad rms
Azimuth-axis tilt	0.02	
Azimuth encoder to true north	0.03	0.01
Optical-elevation-axis nonorthogonality	0.02	
Elevation-azimuth-axis nonorthogonality	0.02	
Depth of null, 10° rms postcomparator error		0.02
Optical-RF-axis collimation	0.02	
RF-elevation-axis nonorthogonality		
Mechanical	0.02	
Electrical (frequency)	0.02	
RF axis to target position	0.02	Eq. (34-2)
Cross-polarization		0.04
Wind	0.02	0.02
Thermal-mechanical	0.02	
rss	0.065	0.05†
rss total	0.082†	

*Described in Secs. 11-3 and 11-4.

†Exclusive of S/N, multipath, and refraction.

TABLE 11-2 Reflector-Tracking-Antenna Elevation Error Budget, 0–85° Elevation

Contributor*	Systematic residuals, mrad rms	Random, mrad rms
Azimuth-axis tilt	0.02	
Elevation encoder to true vertical	0.02	0.01
Depth of null, 10° postcomparator error		0.02
Optical-RF-axis collimation	0.02	
RF-axis mechanical droop	0.02	
RF axis to target position	0.02	Eq. (34-2)
Cross-polarization		0.04
Wind	0.02	0.02
Thermal-mechanical	0.02	
rss	0.051	0.050†
rss total	0.072†	

*Described in Secs. 11-3 and 11-4.

†Exclusive of S/N, multipath, and refraction.

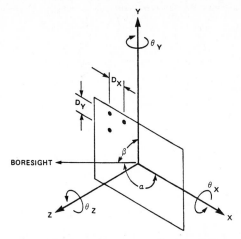

FIG. 11-7 Phased-array coordinate system.

The array face is in the X, Y plane with the Z axis as the normal to the array. The angle α defines the direction of the boresight position, or difference-pattern null, with respect to the X axis, while the angle β defines the direction of the boresight position with respect to the Y axis. On the Z axis, $\alpha = \beta = \pi/2$. The input to command the direction of the boresight position is given as α_i and β_i. The error in boresight direction is then $\alpha - \alpha_i$ and $\beta - \beta_i$.

The signals received in the monopulse difference channels are voltages V_α and V_β, representing the difference between the direction cosines (cos α, cos β) of the boresight axis and the target direction relative to the X and Y axes in the plane of the array. These differences are added to the commanded boresight-direction cosines (cos α_i, cos β_i) along with calibration data to form an estimate of the direction cosines of the target direction relative to the array-normal direction. The direction angles of the target are then combined with the direction of the array normal to estimate the direction of the target in earth-reference coordinates. By using these angles with target ranges, target position may be calculated in a rectilinear system. Predictions are then made as to the next target position. Target direction is transformed back from earth-reference to array-reference coordinates and into the α_i and β_i orders for the next beam position. The orders take into account the frequency of operation and temperature of the array.

Although thermal-noise error is not measured directly during alignment and calibration, the normalized slope of the difference patterns must be measured to estimate that error. Thermal-noise error can be estimated by Eq. (11-2) if it is interpreted to be the standard deviation in the direction cosine of the target direction and the parameters θ_B and K_m are those for the monopulse beam steered normal to the array ($\alpha_i = \beta_i = \pi/2$). In this direction the meaning of Eq. (11-2) is the same for a reflector antenna and for a phase-steered array. This estimate of error can be converted to angle measure by dividing by the square of the cosine of the angle off the array normal ($C^2 = 1 - \cos^2 \alpha - \cos^2 \beta$) for use in Table 11-4. By energy management of the phased array, that is, by making the power available proportional to $1/C^2$, the thermal-noise error is also made independent of scan angle. Other random errors dependent on scan angle are converted to angle measure by dividing by the cosine of the angle off the array normal (C).

TABLE 11-3 Phased-Array-Tracking-Antenna Array Face to Reference Error Budget

Contributor*	Systematic residuals, mrad rms	Random, mrad rms
Array face to true-north reference		
Alignment/calibration	$0.3/\cos\psi$	
Wind	0.1	0.1
Thermal	<0.1	
Tilt		
Alignment/calibration	0.1	
Wind	0.1	0.1
Thermal	<0.1	
Rotation		
Alignment/calibration	<0.1	
Thermal	<0.1	

*Described in Sec. 11-4.

TABLE 11-4 Phased-Array-Tracking-Antenna Array Face to Target-Position Error Budget, $[\pi/6 \le (\alpha \text{ or } \beta) \le 5\pi/6; \cos^2\alpha + \cos^2\beta \le 0.75]$

Contributor*	Systematic residuals, mrad rms	Random, mrad rms
Beam command to boresight		
Position ($\alpha - \alpha_i$, $\beta - \beta_i$)		
Uncorrelated illumination (phase-shifter quantization, phase-shifter roundoff, phase-shifter insertion, rf path insertion, array element position, radome)		$0.2/C$
Precomparator and postcomparator phase (depth of null)	0.1	
Cross-polarization (cross coupling)		$<0.05/C$
Frequency	0.1	
Thermal	0.1	
Boresight position to target	0.1	Eq. $(34\text{-}2)/C^2$
rss	0.20	$0.21/C$†
rss total	$0.20(1 + 1.06/C^2)^{1/2}$†	

*Described in Secs. 11-3 and 11-4.

†Exclusive of S/N, multipath, and refraction.

Just as for the reflector tracking antenna, the gain, frequency, and beam-switching time are specified at levels that support total system requirements with the error budgets to be met under the environment specified. Receiving sidelobe levels in a tactical phased array are generally dictated by anticipated jamming levels rather than by multipath requirements. An illustrative budget for a ground-based tactical system is given in Tables 11-3 and 11-4. The budget is for an array of equal dimensions in X and Y. Otherwise Table 11-4 should be broken into separate α and β tables.

It will be noted that the major error contributor is in the α plane. As indicated in Sec. 11-4, this is due to the use of a north-seeking gyro to provide the reference direction and make the tactical phased array independent of an external reference. The use of an external reference such as employed for the instrumentation reflector tracking system (Sec. 11-5) can reduce this error.

11-3 ELECTRICAL-DESIGN CONSIDERATIONS

Aperture-Distribution Design

A monopulse pattern is a set of two or more patterns, formed simultaneously, that provides both the angular position and the signal-characteristic data from a source. Generally, the aperture distribution for one of these beams, designated the Σ, or sum, beam, will be designed to maximize the gain of the antenna in the principal direction while meeting sidelobe objectives in other directions. The aperture distributions for the other beams, designated Δ, or difference, beams, will be designed to produce a null response along one of two perpendicular planes* intersecting in the boresight-axis direction, to maximize the slope of the response with increasing angle away from the null plane in a direction normal to the plane, and to satisfy gain and sidelobe-level objectives in other directions. The design considerations governing the sum beam, such as gain, beamwidth, and sidelobe level, are similar to those of other antenna requirements. These have been covered adequately elsewhere. In this section we will consider the special requirements of the difference beams in a monopulse cluster.

A measure of performance of the difference pattern is its slope at the central null. This slope may be normalized in at least two ways. The first, the difference slope K, refers only to the difference pattern. It is the slope of the pattern at the boresight null when the excitation is adjusted to radiate unit power relative to the response of a uniform aperture of the same size. In general terms,

$$K = \frac{\dfrac{\partial}{\partial u} \iint d(x,y) e^{-jk(ux+vy)} \, dxdy \, | \, u, v = 0}{\sqrt{A \iint d^2(x,y) \, dxdy}} \qquad \textbf{(11-4)}$$

or

$$K = \frac{-jk \iint x d(x,y) \, dA}{\sqrt{A \iint d^2(x,y) \, dA}} \qquad \textbf{(11-5)}$$

---

*For a phased array, these planes become conical surfaces about perpendicular axes in the array face.

where u (cos α) and v (cos β) are the direction cosines of the observation or target direction, $k = 2\pi/\lambda$, and $d(x,y)$ is the excitation aperture distribution.

The difference slope is maximized when the aperture has an odd linear distribution of illumination amplitude. The exact value of the slope depends upon the shape of the aperture. Two special values of K are of interest for reference purposes. For odd linear difference illumination, the difference slope for a rectangular aperture is

$$K_s = \pi/\sqrt{3} \qquad \text{V/V/SBW}$$

and for a circular aperture is

$$K_s = \pi/2 \qquad \text{V/V/SBW}$$

where K_s is K times the standard beamwidth of the aperture (SBW), in radians (equal to the wavelength divided by the width of the antenna). These maximum-slope functions are seldom used in practical tracking systems because of the high sidelobe levels, -8.3 dB for the rectangular aperture and -11.6 dB for the circular aperture. The aperture distributions (given by Taylor[8] for the sum pattern and by Bayliss[9] for the difference pattern), which maximize gain or slope for a given sidelobe level, are more practical models in designing aperture illumination for a monopulse tracking antenna. Approximate values for the difference slope K_s, derived from the Bayliss distribution for a square aperture, are given in Fig. 11-8.

The second way to normalize the difference slope, generally preferred in the literature of radar systems, is known as the normalized monopulse difference slope K_m, referred to in Sec. 11-2. This difference slope is derived from the difference slope K by normalizing the difference voltage to that of the sum beam and normalizing the angle to the half-power beamwidth (HPBW) of the sum beam. Thus,

$$K_m = K \frac{\theta_B}{\sqrt{\eta_\Sigma}} \qquad\qquad (11\text{-}6)$$

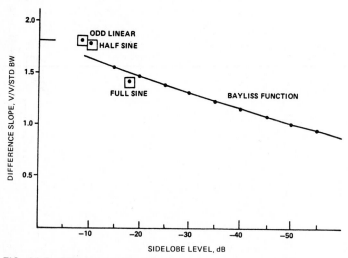

FIG. 11-8 Difference-pattern slope.

where θ_B = HPBW of sum beam

η_Σ = efficiency factor for sum beam, including only aperture illumination, spillover, and blockage

Feed Systems

The feed system distributes the RF excitation over the aperture and establishes the level of excitation at each point in the aperture. The special feature of the monopulse feed is its ability to establish three orthogonal distributions of excitation simultaneously.

A feed is constrained when the excitation proceeds from a beam port through a branching transmission line or waveguide network to the individual elements of the radiating aperture. Space feeds, or optical feeds, are those feed systems that employ a few elements, in a primary feed system, that radiate RF excitation to be intercepted and collimated by a larger secondary aperture. In this subsection, the term *optical feed* is used for this class of feed.

Constrained Feed Systems for Phased Arrays   The simplest kind of monopulse feed system for a tracking array is formed by dividing the array into equal quadrants, as illustrated in Fig. 11-9. The sum of all four quadrants forms the sum beam, and the difference between the sums of the top pair and the bottom pair forms an elevation difference beam. The sum of the left pair is subtracted from the sum of the right pair to form the azimuth difference pattern. A fourth orthogonal beam, sometimes called the Q beam, can be formed by taking the difference of the differences between the quadrants. This beam is usually terminated in a matched load to prevent signals received off axis from being reflected into a difference channel.

This simple feed system does not allow for sidelobe-level optimization or error-slope optimization among the three beams of the monopulse cluster. Large tracking

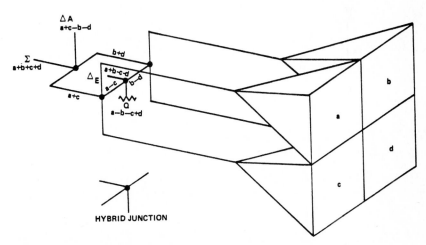

FIG. 11-9   Simple monopulse feed.

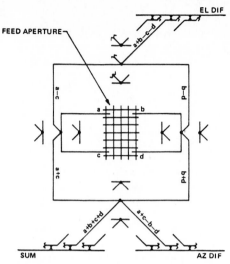

FIG. 11-10 A feed network providing complete independent control of the sum and both difference illumination distributions.

array antennas will usually employ a feed system that will allow independent optimization of the sum-and-difference-beam illumination.

Feed systems that provide sufficient degrees of freedom to allow independent optimization of the monopulse aperture distributions must take full advantage of the symmetry inherent in these distributions. Array elements or subarrays equally displaced from the monopulse axes in the array face all receive the same excitation amplitude for any given function, differing only in the pattern of phase (algebraic sign) for the different beams. A network similar to that shown in Fig. 11-9 can be used to connect symmetrically located elements or subarrays. This circuit is illustrated in Fig. 11-10. The sum ports of these may be combined to provide the array sum illumination distribution. The elevation difference ports may be independently combined to optimize the elevation difference pattern, and similarly the azimuth difference pattern may be independently optimized.

An alternative arrangement that simplifies the feed network with some loss of independence in design of the illumination function is shown in Fig. 11-11. This network allows complete independence for the elevation difference pattern, but the column sum distribution is shared between the sum and azimuth difference distributions. Because the column functions of these two distributions are similar, a compromise design based upon this sharing has little impact on the array pattern.

A row- or column-oriented feed network, subject to the compromise described for the alternative parallel feed, can be implemented with a serial monopulse ladder network as shown in Fig. 11-12. This network can be used in place of the column network or the row network, or both. There does not appear to be a generally applicable choice between these feed types, so that both types must be considered for the requirements of a specific tracking antenna.

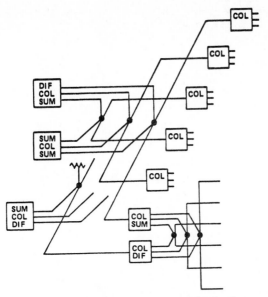

FIG. 11-11   Row-column-array monopulse feed.

Optical Feed Systems   The design of a monopulse feed system for an antenna with optical magnification, such as a reflector, a lens, or a transmission lens array, is complicated by the fact that there are two apertures to consider, the main aperture and the feed aperture. Feed apertures differ for the different beams, and there are fewer degrees of freedom available for aperture-distribution control. Optical monopulse feed systems are excellent examples of the art of design compromise.

The design of a monopulse optical feed, as also the design of a single-beam optical feed, involves a careful balance of aperture illumination efficiency, spillover efficiency, and, in case of a feed for a reflector antenna, blockage. The monopulse feed design is complicated by the need to balance the requirements of the difference beams with those of the sum beam. The optimization of the difference feed differs from that

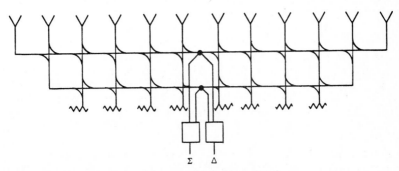

FIG. 11-12   Center-fed serial monopulse ladder network.

FEED EXCITATION WITH $(\frac{aA}{2\lambda F}$ AND $\frac{bB}{2\lambda F})$ FOR MAXIMUM SUM GAIN	SUM		AZ DIF		EL DIF	
	GAIN	SPILLOVER EFFICENCY	SLOPE	SPILLOVER EFFICENCY	SLOPE	SPILLOVER EFFICENCY
El / Az / S AzEl (a) FOUR-HORN	.58	.66	.943	.28	.87	.24
El / Az / S AzEl (b) TWO-HORN DUAL-MODE	.75	.84	1.23	.50	1.00	.31
•.41 El / Az / •.41 S AzEl (c) TWO-HORN TRIPLE-MODE	.75	.83	1.47	.80	1.00	.31
El / Az / S AzEl (d) TWELVE-HORN	.58	.66	1.29	.63	1.22	.62
•.41 El / Az / •.41 S AzEl (e) FOUR-HORN TRIPLE-MODE	.75	.83	1.47	.80	1.36	.78

* RATIO OF MODE AMPLITUDES
El - ELEVATION
Az - AZIMUTH

FIG. 11-13 Monopulse feed characteristics. *(After Ref. 10.)*

for an array. Spillover efficiency for an optical feed is poor for an odd linear distribution, giving it a smaller angular sensitivity than a distribution that results from placing a greater part of the primary (feed-system) difference pattern on the focusing aperture. The optimum size of the difference feed is larger than the optimum size of the sum feed. Hannan[10] suggests that the linear dimension of the optimum difference aperture, in the plane of the difference pattern, is twice that of the sum pattern. Thus, an optimum monopulse feed may have 3 to 4 times the aperture area of an optimum single-beam feed for the same optics. As a result, a monopulse feed will have substantially larger aperture blockage than the optimum single-beam feed.

The performance of a conventional four-horn monopulse feed is poor because it requires a compromise between optimum size for sum-beam performance and for difference-beam performance. Several configurations that provide additional freedom to improve monopulse feed performance include both multibeam and multimode feeds (see Fig. 11-13). The multihorn feeds have network configurations similar to monopulse phased-array networks. For both multibeam and multimode feeds, the increased design complexity requires an increase in the minimum electrical size of the feed, making them more appropriate for large f/D optics.

Figure 11-13, which follows Hannan, represents comparative performance parameters of different feed configurations that are optimized on the basis of a collimating aperture of a rectangular cross section with dimensions A by B, located at a distance F from the feed aperture of dimensions a by b. The spillover efficiency given in this figure is the ratio of the energy radiating from the feed aperture intersecting

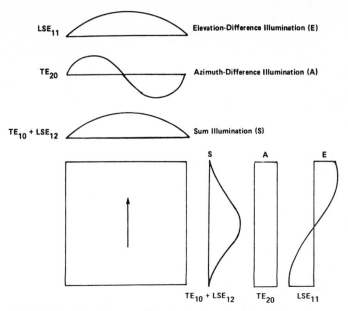

FIG. 11-14   Single-horn multimode feed-aperture distribution.

the collimating aperture to the total energy radiated by the feed in that mode. A major source of the improved performance of the more complex multihorn and multimode feeds is the reduction of spillover loss in the difference beams.

The multimode feed combines the network simplicity that is characteristic of the four-horn feed with the greater design freedom of the multihorn feed. A multimode feed uses several waveguide modes in combination to achieve a desired aperture illumination. A single mode with odd symmetry is used for the monopulse difference function. The sum-beam illumination is usually achieved by using a linear combination of two different modes, each with even symmetry. This has the desirable effect of substantially increasing the primary-pattern beamwidth and thereby improving aperture illumination efficiency at the reflector. Since two modes of different orders will have different velocities of propagation, they can be held in the proper phase relationship at the aperture over only a limited bandwidth. In practice, however, multimode feeds often compare favorably in bandwidth with more complex multihorn feeds.

An example of a dual-polarized multimode design is shown in Fig. 11-14. This feed makes use of two waveguide modes, the TE_{10} and the LSE_{12} modes, to form the sum-beam illumination. The azimuth difference-beam illumination uses the TE_{20}, and the elevation difference-beam illumination uses the LSE_{11} mode. These feed-aperture illumination distributions are illustrated in Fig. 11-14. The bandwidth of the multimode feed is limited by the difference in phase velocity between the TE_{10} and LSE_{12} modes forming the sum distribution. This feed design generates the LSE_{12} mode in phase opposition with the TE_{10}. The length of the tapered horn section is then set by the requirement to provide a 180° differential phase shift between these two modes.

Contributors to Pointing Errors

Defects in an antenna feed system as it affects the sum-beam performance, measured by such factors as gain, beamwidth, and sidelobe level, are treated elsewhere in this handbook. Defects that affect the tracking performance of a monopulse feed system are considered here. Included in this subsection are those errors in the distribution of illumination over the antenna aperture that have relatively little correlation in the aperture. When the error* locations in the aperture are separated by more than a few wavelengths, they often can be considered independently of other errors. Errors that are correlated over the entire part of the aperture, served by one port of the comparator, are treated as precomparator and postcomparator errors. Finally, the dependence of monopulse angle estimation on the polarization of the received signal will be considered. The effect of these illumination errors is an apparent shift in the position of the difference-pattern null, or a boresight shift, and a change in the relative slope of the difference pattern, or a gain shift, which may vary with frequency, time, temperature, or other environmental factors.

Uncorrelated Illumination Errors   The effects of uncorrelated errors must be considered in the design of a tracking antenna, whether for specifying either reflector-surface tolerance or allowable element errors in a phased array. A distribution of phase errors $\xi(x,y)$ across the aperture will produce a voltage V_e that is proportional to

$$\iint \xi(x,y)\ d(x,y)\ dA \approx V_e$$

which to the first order of approximation is in phase with the difference voltage which would exist in the absence of uncorrelated errors. This voltage may be divided by the difference slope in Eq. (11-5) to determine the shift in the apparent position of the difference-pattern null:

$$\Delta u = \frac{\iint \xi(x,y)\ d(x,y)\ dA}{k\iint x d(x,y)\ dA} \tag{11-7}$$

This expression is valid for an arbitrary distribution of any correlation interval across the array, including such periodic phase errors as those resulting from phase quantization errors in some beam-steering equipment.

The specific distribution of phase errors in the difference-pattern illumination for an antenna often is unknown, but we can infer by means of specification or measurement the statistics of such distributions that are averaged over an ensemble of similar antennas. The null shift given in Eq. (11-7) can be treated as a random variable, and the expected value of the null shift is dependent upon the distribution of the expected value of the phase error at each point in the aperture:

$$\overline{\Delta u} = \frac{\iint \overline{\xi}(x,y)\ d(x,y)\ dA}{k\iint x d(x,y)\ dA} \tag{11-8}$$

Usually the phase error, and therefore the null shift, will have a zero mean (expected) value. If the error distribution is such that the rms phase error σ (radians) is the same for every independent uncorrelated region of the antenna, the rms of the null shift (boresight) error is

---

*Error as used here is that phase or amplitude error in the final aperture illumination function which contributes to the beam null shift or pointing error in Tables 11-1, 11-2, and 11-4.

$$\sigma_u = \frac{\sigma}{K\sqrt{N_e}} \qquad (11\text{-}9)$$

where N_e is the effective number of uncorrelated independent regions in the aperture with that phase-error statistic. If the antenna is a phased array and the statistic refers to phase errors that are independent from element to element, then N_e is just the total number of elements in the array. If the phase errors are attributable to some higher organizational level in the array, N_e is the number of components in that level.

Nester[11] has shown that, for practical ranges of error values, the phase error has the predominant effect on null shift. He has also shown the changes in the error slope due to phase and amplitude errors. These are second-order because both the position of the target off boresight and the error in the normalized difference slope are small, independent random variables.

The illumination error distribution associated with the feed system of a phased-array antenna is invariant with scan. This error contribution can be significantly reduced by alignment. The illumination error distribution may also have a component due to a setability error in the phase shifters that is dependent upon scan. This error is difficult to remove by monopulse calibration. Although it might be removed by calibration at the element level, the computational effort makes this approach unattractive for arrays of more than a few elements.

Precomparator and Postcomparator Errors   The accuracy of the monopulse estimation of target location is influenced by both precomparator and postcomparator errors. Precomparator errors affect the symmetry of the patterns, producing a modified sum-pattern error in the difference beam and a modified difference-pattern error in the sum beam. If the modified sum-pattern error is in phase with the difference pattern (resulting from a precomparator phase error), a shift in the angle of the difference null will be observed. If the modified sum-pattern error is in quadrature with the difference pattern, no shift will occur in difference-pattern-null location, but the depth of the null will be reduced. Techniques employed to improve effective null depth make use of the relative phase between the signals from the sum and the difference channels. These techniques make the monopulse performance near the null sensitive to postcomparator phase error in the presence of quadrature precomparator error. The relation between these error sources depends both upon the microwave network employed and upon the technique used in the signal processor to extract angle information.

An example is given in Fig. 11-15 of the phase errors in a monopulse feed system when the difference-pattern error d_t is in quadrature with the difference pattern d. Figure 11-15a illustrates the situation in which no postcomparator phase error exists; d_t is a phasor slowly varying with the angle to the source, while d varies linearly (approximately) with the source angle. In this instance, the ratio

$$\frac{d' \cos\theta}{s} = \frac{d}{s} \qquad (11\text{-}10)$$

is the best estimator of the source angle. In Fig. 11-15b, the effect of the postcomparator phase error ϕ is illustrated. Here, the estimator of source angle

$$\frac{d' \cos(\theta + \phi)}{s} = \frac{d}{s} \frac{\cos\theta\cos\phi - \sin\theta\sin\phi}{\cos\theta} \qquad (11\text{-}11)$$

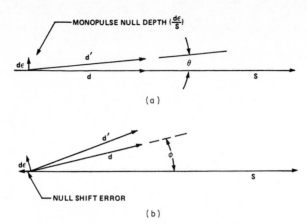

FIG. 11-15 Monopulse comparator phase errors. (*a*) Pre-comparator phase error. (*b*) Precomparator and postcomparator phase error.

will produce a shift in the apparent null position (boresight) such that

$$e_{dn} = \frac{\theta_B}{K_m}\frac{d}{s} = \frac{-\theta_B}{K_m}\frac{d_i}{s}\tan\phi \qquad \text{(11-12)}$$

Precomparator errors that are in phase with the difference pattern produce a null shift error in source-angle prediction. Postcomparator errors produce a null shift error if precomparator errors are in quadrature with the difference pattern. Design or alignment action to reduce these errors depends upon identifying the source of the phase error. Determining the behavior with frequency will often help with this identification. When the alignment has reduced this error to an acceptable level, it may be further reduced by calibrating and correcting the observed angle data. The boresight error due to this source will be a function of frequency.

Cross-Polarization Angle Crosstalk   The normalized output voltage from the two difference channels of a monopulse feed can be related to the displacement of the target from the RF axis of the antenna by the matrix equation

$$\begin{bmatrix} V_u \\ V_v \end{bmatrix} = \begin{bmatrix} K_{uu} & K_{uv} \\ K_{vu} & K_{vv} \end{bmatrix} \begin{bmatrix} \Delta u \\ \Delta v \end{bmatrix} \qquad \text{(11-13)}$$

where the main diagonal terms are the difference slope coefficients from Eq. (11-2), and the cross-diagonal terms are cross-coupling terms due to nonorthogonality between the monopulse null planes (or cones). These coefficients are generally a function of frequency and of the relative polarization of the received signal. If the receiving system does not sense the polarization of the incoming signal, this cross-polarization angle crosstalk represents an uncorrected error in the angle estimate, which can result in loss of track.[12,13]

An example of the difference-voltage equation applied to the elevation-over-azimuth pedestal can be obtained by means of coordinate conversions:

$$\Delta v = \Delta E$$

$$\Delta u = \cos E \, \Delta A$$

giving

$$\begin{bmatrix} V_u \\ V_v \end{bmatrix} = \begin{bmatrix} K_{uu} \cos E & 0 \\ 0 & K_{vv} \end{bmatrix} \begin{bmatrix} \Delta A \\ \Delta E \end{bmatrix} \tag{11-14}$$

when no angle crosstalk occurs and perfect alignment exists between the RF null planes and the pedestal coordinate planes. The angle estimates in pedestal coordinates are obtained by inverting the matrix, giving the relation

$$\begin{bmatrix} \Delta A \\ \Delta E \end{bmatrix} = \begin{bmatrix} 1/(K_{uu} \cos E) & 0 \\ 0 & 1/K_{vv} \end{bmatrix} \begin{bmatrix} V_u \\ V_v \end{bmatrix} \tag{11-15}$$

The cross-polarization angle crosstalk is a consequence of the polarization characteristics of the feed system. It tends to be most pronounced in optical feed systems but is also present in small phased-array antennas. Crosstalk is not observed in large arrays.

A classic explanation of cross-polarization crosstalk is derived by considering the feed for a reflector antenna with polarization characteristics identical to that of a linear electric dipole. The electric vector at the surface of the reflector will be directed along the intersection of the parabolic surface and the plane containing the feed dipole axes and a point on the reflector surface. The copolarization of the antenna is determined by the plane through the feed dipole and the vertex of the reflector. Most of the energy collimated by the reflector is polarized parallel to the principal polarization at the vertex, but each quadrant of the reflector aperture has a cross-polarized component of excitation.[14] During sum-beam excitation, the phase of this cross-polarized energy alternates from quadrant to quadrant, as shown in Fig. 34-16, producing a cross-polarized Q beam. In the principal polarization elevation difference beam, the

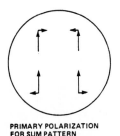

PRIMARY POLARIZATION
FOR SUM PATTERN
_____
CROSS-POLARIZATION
PRODUCES Q-BEAM PATTERN

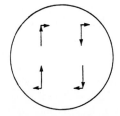

PRIMARY POLARIZATION FOR
AZIMUTH-DIFFERENCE PATTERN
_____
CROSS-POLARIZATION PRODUCES
ELEVATION-DIFFERENCE PATTERN

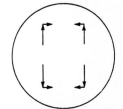

PRIMARY POLARIZATION FOR
ELEVATION-DIFFERENCE PATTERN
_____
CROSS-POLARIZATION PRODUCES
AZIMUTH-DIFFERENCE PATTERN

FIG. 11-16 Cross-polarization crosstalk electric dipole feed on parabolic reflector.

cross-polarization response is that of an azimuth difference pattern. Thus, a cross-polarized source displaced in azimuth would drive the beam in the elevation direction. The same situation occurs in the principal polarization azimuth difference beam, in which the cross-polarized response is that of an elevation difference pattern.

If this feed were to be used as an optical feed for a planar transmission lens array, no such cross-polarization effect would be observed. This source of crosstalk depends upon both the feed polarization and the surface of the focusing element. Koffman[15] has determined the "ideal" feed polarization for each type of surface derived from a conic section by showing that, in general, the strength of the electric dipole polarization should equal the strength of the magnetic dipole polarization, multiplied by the ellipticity of the focusing surface. Thus, a magnetic dipole is optimum for a sphere, an electric dipole is optimum for a plane, and a Huygens source (having equal electric and magnetic dipole strength) is optimum for the paraboloid.

Optical polarization conversion, however, is not the only source of crosstalk; it can arise because of cross coupling in the feed itself. An example of this effect in a five-horn monopulse antenna has been studied by Bridge.[16] The particular feed he investigated was designed for dual polarization and had a direct circuit for measuring the coupling or isolation between the polarizations in the monopulse network.

Multimode feeds are particularly susceptible to cross-polarization crosstalk because the relatively large size of the single aperture can support higher-order cross-polarized modes, and the feed system usually will excite them unless particular care is exercised. Multiple-aperture, single-polarization feeds may exhibit some relatively low-level cross-polarization because of edge diffraction (if we assume that the individual apertures will support only one polarization). This source of angle crosstalk becomes negligible as the size or number of apertures in the feed array becomes large.

11-4 MECHANICAL-DESIGN CONSIDERATIONS

Antenna Mounts

The mount not only must support the antenna and direct it in angle but must accurately provide that direction under dynamic conditions that are introduced by target and/or antenna-mount-platform motion or stress that is introduced by wind. The requirements on the mount are interdependent with the requirements on the antenna. The mechanical bandwidth of the tracking antenna as well as its mount depends upon the resonance characteristics of the antenna structure and its mount as well as upon the coupled resonance. A discussion of bandwidth and design of servo systems to drive tracking-antenna systems is given by Humphrey.[17]

Many reflector tracking antennas use the mount of Fig. 11-21 below, though tracking coverage in this mount is limited near zenith. For a target (see Fig. 11-17) traveling normal to the line of sight and perpendicular to the axis of rotation of a tracking antenna at a ground range R_g and velocity v, the angular velocity of the tracking antenna reaches v/R_g. The acceleration about the azimuth axis of rotation reaches a value of $0.65v^2/R_g^2$ at $\pm 30°$ from the normal. The transfer function that describes the angular velocity and acceleration of the tracking antenna about the azimuth axis has a "pole" when the ground range approaches zero

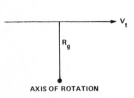

FIG. 11-17 Target geometry.

and the target is near overhead. A loss of track can occur. The three-axis mount described in Skolnik[3] removes the pole at the expense of increasing weight, complexity, and cost.

Even though coverage is sacrificed, a two-axis mount is often chosen for reflector-antenna tracking systems for fixed earth stations or for installation on aircraft or ship platforms. The mount is oriented so that the loss in mission coverage due to the pole is minimized and has little impact on overall operation.

Phased-array-antenna tracking systems that must provide hemispherical coverage for different missions but limited coverage for a single mission may also use the elevation-over-azimuth mount or may use an azimuth-only mount. Dynamic performance requirements are usually limited, but accuracy of positioning must still be maintained.

Mechanical Errors in Reflector Tracking Antennas

This subsection discusses mechanical errors in using, for example, an elevation-over-azimuth mount. The errors are treated as being independent. Though not strictly true, the assumption appears to be valid for the size of errors expected in an accurate system. These formulas may be used as corrective algorithms to the elevation and azimuth angular estimates once the constants have been determined by calibration. More exact expressions for some of the errors will be found in Rozansky.[18] In a new design or in one whose error models are not well known, these algorithms may be used initially, but the final algorithm should be adjusted to match the particular system.

Encoder Errors   The angular position of the antenna is translated into digital or analog data by using angle encoders. Most designs today incorporate digital encoders. The output of these encoders is subject to both systematic and noise errors within the encoder itself; in the data gear train, if used; and in the coupling of the encoder to the mechanical takeoff from the antenna. The state of the art is such that the latter two are usually the greater source of error. If the coupling is rigid, the relative motion between the driving axis and the encoder shaft introduces bending movements in the shaft of the encoder and cyclic errors in the output. Multicyclic errors can be produced by gear trains.

When a flexible coupling between the axis takeoff and the encoder is used, mechanical eccentricities that produce systematic cyclic variations in output data may occur. Backlash produces random errors. Noise in the encoder also produces a random error. With the encoder shaft held fixed, a variation of ± 2 bits in a high-resolution (19-bit) system is not unexpected.

Torquing of setscrews or like fasteners can produce bending moments that may distort the encoder shaft. Rather than adjusting the encoders and creating such an error, provision can be made in the calibration process to introduce the values by which the encoders are offset from the reference positions.

The equation for a combined cyclic and encoder offset error with respect to the reference position is

$$E_{Ac0} = M \cos (A_m - A) + A_0 \qquad \textbf{(11-16)}$$

where M = maximum amplitude of cyclic error
A = azimuth reading of encoder

A_m = azimuth reading of cyclic-error maximum amplitude

A_0 = azimuth reading for external azimuth reference

This equation is also valid for the elevation encoder when the elevation values E, E_m, E_0 are substituted for A, A_m, and A_0. The use of this equation, when M and A_m (or E_m) are measured during the calibration process as a correction algorithm, leads to the level of systematic residual error given in the error budgets in Tables 34-1 and 34-2.

Azimuth-Axis Tilt   Leveling a reflector tracking antenna can be accomplished with three adjustable legs, using two orthogonal gravitational sensors that are parallel to the azimuth plane. An accurate tracking system is usually leveled to within 0.1 to 0.2 mil of the vertical, preferably under cloudy conditions or at night to eliminate temperature gradients. Leveling to a tighter tolerance is not productive, since day-to-day variations of the pedestal vertical, as shown in Fig. 11-22 below, can be greater. In addition, the local vertical may not be the true vertical because of gravitational anomalies, and calibration with respect to the true vertical may be required.

When the azimuth axis is tilted at an angle γ from the vertical reference as shown in Fig. 11-18, errors in both azimuth and elevation that are generated may be approximated by

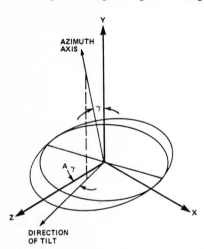

FIG. 11-18   Azimuth-axis tilt.

$$e_{A_\gamma} = \gamma \sin (A_\gamma - A) \tan E \qquad (11\text{-}17)$$

$$e_{E_\gamma} = \gamma \cos (A_\gamma - A)$$

where γ = tilt angle

A_γ = azimuth direction of tilt

A = azimuth encoder output

E = elevation encoder output

Boresight-Telescope Nonorthogonality   The boresight telescope used for system alignment and calibration usually will not be exactly orthogonal to the elevation axis. An error in azimuth, as shown in Fig. 11-19, is produced for the optical axis. Correction can be made through physical adjustment to the body of the telescope or through correction of the data by measuring the angle ζ_0 of nonorthogonality. The error-correction algorithm is approximated by

$$e_{A\zeta 0} = \zeta_0 \sec E \qquad (11\text{-}18)$$

where E is the elevation angle. This equation is needed to derive the RF-axis nonorthogonality to the elevation axis and is not used directly in correcting tracking-antenna azimuth data.

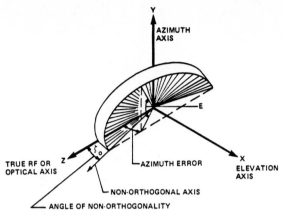

FIG. 11-19 RF- or optical-axis to elevation-axis nonorthogonality error.

RF-Axis Nonorthogonality   The same type of error is produced in azimuth data if the RF axis is not orthogonal to the elevation axis. However, RF-axis-nonorthogonality errors may be a combination of mechanical and electrical boresight shifts, with the latter being a function of frequency of operation.

$$e_{A\zeta r} = \zeta_r \sec E \qquad (11\text{-}19)$$

Steps in the calibration process require that the optical nonorthogonality be determined first and that the RF nonorthogonality as a function of frequency then be determined with the use of a target that is visible both optically and at RF.

Elevation-Axis Nonorthogonality (Skew)   If the elevation axis is nonorthogonal or skewed to the azimuth axis by a small angle ϵ_0, as shown in Fig. 11-20, the resultant azimuth error and correction algorithm is approximated by

$$e_{A_\epsilon} = \epsilon_0 \tan E \qquad (11\text{-}20)$$

for both the optical and the RF axis.

Droop   Droop is an elevation error produced by gravitational force on the boresight telescope and on the RF feed. The errors are at a maximum at 0 and 180° elevation and at a minimum at 90°. They are normally constant with azimuth position. The droop in the telescope can be approximated as a linear equation:

$$e_{Ed_0} = d_0 (E - \pi/2) \qquad (11\text{-}21)$$

where d_0 is the telescope droop constant. The RF boresight-axis droop is approximated by

$$e_{Ed_r} = d_r \cos E \qquad (11\text{-}22)$$

where d_r is the RF feed constant. The telescope constant is developed first during the

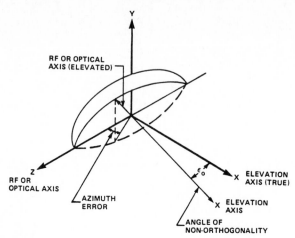

FIG. 11-20 Elevation-axis to azimuth-axis nonorthogonality error.

calibration process and then, with a common target, used to determine the RF-axis droop constant.

Collimation   *Collimation* is defined as "making parallel" or "adjusting accurately the line of sight of [a telescope]." In this chapter, collimation refers to the relationship between the RF boresight axis and the boresight telescope that is used as a reference.

The collimation error between the RF boresight axis and the optical axis is determined by the resolution of the boresight telescope and the optical recording, the depth of the null for the RF difference pattern, multipath error, and the rms phase error of the receiving system. If a common target has a high RF signal-to-noise ratio, optical visibility, reflectivity characteristics that do not depolarize the signal return, high elevation angle, and low velocity and acceleration so that the target can be tracked at or near the RF null, then the collimation error can be reduced to 0.02 mrad rms in each axis. In elevation, the residual error is the error remaining after accounting for mechanical droop, electrical boresight shift, multipath, and the differential offset constant between the optical axis and the RF axis. In azimuth the residual error is the error remaining after accounting for the electrical boresight shift of the RF axis and the nonorthogonality of both the optical axis and the RF axis with respect to the elevation axis.

Thermal Effects   Thermal effects on accuracy primarily are those due to solar radiation producing thermal gradients in the reflector tracking antenna and mount. Differential heating of the mount base will produce an azimuth tilt error. Unequal expansion in the trunion arms supporting the elevation axis can produce an elevation-axis nonorthogonality error. Droop, collimation, and RF-axis-nonorthogonality errors can be accentuated or decreased by gradients in the feed supports. For highly accurate systems, material with low thermal coefficients of expansion should be used. This is particularly true for feed supports. The use of white paint for the entire antenna and mount for nontactical systems and sunshields for the mount will considerably reduce thermal gradients.

Thermal effects are long-term when compared with the time for most instrumentation radar missions. A system designed for rapid premission calibration can reduce the impact of such errors.

Wind   Wind produces both systematic and random errors. A steady-state wind having a velocity V produces a small systematic deflection error:[7]

$$e_{wd} = K_w K_s V_w^2 \qquad (11\text{-}23)$$

where K_w is an aerodynamic constant that relates the torque produced about one of the angular axes to the average wind speed and $V_w K_w$ is a function of the aspect angle of the antenna with respect to wind direction. The mechanical spring constant of the antenna K_s is a measure of the deflection of the antenna about the axis per unit of torque.

Gusty winds produce a small random error with a standard deviation of

$$\sigma_{wd} = 2K_s K_w V_w \sigma_v \qquad (11\text{-}24)$$

where σ_v is a measure of the variation in average wind velocities.

Consideration can be given to the use of a radome to reduce wind-load errors, but a trade-off must be made between radome and wind-induced pointing errors.

Mechanical Errors in Phased-Array Tracking Antennas

Rotation   Angular displacement ($\Delta\theta_z$) about the Z axis or the array normal produces errors in both α and β proportional to $(\beta - \pi/2) \cdot \Delta\theta_z$ and $(\alpha - \pi/2) \cdot \Delta\theta_z$, respectively, where $\Delta\theta_z$ is small. A highly damped tiltmeter, located parallel to the X axis, can be used to level the array initially and automatically monitor for subsequent rotation displacement. High damping is necessary to prevent response of the tiltmeter to vibration and to short-term wind loads but will still permit measurement of permanent shifts.

Tilt   The array is usually tilted backward about the X axis to obtain elevation coverage consistent with tracking demands at the horizon. Angular displacement about the X axis can be measured with a highly damped tiltmeter mounted perpendicularly to that axis with the center of the tiltmeter range at the desired tilt angle. Angular displacement errors about the X axis directly contribute to errors in β, with a minimum amount of coupling into α. Both the rotational and the tilt sensors provide the reference to the local gravitational vertical rather than the true vertical.

Array Face to True North   Alignment of the array-face normal so that its relationship is known with respect to the plane established by the local vertical and true north can be accomplished through a north-seeking gyro designed as an integral subsystem of the array. Gyros are available with an error in pointing of less than 0.3 mrad rms divided by the cosine of the latitude. They require settling times of 5 to 15 min to reach such values.

Thermal Effects   In a planar phased array, a shift in boresight can occur when the overall array-face temperature changes and when there is a uniform differential expansion or contradiction between element distances D_x and D_y. At an α of 45°, a 25-ppm thermal-expansion coefficient and a 40°C differential temperature produce a 1-mrad shift. The formula for the error before correction is

$$e_{\alpha t, \beta t} = L_t \, (T_c - T_0) \cot (\alpha_t, \beta_t) \qquad \textbf{(11-25)}$$

where L_t is the coefficient of expansion, T_c is the calibration temperature, and T_0 is the operational temperature.

A temperature gradient across the face of the array will produce errors dependent on the thermal-gradient pattern. Such errors will be difficult to determine and correct by calibration. Consequently, the thermal design of the array face and structure should minimize such gradients.

An additional thermal error is that due to both absolute temperature and thermal gradients of the feed system. It also will be difficult to determine and correct by calibration. Thermal design must consider the feed architecture and techniques to maintain small thermal gradients between parallel elements in order to minimize differential phase errors.

Wind   The potential for error due to wind and wind gusts is generally greater in a tactical than in an instrumentation tracking-antenna system. The instrumentation system usually has a solid base to which the antenna mount can be rigidly attached, whereas the tactical radar tracking-antenna support structure rests on the ground. The weight of the phased array or wind loads can produce pressure, which in turn compresses or deforms the surface supporting the mount, giving errors about one or more of the array-face axes. Such errors caused by displacement can be read out by tiltmeters and gyros that have been designed as an integral part of the array. In addition to the angular displacement caused by wind loads in θ_X and θ_Z, unless there is means of lateral bracing, angular displacements that are primarily in θ_Y can take place.

In addition to the small permanent shifts, which can be measured by the gyros and tiltmeters, shorter-term shifts due to wind deflection of the mount structure can occur. High and gusty winds can cause large errors, particularly for a system with a low resonant frequency.

11-5   SYSTEM CALIBRATION

Calibration serves two functions: (1) to provide direction for realignment or correction of faults after assembly or after transport in which the antenna may be subject to mechanical stresses; and (2) to provide constants for the algorithms to correct the systematic errors, permitting the system to operate within the accuracy required. To provide maximum accuracy, the system should be calibrated under conditions as near as possible to the conditions under which it will operate. Knowledge of absolute temperature is important for phased-array calibration but is of limited value for reflector-antenna tracking systems. For thermal gradients that distort the antenna and pedestal for reflector tracking antennas used in highly accurate instrumentation systems, it is desirable to calibrate as close to mission time as possible.

The first method of calibration involves direct measurement to determine the errors and then correction of the errors as a function of angular direction and frequency. The second method is the tracking of targets whose trajectories are accurately known, such as satellites, and the use of regression analysis to determine the errors as a function of direction and frequency. This latter approach is generally limited to large power-aperture product systems.

The use of sensors having automatic readout under control of a computer appears to give the fastest, most consistent results and removes the possibility of observer error. When an observer is involved, the design of the readout should reduce the requirements on the observer to a minimum. Physical stress due to temperature extremes, muscular strain due to position, and lighting all contribute to potential observer error.

Frequency of calibration and number of sample measurements depend on user requirements and the stability of the system. It should be noted that in general the larger the number of independent sample points chosen for a particular calibration procedure, the more precisely one may model any particular error. Initially, a tracking-antenna system may require a large number of samples to establish a model or algorithm for the error and a lesser number on subsequent calibrations. A tracking antenna with a specified accuracy of 0.1 mrad (exclusive of thermal noise, glint, multipath, refraction, and clutter) may require calibration for each mission. A tracking antenna with a specified accuracy of 0.5 mrad (exclusive of thermal noise, glint, multipath, refraction, and clutter) may require only leveling and the use of automatic sensors to measure changes.

Calibration for Reflector Tracking Antennas

The order of calibration is chosen to minimize coupling of errors. The order, or technique, used may not be appropriate for all systems, but the principles involved may be applied to develop those required for a specific tracking antenna and user.

Optical-Axis Calibration   Accurate angle encoder readouts with respect to the antenna azimuth and elevation axes of the antenna mount of Fig. 11-21 are fundamental in establishing angular errors in a tracking antenna. A multifaceted optical flat (with 17 facets, for example), temporarily mounted perpendicular to and centered on the axis of rotation and used with an optical collimator, can establish the cyclic errors in the encoder within 0.02 mrad rms. This readout should be performed after installation of the encoder in the mount. Each measurement should be made by rotating about the axis in the same direction to eliminate backlash errors. An encoder readout is made when the reflection of the collimator cross hairs from the optical flat coincides with the collimator cross hairs. Backlash can be determined by reversing the direction of antenna rotation about the axis and reading the encoder at any one of the optical flat positions to establish the backlash random error. The random error due to granularity of the encoder can be read by leaving the antenna stationary and reading the encoder output variation.

The next step in calibration for a reflector antenna after encoder cyclic errors have been established is the use of on-mount levels to bring the azimuth axis parallel to the local vertical. After leveling in one azimuth position, readings are taken at a number of azimuth locations, giving results similar to that of Fig. 11-22.

The error models or algorithms of Sec. 11-4 are used to correct the data output.

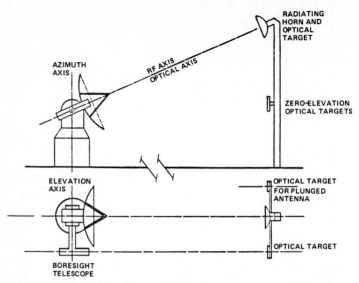

FIG. 11-21 Elevation- over azimuth-axis tracking antenna and boresight tower.

The difference estimates that are furnished by use of these algorithms and true position results in residual errors that are small. Figure 11-22, for example, gives two sets of measurements made on different days. Peak-to-peak error approached 0.4 mil* on March 31 and 0.2 mil on April 3. Correction of the data left a residual error of less than 0.02 mrad for each day. If a second calibration had not been made on April 3, peak errors in excess of 0.1 mil would have been introduced, indicating the need to recalibrate highly accurate systems on a mission basis.

With an antenna capable of plunging, i.e., with an elevation motion greater than 180°, the optical-axis nonorthogonality to the elevation axis may be established by reading the azimuth position of the zero-elevation optical targets of Fig. 11-21 with the antenna in both normal and plunged positions. The difference in azimuth encoder readout should be exactly π rad. One-half of any difference is the optical- to elevation-axis nonorthogonality.

To determine the nonorthogonality or skew constant of the elevation-to-azimuth axis, azimuth encoder readings are then made in the normal and plunged portions of the upper optical targets, taking into account the calibration data from the encoder and the optical-elevation-axis nonorthogonality measurements. The readings should be π rad apart. If not, the skew constant ϵ for the correction algorithm is the difference in reading divided by 2 and then divided by the tan E of the elevation of the upper optical targets. The use of the zero-elevation optical targets precludes the coupling of the optical-elevation-axis-nonorthogonality error into the measurement.

Once the nonorthogonality error of the elevation to azimuth axis has been estab-

---

*This error was taken on a radar using the artillery mil., i.e., ⅟₆₄₀₀ of a circle. This value should not be confused with the milliradian or the artillery-spotting mil (1 yd subtended at 1000 yd). The abbreviation of mil for milliradian is sometimes seen in the literature. Many older systems still utilize the artillery mil.

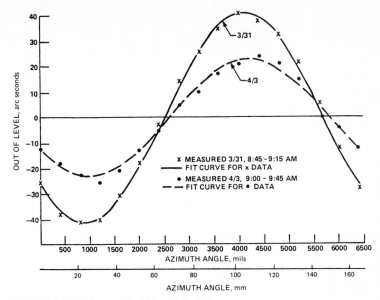

FIG. 11-22   Antenna tilt. *(After Ref. 13.)*

lished, the azimuth encoder offset may also be established by using the upper bore-sight-tower targets as a reference, which usually is surveyed with respect to true north. If the positions of both sets of optical targets and the position of the centerline of the telescope axis and antenna elevation axis are accurately known by an external survey, the droop constant d_0 for the telescope may be determined by the difference in elevation encoder readings between the lower and upper optical targets. Once the droop constant has been established, the elevation encoder offset with respect to the optical axis may be entered into the calibration data.

RF-Axis Calibration   Once the errors have been established for the optical axis and correction made, the relation of the RF axis to the optical axis must be determined. In azimuth, this is relatively simple. Usually there is no azimuth multipath. A bore-sight-tower RF horn radiating a polarized signal and an optical target separated by exactly the same distance as the RF boresight axis of the antenna and the optical axis of the telescope are used, as shown in Fig. 11-22. The RF axis is found at minimum reading from the difference channel in azimuth and elevation. If the depth of the null in either axis does not meet the error-budget requirements, it should be corrected since further measurements will be contaminated by postcomparator phase errors.

The elevation- to azimuth-axis nonorthogonality or skew error is the same for the RF axis as for the optical axis. The azimuth error introduced by RF boresight-axis-to elevation-axis nonorthogonality will be a function of both the mechanical position of the feed and the variation of the boresight axis in azimuth as a function of frequency. Since the skew error is known, the boresight tower can then be used to determine the RF boresight-axis to elevation-axis nonorthogonality as a function of frequency. The test results in Fig. 11-23 show this nonorthogonality to be approximately 0.025 mil and the peak-to-peak variation in azimuth of the boresight axis to be approx-

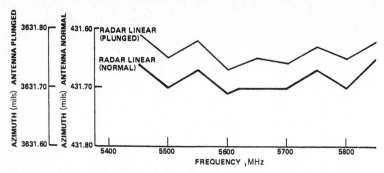

FIG. 11-23 Azimuth RF boresight shift versus frequency. *(After Ref. 13.)*

imately 0.06 mil across the band. As can be seen, the RF to optical-axis collimation error is quite small, <0.01 mil rms.

The multipath, droop in the RF feed, and frequency shift of boresight are combined when measured at a single elevation. The droop is constant at that elevation, but the multipath and frequency shift of the boresight axis are variable. Multipath error may be separated by tracking a balloon-borne metallic sphere; the sphere is centered within the balloon and launched near the tracking antenna. RF tracking is initiated. Corrected data from the radar and optical recording of the boresight-telescope field of view are compared when the elevation angle is above that at which multipath is a factor. A set of 10 frequencies across the band is used for 10 different elevation angles. From this measurement, the boresight shift in elevation with frequency may be determined separately from multipath. Once the droop constant and the boresight shift with frequency have been determined, the RF- to optical-axis offsets may be determined. The remaining error is that due to the collimation error, that is, the limits of telescope resolution, depth of null, and observer.

Off-Axis Calibration Once the electrical boresight axis has been established, off-boresight positions are calibrated. A plot of error versus angle offset is given in Fig. 11-24 in both elevation and azimuth for a single frequency. These measurements were made by using a boresight-tower RF horn radiating a vertically polarized signal. There may be some cross-polarization, multipath, and droop contamination, but the amount should be negligible, since the angular change during measurement was much smaller than the beamwidth.

Star Calibration of the Optical Axis The most accurate method for optical-axis calibration of the antenna mount is that of star calibration.[19] The steps are similar to those used by Meeks,[20] except that the sensitivity of most tracking antennas does not permit direct calibration of the RF axis. A set of stars which is visible (brightness magnitude greater than 3.8) at the time of calibration[21] and distributed uniformly over the celestial sphere between 15 and 70° with respect to the local vertical is chosen as the reference base. Under computer control the azimuth and elevation positions of each star selected are measured by using the optical telescope. A timing error of ±0.2 s contributes about 0.005 mrad rms to the instrumentation error. A telescope with a resolution of 4 s contributes about the same error.

After the measurements have been taken and corrections for refraction made, a

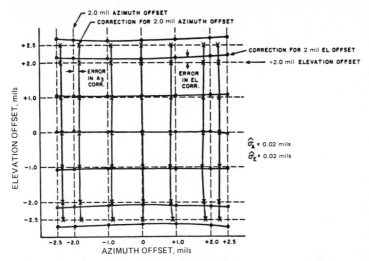

FIG. 11-24 RF off-axis error. *(After Ref. 13.)*

regression analysis will permit determination of the error coefficients for each of the mechanical-error-correction algorithms. The entire process, which may be limited by cloud cover, takes only a few minutes and offers an opportunity to maintain the mechanical system at its highest level of accuracy.

The use of stars as a reference has the advantage of determining true vertical as compared with local vertical—a difference that could be as much as 0.3 mrad in certain locations. It further provides a means of assessing errors as a function of azimuth rather than just the normal and plunged positions. A boresight tower and balloon tracks are still required to establish the relationships of the RF to the optical axis.

Calibration for Phased-Array Tracking Antennas

Calibration techniques for phased arrays depend on the size, configuration, and use of the arrays. If the design is in a limited-scan phased array utilizing a two-axis mount, the mount can be calibrated by using techniques similar to those used for reflector antennas. Once the mount and the face of the array have been established with respect to the azimuth and elevation axis, the array itself may be calibrated by using the mount encoder output and a boresight tower. The same limitation in accuracy by multipath applies here as for the reflector antenna.

If the final phased-array tracking system does not use a mount but the antenna is of a size so that it can be placed on a two-axis mount, then the angular position of the array face to the boresight tower can be established with respect to a theodolite attached to the antenna. A scan-angle order can be given, and the angular position to which the antenna face must be rotated to obtain a difference-pattern null can be established by the theodolite. This angle is then compared with that ordered by the beam-steering controller. An example of the error expected is given in Fig. 11-25. If corrected as a function of frequency, the residual systematic error is less than 0.1 mrad rms.

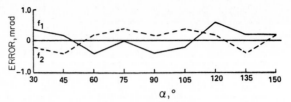

FIG. 11-25 $\alpha - \alpha_i$ error with scan angle.

Another technique that has recently been demonstrated is the use of near-field measurements. The antenna is first mounted in an anechoic chamber. A radiating probe antenna is raster-scanned across the face of the array at a constant distance from the array with the polarization of the probe antenna parallel to that of the array. Measurements are made of amplitude and phase for each of the probe positions at the sum-and-difference output ports. Once the near field is known, the far-field sum-and-difference patterns may be calculated. This technique has been validated by far-field measurements that were made with the same antenna. The near-field technique has the additional advantage of low multipath and radio-frequency interference effects.

The use of the near-field techniques provides data for the sum-and-difference pattern for any one scan angle, frequency, or temperature value. Calculations of the field pattern provide the calibration data for correcting both boresight and off-boresight errors.

A sufficient number of beam positions, both on and off the cardinal planes, must be measured to develop the error-correction algorithms. The same beam positions should be repeated with different frequencies to assure that the beam-steering algorithms in the controller are correct and that errors present in beam position as a function of frequency can be established and corrected. In addition, a set of patterns should also be measured at a different temperature if possible, to develop error-correction algorithms as a function of temperature.

When the antenna is large and cannot be assembled at the factory but must be installed and aligned on site, the position of the array face in the local coordinate system is determined by a survey. By using optical flats and prisms on the antenna face, the orientation of the face with respect to true north, tilt with respect to true vertical, and the rotation about the normal to the array face are determined.

Such an antenna is generally large enough to track such objects as satellites having well-established trajectories. A track of a sufficient number of objects, over the band of operating frequencies, permits a regression analysis that provides for calibration as a function of scan angle and frequency. Scanning about the target establishes the difference-pattern slope. Thermal and wind conditions affecting the array must be recorded concurrently so that the errors produced can be correlated with these conditions and data-correction equations can be developed.

REFERENCES

1 G. M. Kirkpatrick, "Final Engineering Report on Angular Accuracy Improvement," August 1952; reprinted in D. K. Barton, *Monopulse Radar,* Artech House, Inc., Dedham, Mass., 1974.

2 H. Jasik (ed.), *Antenna Engineering Handbook,* 1st ed., McGraw-Hill Book Company, New York, 1961, sec. 25.5.

3 M. I. Skolnik, *Introduction to Radar Systems,* 2d ed., McGraw-Hill Book Company, New York, 1980.

4 D. K. Barton and H. R. Ward, *Handbook of Radar Measurement,* Prentice-Hall, Inc., Englewood Cliffs, N.J., 1969.

5 *IEEE Standard Dictionary of Electrical and Electronics Terms,* 2d ed., Institute of Electrical and Electronics Engineers, New York, 1977.

6 E. B. Wilson, *An Introduction to Scientific Research,* McGraw-Hill Book Company, New York, 1952.

7 D. K. Barton, *Radar System Analysis,* Prentice-Hall, Inc., Englewood Cliffs, N.J., 1964.

8 T. T. Taylor, "Design of Circular Apertures for Narrow Beamwidth and Low Sidelobes," *IRE Trans. Antennas Propagat.,* vol. AP-8, January 1960, pp. 17–22.

9 E. T. Bayliss, "Design of Monopulse Antenna Difference Pattern with Low Sidelobes," *Bell Syst. Tech. J.,* vol. 47, May 1968, pp. 623–650.

10 P. Hannan, "Optimum Feeds for All Three Modes of a Monopulse Antenna in Theory and Practice," *IRE Trans. Antennas Propagat.,* vol. AP-9, no. 5, September 1961, pp. 444–461.

11 W. H. Nester, "A Study of Tracking Accuracy in Monopulse Phased Arrays," *IRE Trans. Antennas Propagat.,* vol. AP-10, no. 3, May 1962, pp. 237–246.

12 M. I. Skolnik (ed.), *Radar Handbook,* McGraw-Hill Book Company, New York, 1970, pp. 21-50, 21-53.

13 R. Mitchell et al., "Measurements and Analysis of Performance of MIPIR (Missile Precision Instrumentation Radar Set AN/FPQ-6)," final rep., Navy Cont. NOW 61-0428d, RCA, Missile and Surface Radar Division, Moorestown, N.J., December 1964.

14 S. Silver (ed.), *Microwave Antenna Theory and Design,* McGraw-Hill Book Company, 1949, p. 419.

15 I. Koffman, "Feed Polarization for Parallel Currents in Reflectors Generated by Conic Sections," *IEEE Trans. Antennas Propagat.,* vol. AP-14, no. 1, January 1966, pp. 37–40.

16 W. M. Bridge, "Cross Coupling in a Five Horn Monopulse Tracking System," *IEEE Trans. Antennas Propagat.,* vol. AP-20, no. 4, July 1972, pp. 436–442.

17 W. M. Humphrey, *Introduction to Servomechanism System Design,* Prentice-Hall, Inc., Englewood Cliffs, N.J., 1973, pp. 234–289.

18 M. Rozansky, "Exact Target Angular Coordinates from Radar Measurements, Corrupted by Certain Bias Errors," *IEEE Trans. Aerosp. Electron. Syst.,* vol. AES-12, no. 2, March 1976, pp. 203–209.

19 J. T. Nessmith, "Range Instrumentation Radars," *IEEE Trans. Aerosp. Electron. Syst.,* vol. AES-12, no. 6, November 1976, pp. 756–766.

20 M. L. Meeks, J. A. Ball, and A. B. Hull, "The Pointing Calibration of Haystack Antenna," *IEEE Trans. Antennas Propagat.,* vol. AP-16, no. 6, November 1968, pp. 746–751.

21 A. Becvar, *Atlas Coeli II: Katalog 1950.0,* Nakladatelství Československé Akademie věd, Prague, 1960.

Chapter 12

Satellite Antennas * +

Leon J. Ricardi

Lincoln Laboratory
Massachusetts Institute of Technology

*This work is sponsored by the Department of the Air Force.

+The United States government assumes no responsibility for the information presented.

The use of satellites in communication and remote-sensing systems has increased tremendously in the past decade or two, resulting in designers who specialize in satellite antennas. The uniqueness of the platform, the environment, and the applications requires antenna designers to interact with several disciplines including mechanics, structural analysis, thermal analysis, surface charging and radiation effects, and communications theory. This chapter will address fundamental spacecraft-stabilization characteristics and the related considerations of antenna design and polarization. Earth-coverage antennas will be described, and the use of higher-gain multiple-beam antennas will be discussed. Finally, a comparison of two types of cross-link antennas is presented. Although these discussions may apply to antennas used in other types of systems, the communications satellite (COMSAT) will be in mind unless otherwise specified.

12-1 SPACECRAFT-STABILIZATION AND FIELD-OF-VIEW CONSIDERATIONS

Not unlike an airplane, an automobile, a ship, and a tower, spacecraft have their unique "platform" characteristics, which determine some of the antenna's more important characteristics. In particular, a spacecraft derives its "stability" from the gyroscopic action produced by rotating all or some of its mass. Consequently, all satellites are designed so that all or a significant part of their mass rotates so as to establish a reference axis, plane, or sphere. In the early years of the satellite age, all satellites were intentionally rotated about the axis with the largest inertial moment; these satellites are commonly referred to as *spinners*. The orientation of this axis was often maintained by applying external forces to the satellite by various means, such as gas propulsion, emission of ionized particles of a solid, and interaction with the earth's magnetic[1] or gravitational fields.

With the development of lubricated bearings that can be operated satisfactorily in the vacuum of space, three-axis stabilization became possible, resulting in a *despun platform* similar to a tower on which the antennas of a microwave land link are mounted. Recently it has become popular to use an *inertia wheel* (i.e., a three-axis gyro) to establish the needed reference system for a three-axis stabilized satellite. Of course, in all cases (except perhaps one-axis stabilization) controlling external forces are applied, in cooperation with pointing and attitude error-sensing devices, to maintain the satellite in the desired orbit or at a particular point in the orbit. The latter function is often referred to as *station keeping*.

Since the prime purpose of a communication satellite is to provide a relay between two or more communication terminals, the earth-bound satellite must have an antenna capable of transmitting signals to and receiving signals from the earth. The degree of stabilization and whether or not the satellite is a spinner determine the fundamental conditions under which it must perform this function. In particular, an antenna on a spinning satellite must have a radiation pattern that is uniform in the plane perpendicular to the spin axis, or it must be designed so that its radiation pattern is "despun" mechanically or electrically. Even an antenna on a three-axis stabilized vehicle must be designed to compensate for satellite attitude and station-keeping errors, again either by modification of its radiation pattern or by physical reorientation of its structure. Clearly, using the tumbling satellite as a platform presents the great-

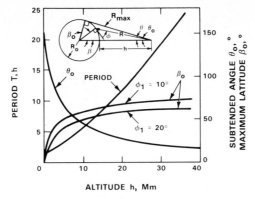

FIG. 12-1 Satellite-earth characteristics.

est challenge to the antenna designer, and the three-axis stabilized station-kept satellite allows for the most sophisticated antenna design.

Communication satellites are often placed in a nearly circular orbit and hence the angle θ_0, subtended by the earth and measured at the satellite, is relatively constant. Since the communications requirement usually involves a relay between widely separated points on the earth, the satellite's field of view (FOV) is often defined by θ_0 and is given by

$$\theta_0 = \sin^{-1}(R_0/R_0 + h) \qquad \textbf{(12-1)}$$

Referring to Fig. 12-1, we see that θ_0 varies from about 145° for a low-attitude satellite to 17.3° for a satellite in a synchronous (i.e., 24-h-period) orbit.

Although most communication satellites make use of the geostationary properties of the equatorial synchronous orbit, other considerations may dictate an elliptical orbit and the concomitant change in θ_0. For an elliptical orbit, the earth is located at one of the foci of the ellipse traced by the orbit and is defined as the barycenter. The eccentricity ϵ of the ellipse is given by

$$\epsilon = [1 - (b/a)^2]^{0.5} \qquad \textbf{(12-2}a\textbf{)}$$

where a and b are the semimajor and semiminor axes of the elliptical orbit, for a circular orbit $a = $ b and $\epsilon = 0$. The orbital period T is independent of ϵ (i.e., as long as the orbit does not intersect the earth), and it is given by

$$T = 2\pi[(R_0 + h)^3/\alpha]^{0.5} \qquad \text{h} \qquad \textbf{(12-2}b\textbf{)}$$

where $R_0 = 6378$ km is the radius of the earth, h is the maximum altitude of the satellite above the earth's surface (measured in kilometers), and $\alpha = 5164 \times 10^{12}$.

An earth terminal located at the edge of the satellite's FOV views the satellite at an elevation angle $\phi = 0°$; that is, the satellite is on the local earth horizon. System performance usually requires $\phi > 10°$; hence the satellite's FOV ϕ_0' is somewhat less than that described by θ_0. The angle θ between the earth-satellite axis and the terminal-satellite direction determines the satellite elevation angle ϕ, as shown in Fig. 12-2. The relation between ϕ and θ is plotted for a satellite at synchronous altitude and at binary multiples of synchronous altitude. The corresponding satellite period T is also indicated for ease of reference. Clearly, $\phi = 0$ when $\theta = \theta_0$ and $\phi = 90°$ when

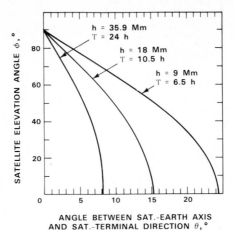

FIG. 12-2   Elevation angle of the satellite measured at the earth terminal.

$\theta = 0$. Using Fig. 12-2, β_0, the maximum latitude of a terminal, was computed by using

$$\beta_0 = 90 - \phi_1 - \theta_1 \qquad (12\text{-}3a)$$

and

$$\theta_1 = \sin^{-1}[R_0 \sin(90 - \phi_1)/(R_0 + h)] \qquad (12\text{-}3b)$$

where ϕ_1 is the minimum value of ϕ and θ_1 is the corresponding value of θ for the $\phi = \phi_1$ (see Fig. 12-2). With reference to Fig. 12-1, β_0 is plotted for $\phi_1 = 10°$ and $20°$ and indicates that satellites in an equatorial orbit cannot operate with terminals located at north or south latitudes in excess of about 70° unless their altitude above the earth exceeds 40 Mm.

Inclined 12-h elliptical and 24-h polar orbits and the equatorial synchronous orbits form what might be considered a primary set of orbits that have most desirable characteristics. The geostationary orbit is most often used by systems with terminals located in the tropical and temperate zones. Satellites in the polar and, most often, the inclined elliptical orbit are used by terminals in the temperate and polar regions. Systems with satellites in both orbits provide worldwide coverage; worldwide service is obtained through the use of satellite-to-satellite links (i.e., cross links) or earth terminals as gateways. Gateway terminals must be in the FOV of at least two satellites. Cross-link-antenna design[2] will be discussed in Sec. 12-7; however, it is of importance to note that θ_0 and h of a critically inclined 12-h elliptical orbit[3] vary as shown in Fig. 12-3. This orbit is often referred to as the Molniya orbit since it was first used by the Russians to provide "commercial" communications capacity to all populated parts of the Soviet Union. For the critically inclined orbit, solar pressure will not cause it to precess. The times indicated in Fig. 12-3 are measured either prior to or after apogee. A pair of satellites spaced about 6 h along the same orbit provide continuous coverage in the temperate regions and one polar region. Note that θ_0 and h vary by approximately 30 percent over the 6-h useful portion of the orbit, that is, during the period from 3 h before to 3 h after the satellite is at its highest altitude above the earth. It is

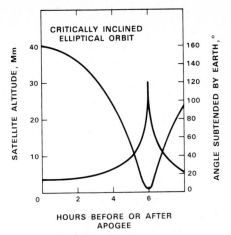

FIG. 12-3 Characteristics of a 12-h elliptical orbit.

the indicated change in θ_0 and the relative direction to the center of the earth that affect antenna-design considerations.

Before considering antenna designs as they interact with various stabilization methods, it is important to point out that satellite-antenna-pattern coverage, as viewed from the angular space conventionally used to describe antenna performance, becomes distorted when radiation-pattern contours are plotted on conventional (e.g., Mercator, gnomonic, etc.) projections of the earth. Presentation of antenna-pattern contours of constant gain on earth maps are referred to as *beam footprints*. The beam-contour coordinates ϕ and θ can be transferred to earth latitude and longitude to determine the beam footprint by first assuming that the satellite's nadir is located at a longitude $= a°$ and at $0°$ latitude. Then use

$$\phi = \cos^{-1} (1 + h/6378) \sin \theta \qquad (12\text{-}4a)$$

$$\beta = 90 - \phi - \theta \qquad (12\text{-}4b)$$

to compute β. The coordinates of a point P on the earth at longitude ζ and latitude ξ are related to θ and ψ through

$$\xi = \sin^{-1} (\sin \beta \cos \psi) \qquad (12\text{-}5a)$$

$$\zeta = a + \tan^{-1} (\tan \beta \sin \psi) \qquad (12\text{-}5b)$$

where ψ is the angle measured from the plane containing the satellite and the earth's polar axis to the plane determined by the earth-satellite axis and the line from the satellite to the point P (Fig. 12-4). When the satellite's nadir is not on the equator, the transformation is given by

$$\xi = \sin^{-1} [\sin (b + \beta) + \cos b \sin \beta (\cos \psi - 1)] \qquad (12\text{-}6)$$

$$\zeta = a + \sin^{-1} (\sin \psi \sin \beta/\cos \xi) \qquad (12\text{-}7)$$

where b is the latitude of the satellite's nadir and a is its longitude.

The foregoing is presented to point out that the satellite's stabilization method

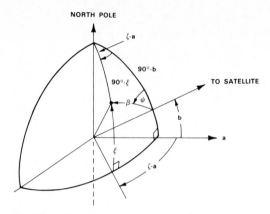

FIG. 12-4 Transformation coordinate system.

and its orbit have a serious impact on its antennas and their FOV. It is beyond the scope of this chapter to discuss these interactive design relationships in detail; however, some designs will be described to indicate methods of accommodating these characteristics.

12-2 DESPUN-ANTENNA SYSTEMS

Many earth satellites are cylindrically or spherically shaped bodies rotating about the axis for which their inertial moment is maximum. If the orientation of this axis is uncontrolled, the antenna has to produce a radiation pattern that is uniform in the plane perpendicular to the spin axis. Alternatively, one could use a directional antenna whose radiation pattern is either despun by varying the excitation of the antenna (i.e., as in the case of an array) or by mechanically rotating the antenna at the same angular rate but in a direction opposite to the satellite's rotation. Clearly if the antenna radiation pattern is not despun and the spin-axis orientation is not controlled (one-axis stabilization), the antenna pattern must be independent of direction. This leads to an *isotropic* antenna and the concomitant directivity and directive gain of 0 dB.

Telstar, the first active communication satellite, employed antennas[5] having a nearly isotropic radiation pattern. The satellite rotated on its axis of maximum moment of inertia and had some attitude control; it was, therefore, a two-axis stabilized spacecraft. It was spherically shaped with an approximate radius of 35 in (889 mm). It had antennas operating at very high frequency (VHF), at 4 and at 6 GHz. All radiation patterns were uniform in a plane perpendicular to the spin axis, providing the necessary despinning function. Except for the surface occupied by the antennas, solar cells covered the satellite's outer surface.

The first Lincoln experimental satellite (LES-1) had two-axis stabilization; the orientation of the spin axis was maintained by force derived from an interaction with the earth's magnetic field. One of the major purposes of LES-1 was to demonstrate the feasibility of electronically despinning the radiation pattern by switching among eight low-directivity (about 7 dB) antennas distributed uniformly over the satellite duodecahedron shape. This experimental satellite also demonstrated the feasibility of a simple magnetic attitude-control system that was subsequently used[6] on LES-4 and enabled the use of antennas with higher directivity (about 14 dB).

With the exception of a VHF antenna on Telstar, communication antennas on these earlier satellites operated in the superhigh-frequency (SHF) range (i.e., 4, 6, and 8 GHz). This enabled the antenna designer to realize moderate antenna gain without significantly compromising the other satellite functions and systems. However, the SHF ground terminals were necessarily large and hence few in number, resulting in communication satellites which served only a few earth terminals or direct users. In the late 1960s, LES-5 and LES-6 were placed[7] in synchronous orbit and operated in the ultra high-frequency (UHF) communications band to service a wide variety of military communication needs because only modest-size earth terminals were required. The antenna systems on LES-5 and LES-6 had approximately the same radiation pattern as Telstar and LES-4 respectively. It is of interest to compare the methods of achieving these radiation patterns.

Telstar employed a large number of open-ended waveguide radiators spaced one-half wavelength on center along the satellite's equator. Two such antenna arrays, one operating at 4.17 GHz and one at 6.39 GHz, were used; each was fed so as to excite each element in phase and with the same signal amplitude. This symmetry guarantees a symmetrical radiation pattern in the plane perpendicular to the satellite's spin axis. Uniformity of the radiation pattern is controlled by the number of individual radiators per wavelength along the satellite's equator and the phase and amplitude equality of the signals exciting each radiator in the array. These antennas had an equatorial-plane radiation pattern that was uniform within ± 1 dB. The feed network[5] provided the desired excitation amplitude and phase within 0.5 dB and 5° respectively.

LES-5 was a cylindrically shaped satellite[7] about one wavelength (λ) long and one wavelength in diameter; that is, it was much smaller than Telstar in terms of their respective operating wavelengths. The communication antenna consisted of two circular arrays. One array was made up of eight full-wave dipoles spaced about $\lambda/2$ circumferentially on the satellite's cylindrical surface and excited in phase with equal-amplitude signals. The other array consisted of two rings of eight half-wave cavity-backed slots colocated with the dipole array. The slots were also excited in phase with equal-amplitude signals but in-phase quadrature with the dipole array so that the combined dipole-slot pair would radiate and receive circularly polarized signals. The antenna radiation pattern was uniform, within ± 1 dB, in planes perpendicular to the satellite's spin axis. Unlike Telstar, LES-5's radiation pattern in a plane containing the spin axis approximated that of a linear array of two $\lambda/2$ dipoles spaced about $\lambda/2$ on center along the satellite's spin axis.

It is of interest to note that the configuration of each satellite (Telstar and LES-5) prohibited placing the antennas at the center of the satellite, where the small single radiator could have been placed to obtain the desired circularly symmetric radiation patterns in the satellite's equatorial plane (i.e., such as that obtained in a plane perpendicular to the axis of a dipole). In fact, most of the volume within the satellite was reserved for other subsystems of the satellite. This required that the antenna be distributed around the outside surface of the satellite and appropriately excited through circuits, made up of transmission lines and n-way power dividers, integrated with the other subsystems located within the satellite's surface. Virtually any one- or two-axis stabilized satellite without a despun antenna must use this general type of antenna configuration to meet the usual requirements dictated by earth-bound stations using a communication satellite. The antenna configuration is often similar to that of public television broadcast antennas mounted on a high tower.

Occasionally a dipole, stub, biconical, or slot antenna that will produce the desired despun radiation pattern can be mounted at one or both poles of the satellite. Alternatively, a somewhat simple antenna can be used when the small volume described by a thick disk coincident with the satellite's equator is available exclusively as the antenna site. In this case, a radial transmission line (i.e., parallel disks excited at their center) exciting a ring of open-ended waveguide radiators[8] represents that case when antenna performance is essentially uncompromised by its necessary integration into the satellite.

Electronically despun antennas are similar in that they usually consist of an array of low- to medium-gain antennas (i.e., 5- to 15-dB directivity) dispersed uniformly around the spin axis of the satellite and excited through a switch network that connects the transmitter or receiver to only those antennas that point most closely toward the center of the earth. An alternative, less common, electronically despun antenna consists of a circular array of low-gain elements simultaneously excited with equal power and phased to produce a unidirectional beam. The phase is varied to despin the beam as the satellite rotates. However, this type of antenna must occupy a complete cylindrical section of the satellite, usually on one end of the spacecraft. Consequently, electronically despun antennas usually consist of an array of several antennas which are sequentially switched to despin the radiation pattern. The details of these antennas are very well described in the literature.[6,7,9]

Before leaving this topic, it is important to discuss mechanically despun antennas used on Intelsat-III[10] and ATS-III[11] (Fig. 12-5a and b). Both of these employ reflector antennas and despin the radiation pattern by counterrotating the reflector (with respect to the satellite sense of rotation) so that its focal axis always points toward the center of the earth. The ATS-III antenna system (Fig. 11-4) consists of two antennas, one operating at 6.26 GHz and one operating at 4.85 GHz. Each antenna employs a parabolic cylinder illuminated by a collinear array of dipoles located on the focal line of the reflector. The latter reflects and converges the incident energy into a beam coincident with the focal axis of the parabola that generates the reflector's surface. Since the radiation pattern of a collinear array of dipoles is uniform in a plane perpendicular to the axis of the array, rotating the reflector about the dipole array rotates the antenna's beam in a plane perpendicular to the axis of the dipole array. The latter is installed coincident with the satellite spin axis so that counterrotation of the reflector results in a despun antenna. A unique feature of this antenna permits the reflector to be ejected if the rotary mechanism fails. Without the reflector the antenna has an omnidirectional radiation pattern permitting continued operation but with about 13-dB reduction in directivity.

Intelsat-III also uses a rotating reflector to despin its radiated beam; however, the reflector surface is flat and oriented at a 45° angle with respect to the spin axis of the satellite. A horn antenna produces a high-directivity (~21-dB) beam pointed along the spin axis of the satellite toward the reflector. The latter redirects the beam perpendicular to, and causes it to rotate about, the satellite spin axis. This antenna, the ATS-III despun, and similar mechanically scanned devices continuously despin the radiation pattern with essentially no variation in the antenna's directive gain. Electronically despun antennas usually scan the radiated beam in a stepwise fashion; this usually produces a significant (from 1- to 3-dB) variation in the antenna's directive gain as it would be measured by a fixed terminal on the earth.

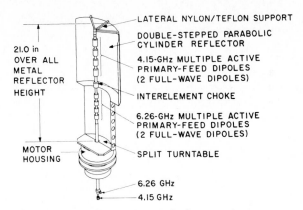

FIG. 12-5a ATS-III mechanically despun antenna.

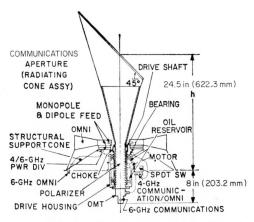

FIG. 12-5b ITS-III mechanically despun antenna.

12-3 ANTENNA POLARIZATION

Before considering three-axis stabilized satellites, it is important to discuss the antenna's polarization and why it plays an important role in satellite-antenna design. Recall that an antenna is defined by the polarization of the electromagnetic (EM) energy that it radiates. It is important to measure this polarization in the far zone of the antenna, that is, at distances sufficiently far from the antenna so that a further increase in this distance will not change the measured polarization. A distance $R = 2D^2/\lambda$ is customarily chosen as adequate for measuring the antenna's polarization and directive gain, where D is the antenna-aperture size and λ is the operating wavelength. The electric field direction defines the polarization of the EM energy.

Although essentially all polarization properties of EM waves play a role in satellite-antenna design, let us review those which are most important. For example, a linearly polarized (LP) antenna such as a dipole, oriented with its axis vertical (with respect to the earth's surface), will radiate and receive vertically polarized signals. Conversely, it will neither radiate nor receive horizontally polarized signals. This phenomenon is commonly referred to by stating that an antenna will not radiate or receive cross-polarized signals, or that orthogonally polarized signals are rejected. This statement is not limited to LP antennas; circular and elliptically polarized EM waves and antennas have copolarized and cross-polarized properties identical to those of LP waves and antennas.[12] Circularly polarized (CP) waves have a right-hand sense (i.e., RHCP) if the electric field vector rotates in a clockwise sense as the wave is propagating away from the observer. The electric field vector of a left-hand circularly polarized (LHCP) wave rotates in a counterclockwise sense for receding waves. Changing both the direction of propagation (i.e., receding to approaching) and the sense of rotation (i.e., clockwise to counterclockwise) does not alter the polarization. Elliptically polarized waves result when the strength of the electric field varies as its direction rotates. In summary, virtually all antennas are, in fact, elliptically polarized, but the ellipticity is such that referring to them as CP or LP is an adequate descriptor. The important point is that copolarized antennas couple well to one another and cross-polarized antennas tend to reject one another's signals.

Now consider earth-satellite signal links when the frame of reference (i.e., vertical and horizontal) of the earth station will not, in general, coincide with the frame of reference (i.e., north and south) of the satellite. Since the satellite usually serves many users simultaneously and its antenna can assume only one polarization at any instant of time, it follows that when LP antennas are used, the earth station must adjust its frame of reference to coincide with the satellite's frame of reference. Although this is possible, it is far simpler to use CP satellite and earth-terminal antennas and remove the need to align them in order to maximize coupling between them. Consequently, it is not surprising that most satellite antennas are circularly polarized.

When an LP satellite antenna is used, orientation of the associated EM waves is altered as they propagate through the earth's ionosphere,[13,14] a phenomenon often referred to as the Faraday rotation effect. This rotation of LP waves is usually negligible (less than a few degrees) at frequencies above a few GHz. However, at frequencies below 1 GHz Faraday rotation effects can rotate the wave polarization more than $360°$. Fortunately, the polarization of a CP wave is not altered by the Faraday rotation effect. Change in polarization due to transverse "static" magnetic fields along the propagation path is much smaller; therefore, circular polarization is preferred because CP waves propagate through the ionosphere with no essential change in polarization.

Most spacecraft and earth-terminal antennas are shared by the associated transmitter and receiver. Consequently, diplexer (or duplexer for a radar system) filtering is required. The use of antennas that are orthogonally polarized for transmitted and received radiation enhances the isolation between the transmitter and the receiver. For this and the foregoing reasons it is customary for satellite antennas to be opposite-sense circularly polarized for simultaneous transmit and receive functions.

12-4   THREE-AXIS STABILIZED SATELLITES

It is quite common to station-keep satellites whose attitude is completely controlled (i.e., orientation of the satellite's three principal axes is controlled). The satellite then takes on many of the characteristics of a relay station installed on a tower between two earth terminals. In particular, high-gain directional antennas can be used to increase the link's data rate, or reliability, and decrease its vulnerability to external noise sources. The antenna can become much more sophisticated than antennas mounted on a satellite whose total mass is spinning. Suffice it to say, the antennas can be as common as a narrow-beam paraboloidal reflector mounted on a two-axis pedestal or as complicated and unique as the multiple-frequency, multiple-beam antenna[15,16] that makes the National Aeronautics and Space Administration's ATS-6 such a unique satellite.

Three-axis stabilization is achieved by despinning a "platform," or part of the spacecraft, or by including an inertia wheel as an integral part of the satellite. The gyroscopic action of the spinning spacecraft or inertia wheel can maintain the satellite's attitude within about 0.1° of a given frame of reference. The addition of a propulsion device keeps the spacecraft at a prescribed location in space.

12-5 EARTH-COVERAGE ANTENNA

Probably the most common spacecraft antennas are those that have a broad pattern with approximately the same directive gain over the earth as it is viewed from the satellite. These are called *earth-coverage antennas* and vary from simple dipole arrays to horn antennas, shaped-lens antennas, or shaped-reflector antennas. For low-altitude satellites, the angle θ_0 subtending the earth is relatively large (see Fig. 12-1); consequently the associated antenna may be a dipole, a helix, a log-periodic dipole, a backfire antenna, or a spiral antenna. Higher-altitude satellites usually use horn-type earth-coverage antennas. In all cases, however, the desired antenna pattern $P_0(\theta)$ should provide the same signal strength at its output port (terminal) as a constant-power signal source is moved on the surface of the earth within the satellite's FOV. At frequencies below about 5 GHz atmospheric attenuation is negligible, and $P(\theta)$ depends only on the variation in path length R (Fig. 12-1) from the satellite to the points on the earth's surface that are within the satellite's FOV. At frequencies above 10 GHz atmospheric attenuation becomes significant, and $P(\theta)$ should be adjusted accordingly. Atmospheric attenuation A depends on absorption by the water-vapor and oxygen molecules and particulate scattering principally due to rain. Total attenuation due to atmospheric effects is, to a first order, proportional to the path length L_a through the atmosphere. Since the height h_a of the atmosphere in the vicinity of an earth terminal is nearly constant, atmospheric attenuation is approximately inversely proportional to the cosine of the satellite elevation angle ϕ (Fig. 12-1).

The desired earth-coverage-antenna pattern $P(\theta)$ must equalize the change in R and A over the FOV; hence

$$P(\theta) \propto (R(\theta)/h)^2 A(\theta) \qquad (12\text{-}8)$$

To determine $A(\theta)$ refer to Fig. 12-6 and note that the path length L_a through the earth's atmosphere is given by

$$L_a = [(R_0 + h_a)^2 + R_0^2 - 2R_0(R_0 + h_a)\cos\alpha]^{1/2} \qquad (12\text{-}9)$$

$$L_a \approx h_a[1 + 4(R_0/h_a)^2 \sin^2(\alpha/2)]^{1/2} \qquad (12\text{-}10a)$$

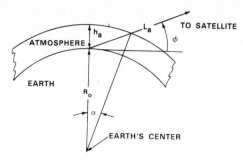

FIG. 12-6   Atmospheric path L_a.

where R_0/h_a is assumed $\gg 1$ and

$$\alpha \approx 90° - \phi - \sin^{-1}(\cos \phi) \qquad (\mathbf{12\text{-}10}b)$$

Hence,

$$A(\theta) = e^{A_0(1 - L_a/h_a)} \qquad (\mathbf{12\text{-}11})$$

where A_0 is the attenuation when $\phi = 90°$. It remains to relate θ and ϕ and determine $R(\theta)$.

Referring to Fig. 12-1, note that

$$R = [R_0^2 + (R_0 + h)^2 - 2R_0(R_0 + h) \cos \beta]^{1/2} \qquad (\mathbf{12\text{-}12})$$

which reduces to

$$R = h\{1 + 4[(R_0/h)^2 + R_0/h] \sin^2 (\beta/2)\}^{1/2} \qquad (\mathbf{12\text{-}13}a)$$

where

$$\beta = 90 - \theta - \phi \qquad (\mathbf{12\text{-}13}b)$$

and

$$\phi = \cos^{-1}[(1 + h/R_0) \sin \theta] \qquad (\mathbf{12\text{-}13}c)$$

The directivity of an antenna with a radiation pattern given by Eq. (12-8) can be calculated from

$$D = \frac{4\pi}{2\pi \int_0^{\theta_1} P(\theta) \sin \theta d\theta} \qquad (\mathbf{12\text{-}14})$$

where θ_1 is the edge of the coverage zone and $P(\theta) = 0$ for θ greater than θ_1. Note that $P(\theta)$ is identical in all planes containing the $\theta = 0$ axis and that θ_1 is given by

$$\theta_1 = \sin^{-1}[R_0 \cos \phi_0/(R_0 + h)] \qquad (\mathbf{12\text{-}15})$$

where ϕ_0 is the minimum satellite elevation angle within the desired FOV.

Having determined the optimum earth-coverage-antenna pattern $P(\theta)$, it is of interest to calculate the corresponding antenna directivity and directive gain and define a figure of merit F_e for assessing the performance of a specific earth-coverage antenna.

Toward this end, consider an earth-coverage antenna on a synchronous satellite (i.e., $h = 3.9$ Mm). Operational experience indicates that $\phi_0 > 20°$ prevents intolerable atmospheric-related diffraction effects and does not include intolerable atmospheric attenuation (i.e., atmospheric attenuation less than 1.5 dB at 10 GHz and less than 3 dB at 45 GHz) at the higher frequencies. For $\phi_0 = 19°$, Eq. (12-15) gives θ_0 = 8.2. By using these values for θ_0 and ϕ_0, the directivity of an optimum earth-coverage pattern is shown in Fig. 12-7 for $A_0 = 0$ and 1.0 dB. Note that increasing the atmospheric attenuation, in the zenith direction, from 0 to 1 dB changes the directivity of the ideal earth-coverage pattern by about 0.4 dB. However, the directivity D (8.2°) toward the edge of the coverage area increases by 1.4 dB in order to overcome the increase in atmospheric attenuation.

A shaped earth-coverage pattern is shown to indicate a feasible approximation to the ideal pattern. Finite antenna-aperture size will permit only an approximation to the cusp (at $\theta = \theta_0$) in the ideal pattern. Nevertheless, pattern shaping that increases $D(-\theta_0)$ will enhance the overall system performance even if $D(0)$ is reduced by 2 dB, as indicated by the shaped pattern (Fig. 12-7).

A typical conical-horn pattern is shown (Fig. 12-7) to indicate a primitive yet commonly used earth-coverage antenna. Note that for the shaped pattern $D(\theta_0)$ is about 4 dB greater than for the horn pattern. On the other hand, $D(0)$ is about 1 dB greater for the horn than for the shaped pattern. This conflict in performance points out the need for a figure of merit F_e that will aid in assessing the performance of an earth-coverage antenna. For this reason, it is suggested that F_e be set equal to the maximum difference in the antenna directivity $D_a(\theta)$ and the directivity $D(\theta)$ of the ideal antenna pattern. That is,

$$F_e = \max[D_a(\theta) - D(\theta)] /_{\theta=0}^{\theta=\theta_0} \quad (12\text{-}16)$$

where $D(\theta)$ and $D_a(\theta)$ are expressed in decibels. Therefore, F_e is negative with a maximum possible value of zero.

To estimate the maximum possible F_e, assume that the antenna aperture is

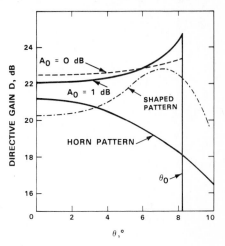

FIG. 12-7 Directive gain of an earth-coverage antenna on a synchronous satellite.

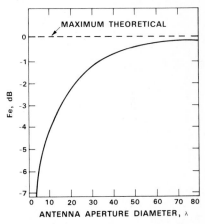

FIG. 12-8 Earth-coverage-antenna figure of merit.

inscribed in a sphere of radius a and that a set of spherical waves, with wave number $N < 2\pi a/\lambda$, is used to approximate $P(\theta)$. The directivity $D_0(\theta)$ of the pattern generated by the finite set of spherical waves is then used to compute $D_a(\theta)$ and F_e. Since the synthesis procedure guarantees a least-mean-square fit to $P(\theta)$ and the antenna-aperture excitation is prevented from exciting supergain waves[17] (i.e., $N < 2\pi a/\lambda$), the computed F_e is maximum for the given-size antenna aperture $D = 2a$. The maximum value of F_e shown in Fig. 12-8 was calculated by using this procedure.

12-6 SPOT-BEAM COVERAGE

Satellite antennas designed to provide high gain to a point within view of the satellite are required to have directivity greater than that of a simple fixed-beam earth-coverage antenna. In the case of a satellite in synchronous orbit, the theoretical maximum directivity of an earth-coverage antenna is about 24 dB; practical considerations limit its directivity to about 20 dB. Hence, for a directivity greater than 20 dB and an earth FOV, the synchronous-satellite antenna must produce a beam that scans or steps over the earth or the FOV. If a sequential raster scan is desired, the antenna might be steered mechanically or electronically over the FOV. However, for pseudo-random coverage of the FOV, step scan at microsecond rates is usually preferred. When step scan is preferred, the FOV is covered by several overlapping beam footprints. The minimum gain provided by N beams, designed to cover the FOV, is required to estimate system performance and is the subject of this section.

Since the cross section of a high-gain antenna beam is usually circular, the following analysis assumes that N beams with circular cross section are arranged in a triangular grid with beam-axis-to-beam-axis angular separation θ_b (see Fig. 12-9). A circular FOV is assumed; however, this analytic method can be readily extended to

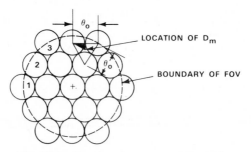

FIG. 12-9   Spot-beam coverage of the earth.

any FOV with a noncircular boundary. The analysis considers the black area in Fig. 12-9 and assumes that the directivity variation over this shaded area is replicated over the entire FOV. An edge-correction effect will be discussed following detailed consideration of the black area.

The analysis assumes a gaussian-shaped beam and determines θ_b and beam directivity D_0 that maximize the minimum directivity D_m over the black area. It follows that D_m is also the maximum value of the minimum directivity over the entire FOV with the possible exception of some small areas along the boundary of the FOV.

Since the useful directivity D of a beam will be within 5 dB of the beam directivity D_0, the directivity of all beams can be adequately represented by the function

$$D = D_0 e^{-\alpha(2\theta/\theta_1)^2} \tag{12-17}$$

where θ_1 = the half-power beamwidth (HPBW) of the beam. It is further assumed that when θ_1 is expressed in radians,

$$D_0 = \eta \frac{4\pi}{\theta_1^2} \tag{12-18}$$

where η ranges from about ½ to ¾. (Experimental data indicate that the antenna efficiency of a center-fed parabola is about 93η percent.) By noting that $D = D_0/2$ when $\theta = \theta_1/2$ and using Eq. (12-17),

$$\alpha = \ln 2 = 0.693 \qquad \textbf{(12-19)}$$

and

$$D = D_0 e^{-2.77(\theta/\theta_1)^2} \qquad \textbf{(12-20)}$$

Using Eqs. (12-18) and (12-20) gives an alternative expression for D, namely,

$$D = D_0 e^{-0.22\theta^2(D_0/\eta)} \qquad \textbf{(12-21)}$$

Over the FOV (not including the boundary), D_m occurs at the center of the equilateral triangle formed by the center of three adjacent beam footprints. The angle θ_m between the axis of the three adjacent beams and the direction of minimum direction gain is given by

$$\theta_m = \frac{\theta_b}{\sqrt{3}} \qquad \textbf{(12-22)}$$

Hence, from Eqs. (12-22) and (12-21)

$$D_m = D_0 e^{-0.074\theta_b^2(D_0/\eta)} = D_0 e^{-BD_0} \qquad \textbf{(12-23)}$$

Differentiating D_m with respect to D_0 gives

$$\frac{\partial D_m}{\partial D_0} = (1 - BD_0)e^{-BD_0} \qquad \textbf{(12-24)}$$

where $B = 0.074\theta_0^2/\eta$. From Eq. (12-24) D_m is maximum when

$$D_0 = \frac{\eta}{0.074\theta_0^2} \qquad \textbf{(12-25)}$$

$D_{\max}$, the maximum value of D_m, is given by

$$D_{\max} = D_0 e^{-1} \qquad \textbf{(12-26)}$$

Expressing Eq. (12-26) in decibels gives

$$D_{\max} = D_0 \text{ (dB)} - 4.34 \text{ dB} \qquad \textbf{(12-27)}$$

Hence, for maximum gain over the FOV, the crossover level between three adjacent beams is -4.34 dB with respect to D_0, the directivity of a beam.

By returning to Eq. (12-21), the crossover level D_2 between two adjacent beams is given by

$$D_2 = D_0 e^{-0.22(D_0/\eta)(\theta_b/2)^2} \qquad \textbf{(12-28)}$$

Substituting Eq. (12-25) into Eq. (12-28) gives

$$D_2 = D_0 e^{-3/4}$$

or

$$D_2 \text{ (dB)} = D_0 \text{ (dB)} - 3.25 \text{ dB} \qquad \textbf{(12-29)}$$

Hence the crossover level between two adjacent beams is 1.1 dB higher than the crossover level between three adjacent beams.

Statistical Distribution of D

Having knowledge of D_m is not always satisfying to the system designer because it represents a worst case. Whenever the estimated performance indicates that D_m may not be large enough, it is important to address the probability that the worst case will occur. With this probability placed in perspective with the foregoing analysis, it is important to determine the probability that $D > D' \geq D_m$. Toward this end, consider an enlargement (Fig. 12-10) of the black area in Fig. 12-9. The probability that $D > D_2$ is given by

$$P(D \geq D_2) = \frac{2(\pi/12)(\theta_0/2)^2}{(\theta_b/2)(\theta_b/2\sqrt{3})} = 0.907 \qquad (\textbf{12-30})$$

Hence less than 10 percent of the FOV has a directivity less than $D_2 = D_0 - 3.25$ dB. The probability that $D > D'$ can be found for $D' > D_2$ [i.e., $\theta_1 \leq \theta_b/2$)] by first finding the probability that $\theta < \theta_1$. That is,

$$P(\theta < \theta_1) = P(D > D') = 3.628 \left(\frac{\theta'}{\theta_b}\right)^2 \qquad (\textbf{12-31})$$

From Eq. (12-21)

$$D' = D_0 e^{-0.22(D_0/\eta)\theta'^2} \qquad (\textbf{12-32})$$

Solving Eq. (12-32) for θ'^2, substituting in Eq. (12-31), and using Eq. (12-25) give

$$P(D \geq D') = 1.22 \ln\left(\frac{D_0}{D'}\right) \qquad (\textbf{12-33})$$

By using Eq. (12-33) and a linear interpolation for $\theta_m > \theta > \theta_0$, the probability that $D \geq D'$ is given in Fig. 12-11.

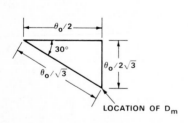

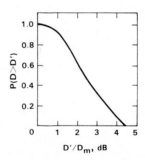

FIG. 12-10 Characteristic area of spot-beam coverage.

FIG. 12-11 Distribution of directivity of a multiple-beam antenna.

The foregoing derives $D_{\max}$ in terms of D_0. It remains to determine D_0 in terms of N, the number of beams required to cover the FOV. By assuming a hexagonal FOV, N is given by

$$N = 1 + \sum_{m=1}^{(M-1)/2} 6m \qquad (\textbf{12-34})$$

where M, the maximum number of beams in a row (Fig. 12-9), is given by

$$M = \frac{\psi}{\theta_b} + 1 \qquad (\textbf{12-35})$$

and ψ is the angle subtended by the major diagonal of the FOV and is measured at the antenna. From Eqs. (12-25) and (12-35)

$$M = \text{int}(1 + \psi\sqrt{0.07D_0/\eta}) \qquad (\textbf{12-36})$$

hence,

$$D_0 = \left(\frac{M-1}{\psi}\right)^2 \frac{\eta}{0.074} \qquad (\textbf{12-37})$$

For a synchronous satellite, the earth subtends an angle $\psi = 17.2°$ (0.3 rad); therefore, from Eq. (12-37) the maximum directivity of a multiple-beam antenna designed to cover the earth FOV with spot beams is given by

$$D_0 = 150(M-1)^2\eta \qquad (\textbf{12-38})$$

In accordance with Eq. (12-27) the corresponding minimum gain $D_m = 55(M-1)^2\eta$.

The foregoing derivation considered only those minimum-directivity points within the FOV. Usually the desired FOV does not conform to that described by the outer boundary of the footprints of the beams that cover the edge of the FOV. For example, for a circular FOV, arranging the beam footprints on a triangular grid results in the minimum number of beams to cover the circular area, but the resultant array is not hexagonal. To determine the number of beams required and their configuration, refer to Fig. 12-9 and note the location of D_m. Since the hexagonal array has sixfold symmetry, a point between beams 2 and 3 determines the edge of the FOV where $D = D_m$. As more beams are added to the array (i.e., as the FOV gets larger or as larger values of D_m are desired), the hexagonal grid gives poorer coverage along the edge of the FOV. Improved coverage is obtained by adding beams just at those points along the edge of the FOV where $D < D_m$ rather than by completing another ring of the hexagonal array of beams. The maximum directivity D_0 and the minimum directivity D_m, for a set of N beams arranged on a triangular grid and designed to cover the earth's disk from a synchronous satellite, are given in Table 12-1. The number of beams M in a diagonal of the hexagon pattern from which the set of N beams is derived is also given. For example, the 31-beam array was obtained by adding 2 beams to each side of a 19-beam array; hence $M = 5$ (for 19 beams) and $N = 19 + 2 \times 6 = 31$. The beam spacing θ_b is obtained from K in Table 35-1 and the relationship

$$\theta_b = 17.2°/K \qquad (\textbf{12-39})$$

In summary, coverage of an area or FOV by exciting any one of N beams can be achieved by arranging the beams in a triangular grid with angular spacing θ_b between adjacent beams. The minimum directivity D_m over the FOV is maximized when $\theta_b = 3.67\sqrt{\eta/D_0}$; D_m is 4.3 dB less than D_0, the directivity of a single beam. If adjacent beams are excited simultaneously,[18,19] D_m can increase by more than 2 dB.

TABLE 12-1 Estimated Directivity, decibels

M	N	K	D_0	D_{min}
			$\eta = 0.5$	
3	7	2.30	26.0	21.7
3	13	3.06	28.4	24.1
5	19	4.16	31.1	26.8
5	31	5.04	32.8	28.5
9	37	5.76	34.0	29.7
9	43	6.10	34.5	30.2
9	55	7.02	35.7	31.4
11	61	7.36	36.1	31.8
11	73	8.08	36.9	32.6
11	85	9.02	37.8	33.5
11	91	9.22	38.0	33.7
11	97	9.44	38.3	34.0
13	109	10.26	39.0	34.7
13	121,127	11.0	39.6	35.3
15	139	11.36	39.9	35.6

12-7 CROSS-LINK ANTENNAS

Communication links between satellites are becoming more common for a number of reasons. Where initially satellites were used to relay communications between two earth terminals worldwide or even continent to continent, communications require more than a single satellite relay station. If communication between satellites is carried out through an earth station (i.e., a gateway), the round-trip delay, the vulnerability of the earth station, and the inherent insecurity of the earth-satellite link versus that of a satellite cross link motivate and often justify the use of cross links. Consequently, this section is addressed to a description of cross-link-antenna characteristics and some considerations essential to their design.

Although satellite cross links can operate at other frequencies (e.g., UHF), a 1-GHz band at 60 GHz (designated by international agreement) set aside for satellite-to-satellite communications is most attractive. Incidentally, operating a radiating system at frequencies different from those designated by international agreement and national law is acceptable to all provided there is no interference with those who have been allocated a frequency band. Consequently, it is wise to choose an appropriate frequency band within the designated frequency bands and obtain a frequency allocation.

The 60-GHz band has much more bandwidth than lower-frequency bands and can accommodate very-high-data-rate communications. In addition, the earth's atmosphere serves to reduce interference of human origin if not to eliminate it. For example, the attenuation of EM waves at frequencies between 55 and 65 GHz is on the order of 10 dB/km (see Fig. 12-12) at the earth's surface and decreases with increasing altitude above mean sea level. Wave attenuation becomes negligible at altitudes

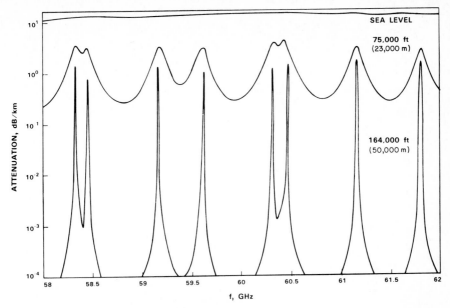

FIG. 12-12 Atmospheric attenuation of EM waves.

higher than 500 km (310 mi). Consequently, the cross links between satellites at altitudes greater than 500 km are shielded from signals or noise generated on or near the earth's surface. Since airplanes fly at altitudes of less than 25 km (82,000 ft), even high-flying platforms can experience substantial atmospheric attenuation unless the spacecraft is near the aircraft's zenith direction.

The natural shield provided by the earth's atmosphere is enhanced by the directivity of cross-link antennas. The short wavelength permits directive ($\sim 1°$-HPBW) beams from a relatively small- (~ 0.5-m-) diameter antenna aperture. Thus, bandwidth, wave attenuation, and antenna directivity considered together with the difficulty with which high-power RF signals can be generated lead to 60 GHz as a preferred cross-link frequency band. Similar reasoning leads to the conclusion[20] that extremely high frequency (EHF) is the preferred operating frequency of military (and perhaps nonmilitary) satellite communication systems.

Having chosen a frequency of operation, the designer must determine the type of antenna and its gain. If it is assumed that the antenna will be steered in the direction of the intended satellite station, it is necessary to develop a trade-off algorithm capable of evaluating an array, a paraboloid, or some other design. For the purposes of this chapter, a conventional paraboloid will be compared with an array of elements, each of which has a beam pattern that defines the desired FOV. The paraboloid is steered by pointing it in the desired direction; the array is steered by appropriate phasing. Furthermore, the array aperture is square with side d, and the paraboloid has a circular aperture with diameter d. The number of elements in the array depends on the FOV. If the elements are assumed to be horn antennas arranged on a square grid, the number of elements versus antenna size is given in Fig. 12-13 for various angular (α) fields of view.

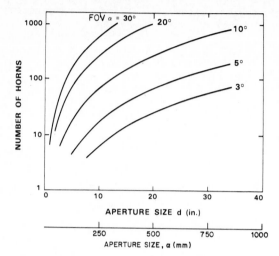

FIG. 12-13   Number of array elements.

By assuming a receiver noise temperature of 1500 K, a 64-Mm (40,000-mi) link between satellites, an energy-per-bit-to-noise power-density ratio $E_b/N_0 = 13$ dB, and an antenna system efficiency of 10 percent for the array and 20 percent for the paraboloid, the antenna and transmitter weight and power were estimated as a function of the aperture size and the data rate. The results shown in Figs. 12-14 and 12-15 indicate that the paraboloid requires about 40 percent more bus power but can be substantially lighter than the array. However, it is definitely clear (Fig. 12-14) that the paraboloid both is lighter and requires much less transmitter power than the array when the data rate exceeds about 10,000 bits per second (b/s). It is further clear that

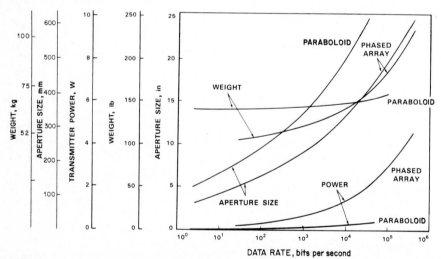

FIG. 12-14   Phased-array and paraboloid characteristics.

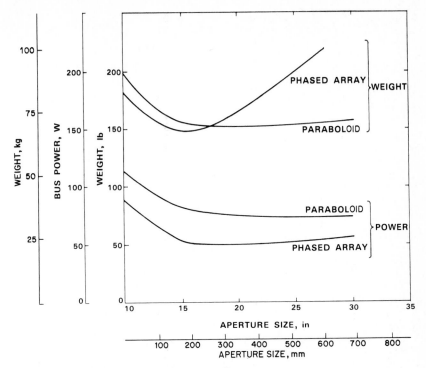

FIG. 12-15 Weight and aperture size of 10^5 b/s system.

if an aperture size of about 0.5 m is acceptable, the paraboloid is probably the best choice for a cross-link antenna.

REFERENCES

1 K. T. Alfriend, "Magnetic Attitude Control Systems for Dual Spin Satellites," *AIAA J.*, vol. 13, no. 6, June 1975, pp. 817–822.
2 W. C. Cummings, "Satellite Crosslinks," Tech. Note 1978-25, MIT Lincoln Laboratory, Aug. 4, 1978.
3 Y. L. Lo, "Inclined Elliptical Orbit and Associated Satellite Field of View," Tech. Note 1978-62, MIT Lincoln Laboratory, Sept. 26, 1979.
4 G. A. Korn and T. M. Korn, *Mathematical Handbook for Scientists and Engineers,* McGraw-Hill Book Company, New York, 1961, par. 3.1-12.
5 J. T. Bangert et al., "The Spacecraft Antenna," *Bell Syst. Tech. J.*, July 1963.
6 J. B. Rankin et al., "Multifunction Single Package Antenna System for Spin-Stabilized Near-Synchronous Satellite," *IEEE Trans. Antennas Propagat.*, vol. AP-17, July 1969, pp. 435–442.
7 M. L. Rosenthal et al., "VHF Antenna Systems for a Spin-Stabilized Satellite," *IEEE Trans. Antennas Propagat.*, vol. AP-17, July 1969, pp. 443–451.
8 W. F. Croswell et al., "An Omnidirectional Microwave Antenna for Use on Spacecraft," *IEEE Trans. Antennas Propagat.*, vol. AP-17, July 1969, pp. 459–466.

9 E. Norsell, "SYNCOM," *Astronaut. Aerosp. Eng.,* September 1963.

10 E. E. Donnelly et al., "The Design of a Mechanically Despun Antenna for Intelsat-III Communications Satellite," *IEEE Trans. Antennas Propagat.,* vol. AP-17, July 1969, pp. 407–414.

11 L. Blaisdell, "ATS Mechanically Despun Communications Satellite Antenna," *IEEE Trans. Antennas Propagat.,* vol. AP-17, July 1969, pp. 415–427.

12 R. E. Collin and F. J. Zucker, *Antenna Theory,* part I, McGraw-Hill Book Company, New York, 1969, pp. 106–114.

13 J. A. Wick, "Sense Reversal of Circularly Polarized Waves on Earth-Space Links," *IEEE Trans. Antennas Propagat.,* vol. AP-15, November 1967, pp. 828–829.

14 I. J. Kantor, D. B. Rai, and F. DeMendonca, "Behavior of Downcoming Radio Waves Including Transverse Magnetic Field," *IEEE Trans. Antennas Propagat.,* vol. AP-19, March 1971, pp. 246–254.

15 A. Kampinsky et al., "ATS-F Spacecraft: A EMC Challenge," 16th EM Compatibility Symp., San Francisco, Calif., July 16–18, 1974.

16 V. J. Jakstys et al., "Composite ATS-F&G Satellite Antenna Feed," *Seventh Inst. Electron. Eng. Ann. Conf. Commun. Proc.,* Montreal, June 1971.

17 R. Harrington, *Time Harmonic Fields,* McGraw-Hill Book Company, New York, 1961, pp. 307–311.

18 L. J. Ricardi et al., "Some Characteristics of a Communications Satellite Multiple-Beam Antenna," Tech. Note 1975-3, DDC AD-A006405, MIT Lincoln Laboratory, Jan. 28, 1975.

19 A. R. Dion, "Optimization of a Communication Satellite Multiple-Beam Antenna," Tech. Note 1975-39, DDC AD-A013104/5, MIT Lincoln Laboratory, May 27, 1975.

20 W. C. Cummings et al., "Fundamental Performance Characteristics That Influence EHF MILSATCOM Systems," *IEEE Trans. Com.,* vol. COM-27, no. 10, October 1979.

Chapter 13

Earth Station Antennas

James H. Cook, Jr.
Scientific-Atlanta, Inc.

13-1 INTRODUCTION AND GENERAL CHARACTERISTICS

An earth-station-antenna system consists of many component parts such as the receiver, low-noise amplifier, and antenna. All the components have an individual role to play, and their importance in the system should not be minimized. The antenna, of course, is one of the more important component parts since it provides the means of transmitting signals to the satellite and/or collecting the signal transmitted by the satellite. Not only must the antenna provide the gain necessary to allow proper transmission and reception, but it must also have radiation characteristics which discriminate against unwanted signals and minimize interference into other satellite or terrestrial systems. The antenna also provides the means of polarization discrimination of unwanted signals. The operational parameters of the individual communication system dictate to the antenna designer the necessary electromagnetic, structural, and environmental specifications for the antenna.

Antenna requirements can be grouped into several major categories, namely, electrical or radio-frequency (RF), control-system, structural, pointing- and tracking-accuracy, and environmental requirements, as well as miscellaneous requirements such as those concerning radiation hazards, primary-power distribution for deicing, etc. Only the electrical or RF requirements will be dealt with in this chapter.

The primary electrical specifications of an earth station antenna are gain, noise temperature, voltage standing-wave ratio (VSWR), power rating, receive-transmit group delay, radiation pattern, polarization, axial ratio, isolation, and G/T (antenna gain divided by the system noise temperature). All the parameters except the radiation pattern are determined by the system requirements. The radiation pattern should meet the minimum requirements set by the International Radio Consultative Committee (CCIR) of the International Telecommunications Union (ITU) and/or national regulatory agencies such as the U.S. Federal Communications Commission (FCC).

Earth station antennas operating in international satellite communications must have sidelobe performance as specified by INTELSAT standards or by CCIR Recommendation 483 and Report 391-2 (see Fig. 13-1).

The CCIR standard specifies the pattern envelope in terms of allowing 10 percent of the sidelobes to exceed the reference envelope and also permits the envelope to be adjusted for antennas whose aperture is less than 100 wavelengths (100λ). The reference envelope is given by

$$G = [52 - 10 \log (D/\lambda) - 25 \log \theta] \quad \text{dBi} \qquad D \leq 100\lambda$$

$$= (32 - 25 \log \theta) \quad \text{dBi} \qquad D > 100\lambda$$

This envelope takes into consideration the limitations of small-antenna design and is representative of measured patterns of well-designed dual-reflector antennas.

Earth station antennas can be grouped into two broad categories: single-beam antennas and multiple-beam antennas. A single-beam earth station antenna is defined as an antenna which generates a single beam that is pointed toward a satellite by means of a positioning system. A multiple-beam earth station antenna is defined as an antenna which generates multiple beams by employing a common reflector aperture with multiple feeds illuminating that aperture. The axes of the beams are determined by the location of the feeds. The individual beam identified with a feed is pointed toward a satellite by positioning the feed without moving the reflector.

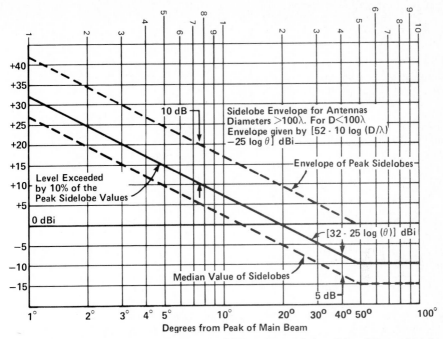

FIG. 13-1   Sidelobe envelope as defined by CCIR recommendation 433 and Report 391-2.

13-2   SINGLE-BEAM EARTH STATION ANTENNAS

Single-beam antenna types used as earth stations are paraboloidal reflectors with focal-point feeds (prime-focus antenna), dual-reflector antennas such as the Cassegrain and gregorian configurations, horn reflector antennas, offset-fed paraboloidal antennas, and offset-fed multiple-reflector antennas. Each of these antenna types has its own unique characteristics, and the advantages and disadvantages have to be considered when choosing one for a particular application.

Axisymmetric Dual-Reflector Antennas

The predominant choice of designers of earth station antennas has been the dual-reflector Cassegrain antenna. Cassegrain antennas can be subdivided into three primary types:

1   The classical Cassegrain geometry[1,2] employs a paraboloidal contour for the main reflector and a hyperboloidal contour for the subreflector (Fig. 13-2). The paraboloidal reflector is a point-focus device with a diameter D_p and a focal length f_p. The hyperboloidal subreflector has two foci. For proper operation, one of the two foci is the real focal point of the system and is located coincident with the phase center of the feed; the other focus, the virtual focal point, is located coincident with the focal

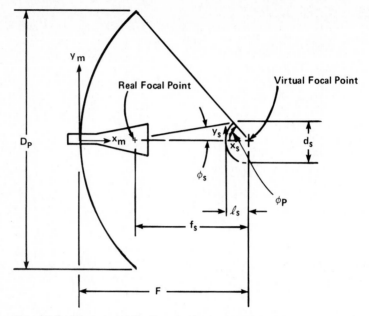

FIG. 13-2   Geometry of the Cassegrain antenna system.

point of the main reflector. The parameters of the Cassegrain system are related as follows:

$$\phi_p = 2 \tan^{-1}(0.25 D_p/F_p) \tag{13-1}$$

$$f_s/d_s = 0.5(\cot \phi_p + \cot \phi_s) \tag{13-2}$$

$$\ell_s/f_s = 0.5(1 - \{\sin [0.5(\phi_p - \phi_s)]/\sin [0.5(\phi_p + \phi_s)]\}) \tag{13-3}$$

In a typical design, the parameters f_p, D_p, f_s, and ϕ_s are chosen, and the remaining three parameters are then calculated.

The contours of the main reflector and subreflector are given by

$$\text{Main reflector: } y_m^2 = 4F_p x_m \tag{13-4}$$

$$\text{Subreflector: } (y_s/b)^2 + 1 = (x_s/a + 1)^2 \tag{13-5}$$

where
$$a = (f_s/2e) \qquad b = a\sqrt{e^2 - 1}$$
$$e = \sin [0.5(\phi_p + \phi_s)]/\sin [0.5(\phi_p - \phi_s)]$$

The quantities a, b, and e are half of the transverse axis, half of the conjugate axis, and the eccentricity parameters of the hyperboloidal subreflector respectively.

2   The geometry of this type consists of a paraboloidal main reflector and a special-shaped quasi-hyperboloidal subreflector.[3,4,5,6,7] The geometry in Fig. 13-2 is appropriate for describing this antenna. The main difference between this design and the classical Cassegrain above is that the subreflector is shaped so that the overall efficiency of the antenna has been enhanced. This technique is especially useful with antenna diameters of approximately 60 to 300 wavelengths. The subreflector shape may be solved for by geometrical optics (GO) or by diffraction optimization, and then, by comparing the required main-reflector surface to a paraboloidal surface, the best-fit paraboloidal surface is found. Aperture efficiencies of 75 to 80 percent can be realized by a GO design and efficiencies of 80 to 95 percent by diffraction optimization of the subreflector (Fig. 13-3).

3 This type is a generalization of the Cassegrain geometry consisting of a special-shaped quasi-paraboloidal main reflector and a shaped quasi-hyperboloidal subreflector.[9,10,11,12] Green[8] observed that in dual-reflector systems with high magnification—essentially a large ratio of main-reflector diameter to subreflector diameter—the distribution of energy (as a function of angle) is largely controlled by the subreflector curvature. The path length or phase front is dominated by the main reflector (see Fig. 13-4). Kinber[9] and Galindo[10,11] found a method for simultaneously solving for the main-reflector and subreflector shapes to obtain an exact solution for both the phase and the amplitude distributions in the aperture of the main reflector of an axisymmetric dual-reflector antenna. Their technique, based on geometrical optics, is highly mathematical and involves solving two simultaneous, nonlinear, first-order, ordinary differential equations. Figure 13-5 gives the geometry showing the path of a single ray. The feed phase center is located as shown, and the feed is assumed to have a power radiation pattern $I(\theta_1)$. The parameters α and β represent respectively the distance of the feed phase center from the aperture plane and the distance between

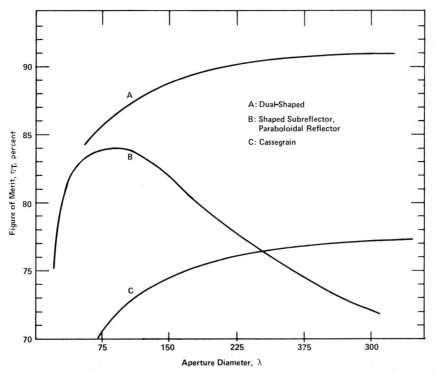

FIG. 13-3 Antenna of figure of merit versus aperture diameter.

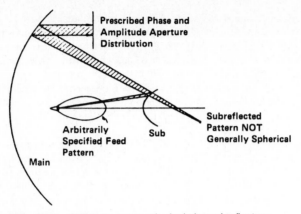

FIG. 13-4 Circularly symmetric dual-shaped reflectors.

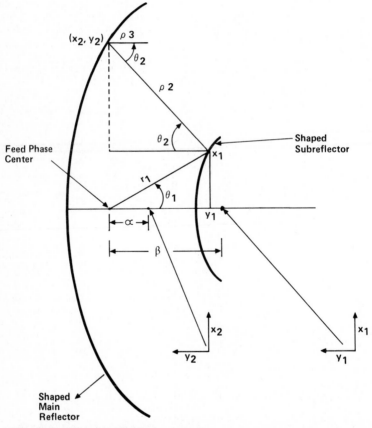

FIG. 13-5 Dual-shaped reflector geometry.

the feed phase center and the back surface of the subreflector. The constraints to the dual-reflector system are as follows:

a The phase distribution across the main-reflector aperture plane will be uniform, or

$$r_1 + p_2 + p_3 + C_p(\theta_1) = \text{constant} \qquad (13\text{-}6)$$

for $0 \leq \theta_1 \leq \theta_{1\text{max}}$. $C_p(\theta_1)$ represents the phase distribution across the primary-feed radiation pattern in units of length.

b The feed energy, or ray bundles intercepted and reflected by the subreflector, is conserved and redistributed according to a specified aperture distribution, or

$$I(\theta_1) \sin(\theta_1)\ d\theta_1 = C \cdot I(X_2)X_2\ dX_2 \qquad (13\text{-}7)$$

where $I(X_2)$ represents the power radiation distribution across the main-reflector aperture and C represents a constant which is determined by applying the conservation-of-power principle.

$$\int_0^{\theta_{1\text{max}}} I(\theta_1) \sin \theta_1\ d\theta_1 = C \int_{X_{2\text{min}}}^{X_{2\text{max}}} I(X_2)X_2\ dX_2 \qquad (13\text{-}8)$$

The lower limit of integration over the main reflector can be arbitrarily chosen so that only an annular region of the main reflector is illuminated.

c Snell's law must be satisfied at the two reflecting surfaces, which yields

$$\frac{dY_1}{dX_1} = \tan\left(\frac{\theta_1 - \theta_2}{2}\right) \qquad (13\text{-}9)$$

$$\frac{dY_2}{dX_2} = -\tan\left(\frac{\theta_2}{2}\right) \qquad (13\text{-}10)$$

Solving Eqs. (36-7) through (36-10) simultaneously results in a nonlinear, first-order differential equation of the form

$$\frac{dY_1}{dX_2} = f(\theta_1, \theta_2, \alpha, \beta, \text{etc.}) \qquad (13\text{-}11)$$

which leads to the cross sections of each reflector when subject to the boundary condition $Y_1 (X_2 = X_{2\text{max}}) = 0$, where X_2 is the independent variable. Equation (13-11) can be solved numerically by using an algorithm such as a Runge-Kutta, order 4.

The above procedure is based on GO, but it is evident that the assumptions of GO are far from adequate when reflectors are small in terms of wavelengths. An improvement in the design approach is to include the effects of diffraction. Clarricoats and Poulton[7] reported a gain increase of 0.5 dB for a diffraction-optimized design over the GO design with a 400λ-diameter main reflector and 40λ-diameter subreflector.

Prime-Focus-Fed Paraboloidal Antenna

The prime-focus-fed paraboloidal reflector antenna is also often employed as an earth station antenna. For moderate to large aperture sizes, this type of antenna has excellent sidelobe performance in all angular regions except the spillover region around the edge of the reflector. The CCIR sidelobe specification can be met with this type of antenna. The reader is referred to Ref. 27 for more information on paraboloidal antennas.

Offset-Fed Reflector Antennas

The geometry of offset-fed reflector antennas has been known for many years, but its use has been limited to the last decade or so because of its difficulty in analysis. Since the advent of large computers has allowed the antenna engineer to investigate theoretically the offset-fed reflector's performance, this type of antenna will become more common as an earth station antenna.

The offset-fed reflector antenna can employ a single reflector or multiple reflectors, with two-reflector types the more prevalent of the multiple-reflector designs. The offset front-fed reflector, consisting of a section of a paraboloidal surface (Fig. 13-6), minimizes diffraction scattering by eliminating the aperture blockage of the feed and feed-support structure. Sidelobe levels of $(29 - 25 \log \theta)$ dBi can be expected from this type of antenna (where θ is the far-field angle in degrees) with aperture efficiencies of 65 to 80 percent. The increase in aperture efficiency as compared with an axisymmetric prime-focus-fed antenna is due to the elimination of direct blockage. For a detailed discussion of this antenna, see C. A. Mentzer.[13]

Offset-fed dual-reflector antennas exhibit sidelobe performance similar to that of the front-fed offset reflector. Two offset-fed dual-reflector geometries are used for

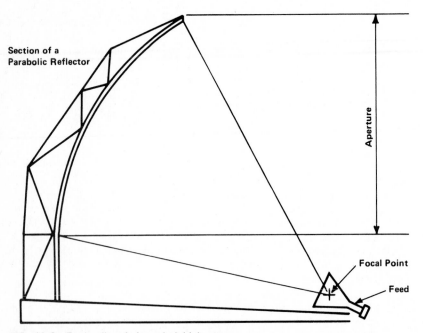

FIG. 13-6   Basic offset-fed paraboloidal antenna.

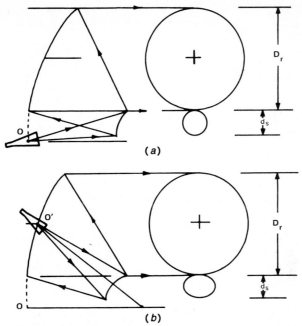

FIG. 13-7 Offset dual-reflector geometries. (*a*) Double-offset geometry (feed phase center and paraboloidal vertex at O). (*b*) Open cassegrainian geometry (feed phase center located at O; paraboloidal vertex, at O).

earth station antennas: the double-offset geometry shown in Fig. 13-7*a* and the open Cassegrain geometry introduced by Cook et al.[14] of Bell Laboratories and shown in Fig. 13-7*b*. In the double-offset geometry, the feed is located below the main reflector, and no blocking of the optical path occurs. In contrast, the open Cassegrain geometry is such that the primary feed protrudes through the main reflector; thus it is not completely blockage-free. Nevertheless, both of these geometries have the capability of excellent sidelobe and efficiency performance.

The disadvantage of offset-geometry antennas is that they are asymmetric. This leads to increased manufacturing cost and also has some effect on electrical performance. The offset-geometry antenna, when used for linear polarization, has a depolarizing effect on the primary-feed radiation and produces two cross-polarized lobes within the main beam in the plane of symmetry. When it is used for circular polarization, a small amount of beam squint whose direction is dependent upon the sense of polarization is introduced. The beam squint is approximately given by[15] $\psi_s =$ arc sin $[\lambda \sin (\theta_0)/4\pi F]$, where θ_0 is the offset angle, λ is the free-space wavelength, and F is the focal length.

During the past few years, considerable analysis has been performed by Galindo-Israel, Mittra, and Cha[16] for this geometry for high-aperture-efficiency applications. Their analytical techniques are reported to result in efficiencies in the 80 to 90 percent range. Mathematically, the offset geometry, when formulated by the method used in designing axisymmetric dual reflectors, results in a set of partial differential equations for which there is no exact GO solution. The method of Galindo-Israel et al. is to solve the resultant partial differential equations as if they were total differential equations. Then, by using the resultant subreflector surface, the main-reflector surface is perturbed until a constant aperture phase is achieved.

13-3   MULTIPLE-BEAM EARTH STATION ANTENNAS

During the past few years there has been an increasing interest in receiving signals simultaneously from several satellites with a single antenna. This interest has prompted the development of several multibeam-antenna configurations which employ fixed reflectors and multiple feeds. The antenna engineering community, of course, has been investigating multibeam antennas for many years. In fact, in the middle of the seventeenth century, Christian Huygens and Sir Isaac Newton first studied the spherical mirror, and the first use of a spherical reflector as a microwave antenna occurred during World War II. More recently, the spherical-reflector, the torus-reflector, and the dual-reflector geometries, all using multiple feeds, have been offered as antennas with simultaneous multibeam capability. Chu[17] in 1969 addressed the multiple-beam spherical-reflector antenna for satellite communication, Hyde[18] introduced the multiple-beam torus antenna in 1970, and Ohm[19] presented a novel multiple-beam Cassegrain-geometry antenna in 1974. All three of these approaches, as well as variations of scan techniques for the spherical reflector, are discussed below.

Spherical Reflector

The properties, practical applications, and aberrations of the spherical reflector are not new to microwave-antenna designers. The popularity of this reflector is primarily due to the large angle through which the radiated beam can be scanned by translation and orientation of the primary feed. This wide-angle property results from the symmetry of the surface. Multiple-beam operation is realized by placing multiple feeds along the focal surface. In the conventional use of the reflector surface, the minimum angular separation between adjacent beams is determined by the feed-aperture size. The maximum number of beams is determined by the percentage of the total sphere covered by the reflector. In the alternative configuration described below, these are basically determined by the f/D ratio and by the allowable degradation in the radiation pattern.

In the conventional use of the spherical reflector, the individual feed illuminates a portion of the reflector surface so that a beam is formed coincident to the axis of the feed. The conventional multibeam geometry is shown in Fig. 13-8. All the beams have similar radiation patterns and gains, although there is degradation in performance in comparison with the performance of a paraboloid. The advantage of this antenna is that the reflector area illuminated by the individual feeds overlaps, reducing the surface area for a given number of beams in comparison with individual single-beam antennas.

The alternative multibeam-spherical-reflector geometry is shown in Fig. 13-9. For this geometry, each of the feed elements points toward the center of the reflector,

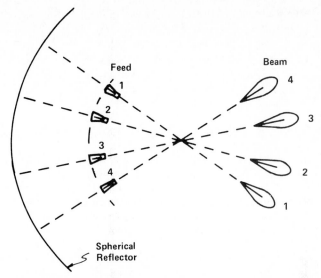

FIG. 13-8 Conventional spherical multibeam antenna using extended reflector and multiple feeds.

with the beam steering accomplished by the feed position. This method of beam generation leads to considerable increase in aberration, including coma; therefore, the radiation patterns of the off-axis beams are degraded with respect to the on-axis beam. This approach does not take advantage of the spherical-reflector properties that exist in the conventional approach. In fact, somewhat similar results could be achieved with a paraboloidal reflector with a large f/D.

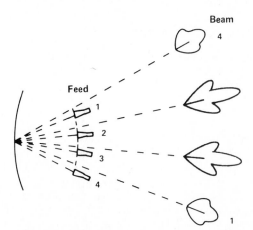

FIG. 13-9 Alternative spherical multibeam antenna using minimum reflector aperture with scanned beam feeds.

Torus Antenna

The torus antenna is a dual-curvature reflector, capable of multibeam operation when it is fed with multiple feeds similar to those of the conventional spherical-reflector geometry. The feed-scan plane can be inclined to be in the orbital-arc plane, allowing the use of a fixed reflector to view geosynchronous satellites. The reflector has a circular contour in the scan plane and a parabolic contour in the orthogonal plane (see Fig. 13-10). It can be fed in either an axisymmetric or an offset-fed configuration. Offset geometry for use as an earth station antenna has been successfully demonstrated by COMSAT Laboratories.[20] The radiation patterns meet a (29-25 log θ)-dBi envelope.

The offset-fed geometry results in an unblocked aperture, which gives rise to low wide-angle sidelobes as well as providing convenient access to the multiple feeds.

The torus antenna has less phase aberration than the spherical antenna because of the focusing in the parabolic plane. Because of the circular symmetry, feeds placed anywhere on the feed arc form identical beams. Therefore, no performance degradation is incurred when multiple beams are placed on the focal arc. Point-focus feeds may be used to feed the torus up to aperture diameters of approximately 200 wavelengths. For larger apertures, it is recommended that aberration-correcting feeds be used.

The scanning or multibeam operation of a torus requires an oversized aperture to accommodate the scanning. For example, a reflector surface area of approximately 214 m^2 will allow a field of view (i.e., orbital arc) of 30° with a gain of approximately 50.5 dB at 4 GHz (equivalent to the gain of a 9.65-m reflector antenna). This surface area is equivalent to approximately three 9.65-m antennas.

Offset-Fed Multibeam Cassegrain Antenna

The offset-fed multibeam Cassegrain antenna is composed of a paraboloidal main reflector, a hyperbolic oversized subreflector, and multiple feeds located along the scan

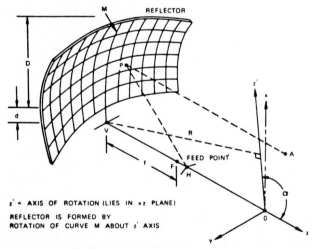

FIG. 13-10 Torus-antenna geometry. (*Copyright 1974*, COMSAT Technical Review. *Reprinted by permission.*)

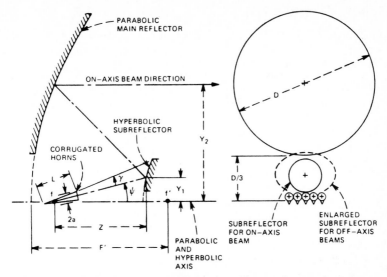

FIG. 13-11 Geometry of the offset-fed multibeam Cassegrain antenna. *(Copyright 1974, American Telephone and Telegraph Company. Reprinted by permission.)*

axis, as shown in Fig. 13-11. The offset geometry essentially eliminates beam blockage, thus allowing a significant reduction in sidelobes and antenna noise temperature. The Cassegrain feed system is compact and has a large focal-length-to-diameter ratio (f/D), which reduces aberrations to an acceptable level even when the beam is moderately far off axis. The low-sidelobe performance is achieved by using a corrugated feed horn which produces a gaussian beam.

A typical antenna design consisting of a 10-m projected aperture would yield half-power beamwidths (HPBWs) and gain commensurate with an axisymmetric 10-m antenna, 0.5° HPBW, and 51-dB gain at 4 GHz. The subreflector would need to be approximately a 3- by 4.5-m elliptical aperture. The feed apertures would be approximately 0.5 m in diameter. The minimum beam separation would be less than 2°, which is more than sufficient to allow use with synchronous satellites with orbit spacings of 2° or greater. For the ±15° scan, the gain degradation would be approximately 1 dB, and the first sidelobe would be approximately 20 to 25 dB below the main-beam peak.

13-4 ANGLE TRACKING

Automatic tracking of satellite position is required for many earth station antennas. Monopulse, conical-scan, and sequential-lobing techniques described in Chap. 11 are applicable for this purpose. Two other types of tracking techniques are also commonly employed: single-channel monopulse (pseudo monopulse) and steptrack. The single-channel monopulse is a continuous angle-tracking scheme, and steptrack is a time-sequencing, signal-peaking technique.

Single-Channel Monopulse

This technique utilizes multiple elements or modes to generate a reference signal, an elevation error signal, and an azimuth error signal. The two error signals are then combined in a time-shared manner by a switching network which selects one of two phase conditions (0° and 180°) for the error signal. The error signal is then combined with the reference signal, with the resultant signal then containing angle information. This allows the use of a single receiver for the tracking channel. This technique is equivalent to sequential lobing in which the lobing is done electrically and can be adjusted to any desired fixed or variable scan rate.

Steptrack

Steptracking is a technique employed primarily for maintaining the pointing of an earth station's antenna beam toward a geosynchronous satellite. Since most geosynchronous orbiting satellites have some small angular box of station keeping, a simple peaking technique can be used to keep the earth station beam correctly pointed. The signal-peaking, or steptrack, routine is a software technique that maneuvers the antenna toward the signal peak by following a path along the steepest signal-strength gradient or by using some other algorithm that accomplishes the same result.[21,22] Any method of direct search for maximization of signal is applicable; one such method calculates the local gradient by determining the signal strength at three points (A, B, C; see Fig. 13-12). From the signal strength at points A, B, and C, an angle α_2, representing the unit's gradient vector angle relative to the azimuth axis, is computed. Once C_1 has been determined, A_2 is set equal to C_1 and another angle θ_2 is defined by $\theta_2 = \theta_1 + \alpha_2$ ($\theta_1 = 0$). The angle is used to relate pedestal movements to a fixed-pedestal coordinate system. The procedure is repeated a number of times until the antenna boresight crosses throughout the peak. At this time the step size is reduced

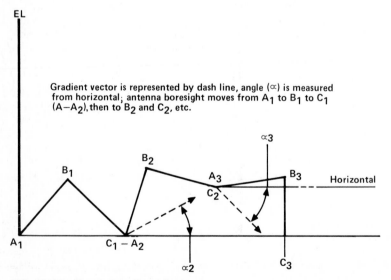

FIG. 13-12 Signal-strength gradients.

by one-half, and for each time that the peak is crossed it is again reduced by one-half until the peaking resolution is accomplished.

13-5 POLARIZATION

Many satellite communication systems are dual-polarized (frequency reuse) and are therefore susceptible to interference from the intended cross-polarized signals when the medium of propagation is such that the ellipticity of the signals is changed. This condition can occur during atmospheric conditions such as rain and Faraday rotation in the ionosphere. Therefore, in order to maintain sufficient polarization discrimination, special precautions must be taken in regard to the polarization purity of dual-polarization antennas; indeed, adaptive polarization-correcting circuits may be necessary.[23,24]

Several types of polarization-discrimination-enhancement schemes may be used. They include:

1 Simple rotation of the polarization major-ellipse axis to correct for rotation due to the ionosphere (applicable for linear polarizations). Transmit and receive rotations are in opposite directions.

2 Suboptimal correction of depolarizations; ellipticity correction with respect to one of the signals but not orthogonalizing the two signals (differential phase).

3 Complete adaptive correction including orthogonalization (differential phase and amplitude).

The polarization-enhancement schemes may be implemented at the RF frequencies, or they may follow a down-conversion stage and be implemented at intermediate frequencies (IF). In either case the circuitry must operate over the full bandwidth of the communication channel. Kreutel et al.[24] treat the implementation of RF frequencies in detail. The reader is also referred to Gianatasio[25,26] for both RF and IF circuitry and experiments to verify performance of the circuitry.

REFERENCES

1 P. W. Hannah, "Microwave Antennas Derived from the Cassegrain Telescope," *IRE Trans. Antennas Propagat.*, vol. AP-9, March 1961, pp. 140–153.

2 P. A. Jensen, "Designing Cassegrain Antennas," *Microwave J.*, December 1962, pp. 10–16.

3 G. W. Collins, "Shaping of Subreflectors in Cassegrainian Antennas for Maximum Aperture Efficiency," *IEEE Trans. Antennas Propagat.*, vol. AP-21, no. 3, May 1973, pp. 309–313.

4 S. Parekh, "On the Solution of Best Fit Paraboloid as Applied to Shaped Dual Reflector Antennas," *IEEE Trans. Antennas Propagat.*, vol. AP-28, no. 4, July 1980, pp. 560–562.

5 J. W. Bruning, "A 'Best Fit Paraboloid' Solution to the Shaped Dual Reflector Antenna," Symp. USAF Antenna R&D, University of Illinois, Urbana, Nov. 15, 1967.

6 P. J. Wood, "Reflector Profiles for the Pencil-Beam Cassegrain Antenna," *Marconi Rev.*, vol. 35, no. 185, 1972, pp. 121–138.

7   P. J. B. Clarricoats and G. T. Poulton, "High-Efficiency Microwave Reflector Antennas— A Review," *IEEE Proc.,* vol. 65, no. 10, October 1977, pp. 1470–1504.

8   K. A. Green, "Modified Cassegrain Antenna for Arbitrary Aperture Illumination," *IEEE Trans. Antennas Propagat.,* vol. AP-11, no. 5, September 1963.

9   B. Y. Kinber, "On Two Reflector Antennas," *Radio Eng. Electron. Phys.,* vol. 6, June 1962.

10 V. Galindo, "Design of Dual Reflector Antenna with Arbitrary Phase and Amplitude Distributions," PTGAP Int. Symp., Boulder, Colo., July 1963.

11 V. Galindo, "Synthesis of Dual Reflector Antennas," Elec. Res. Lab. Rep. 64-22, University of California, Berkeley, July 30, 1964.

12 W. F. Williams, "High Efficiency Antenna Reflector," *Microwave J.,* July 1965, pp. 79–82.

13 C. A. Mentzer, "Analysis and Design of High Beam Efficiency Aperture Antennas," doctoral thesis 74-24, 370, Ohio State University, Columbus, 1974.

14 J. S. Cook, E. M. Elam, and H. Zucker, "The Open Cassegrain Antenna: Part I. Electromagnetic Design and Analysis," *Bell Syst. Tech. J.,* September 1965, pp. 1255–1300.

15 A. W. Rudge and N. A. Adatia, "Offset-Parabolic-Reflector Antennas: A Review," *IEEE Proc.,* vol. 66, no. 12, December 1978, pp. 1592–1611.

16 V. Galindo-Israel, R. Mittra, and A. Cha, "Aperture Amplitude and Phase Control of Offset Dual Reflectors," USNC/URSI & IEEE Antennas Propagat. Int. Symp., May 1978.

17 T. S. Chu, "A Multibeam Spherical Reflector Antenna," *IEEE Antennas Propagat. Int. Symp.: Program & Dig.,* Dec. 9, 1969, pp. 94–101.

18 G. Hyde, "A Novel Multiple-Beam Earth Terminal Antenna for Satellite Communication," *Int. Conf. Comm. Rec.,* Conf. Proc. Pap. 70-CP-386-COM, June 1970, pp. 38-24–38-33.

19 E. A. Ohm, "A Proposed Multiple-Beam Microwave Antenna for Earth Stations and Satellites," *Bell Syst. Tech. J.* vol. 53, October 1974, pp. 1657–1665.

20 G. Hyde, R. W. Kreutei, and L. V. Smith, "The Unattended Earth Terminal Multiple-Beam Torus Antenna," *COMSAT Tech. Rev.,* vol. 4, no. 2, fall, 1974, pp. 231–264.

21 M. J. D. Powell, "An Efficient Method for Finding the Minimum of a Function of Several Variables without Calculating Derivatives," *Comput. J.,* vol. 7, 1964, p. 155.

22 J. Kowalik and M. R. Osborne, *Methods for Unconstrained Optimization Problems,* American Elsevier Publishing Company, Inc., New York, 1968.

23 T. S. Chu, "Restoring the Orthogonality of Two Polarizations in Radio Communication Systems, I and II," *Bell Syst. Tech. J.,* vol. 50, no. 9, November 1971, pp. 3063–3069; vol. 52, no. 3, March 1973, pp. 319–327.

24 R. W. Kreutel, D. F. DiFonzo, W. J. English, and R. W. Gruner, "Antenna Technology for Frequency Reuse Satellite Communications," *IEEE Proc.,* vol. 65, no. 3, March 1977, pp. 370–378.

25 A. J. Gianatasio, "Broadband Adaptively-Controlled Polarization-Separation Network," *Seventh Europ. Microwave Conf. Proc.,* Copenhagen, September 1977, pp. 317–321.

26 A. J. Gianatasio, "Adaptive Polarization Separation Experiments," final rep., NASA CR-145076, November 1976.

27 K. S. Kelleher and G. Hyde, "Reflector Antennas," in R. C. Johnson and H. Jasik (eds.), *Antenna Engineering Handbook 2/e,* McGraw-Hill Book Company, New York, 1984, Chap. 17.

Chapter 14

Aircraft Antennas

William P. Allen, Jr.
Lockheed-Georgia Company

Charles E. Ryan, Jr.
Georgia Institute of Technology

14-1 INTRODUCTION

The design of antennas for aircraft differs from other applications. Aircraft antennas must be designed to withstand severe static and dynamic stresses, and the size and shape of the airframe play a major role in determining the electrical characteristics of an antenna. For this latter reason, the type of antenna used in a given application will often depend on the size of the airframe relative to the wavelength. In the case of propeller-driven aircraft and helicopters, the motion of the blades may give rise to modulation of the radiated signal. Triboelectric charging of the airframe surfaces by dust or precipitation particles, known as *p* static, gives rise to corona discharges, which may produce extreme electrical noise, especially when the location of the antenna is such that there is strong electromagnetic coupling to the discharge point. In transmitting applications, corona discharge from the antenna element may limit the power-handling capacity of the antenna. Also, special consideration must frequently be given to the protection of aircraft antennas from damage due to lightning strikes.

14-2 LOW-FREQUENCY ANTENNAS

The wavelengths of frequencies below about 2 MHz are considerably larger than the maximum dimensions of most aircraft. Because of the inherently low radiation efficiency of electrically small antennas and the correspondingly high radio-frequency (RF) voltages required to radiate significant amounts of power, nearly all aircraft radio systems operating at these lower frequencies are designed so that only receiving equipment is required in the aircraft.

Radiation patterns of aircraft antennas in this frequency range are simple electric or magnetic dipole patterns, depending upon whether the antenna element is a monopole or a loop. In considering first electric dipole antennas, it can be shown with reference to Fig. 14-1 that while the pattern produced by a small monopole antenna will always be that of a simple dipole regardless of location, the orientation of the equivalent-dipole axis with respect to the vertical will depend upon location. The figure shows the electric field fringing produced by an airframe for incident fields polarized in the three principal directions: vertical, longitudinal, and transverse. A small antenna element placed on the airframe would respond to all three of these field components, indicating that the dipole moment of the antenna-airframe combination has projections in all three directions.

The sensitivity of low-frequency (LF) antennas is customarily expressed in either of two ways, depending upon whether the antenna is located on a relatively flat portion of the airframe such as the top or bottom of the fuselage or at a sharp extremity such as the tip of the vertical stabilizer. In the first case it can be assumed (at least when the antenna element is small relative to the surface radii of curvature) that the antenna performs as it would on a flat ground plane, except that the incident-field intensity which excites it is greater than the free-space incident-field intensity because of the field fringing produced by the airframe (Fig. 14-1). The effect of the airframe on antenna sensitivity is hence expressed by the ratio of the local-field intensity on the airframe surface to the free-space incident-field intensity. For a vertically polarized incident field (which is the case of primary importance in LF receiving-antenna

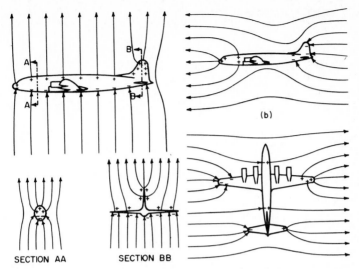

FIG. 14-1 Low-frequency field fringing due to the airframe.

design) this ratio is designated as F_v and is called the *curvature factor for vertical polarization*. Survey data giving the equivalent-dipole tilt angle and the factor F_v for top and bottom centerline locations on a typical airframe are shown in Fig. 14-2.

Effective height and capacitance data for antennas installed in locations for which this method is applicable are usually obtained by measurements or calculations for the antenna on a flat ground plane. The effective height (for vertically polarized signals) of the antenna installed on the airframe is then estimated by multiplying the flat-ground-plane effective height by the factor F_v appropriate to the installation, while the capacitance may be assumed to be the same as that determined with the antenna on a flat ground plane. The presence of a fixed-wire antenna on the aircraft may have a significant effect on F_v. Because of the shielding effect due to a wire, LF antennas are seldom located on the top of the fuselage in aircraft which carry fixed-wire antennas.

Flat-ground-plane data for a T antenna are shown in Fig. 14-3. The capacitance curves shown apply to an antenna made with standard polyethylene-coated wire [0.052-in- (1.32-mm-) diameter conductor and 0.178-in- (4.52-mm-) diameter polyethylene sheath]. Two antennas of this type are frequently located in close proximity on aircraft having dual automatic-direction-finder (ADF) installations. The effects of a grounded T antenna on h_e and C_a of a similar nearby antenna are shown in Fig. 14-4. To determine the significance of capacitance interaction it is necessary to consider the Q's of the input circuits to which the antennas are connected and the proximity of the receiver frequencies.

Sensitivity data for two flush antennas and a low-silhouette antenna consisting of a relatively large top-loading element and a short downlead are shown in Figs. 14-5, 14-6, and 14-7 respectively. The antenna dimensions shown in these figures are not indicative of antenna sizes actually in use but rather are the sizes of the models used in measuring the data. Both h_e and C_a scale linearly with the antenna dimensions.

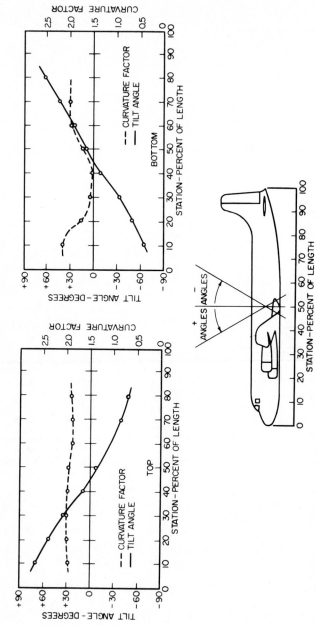

FIG. 14-2 Low-frequency-antenna survey data for a DC-6 aircraft.

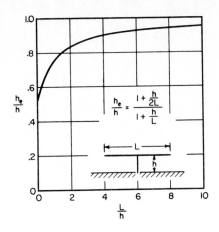

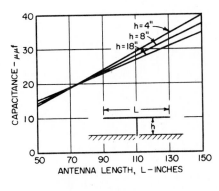

FIG. 14-3 Effective height and capacitance of a T antenna.

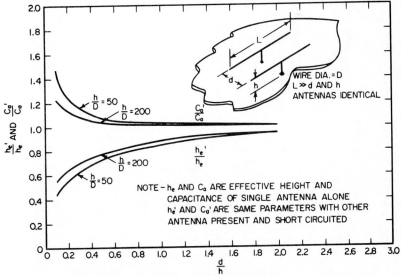

FIG. 14-4 Shielding effect of one T antenna on another.

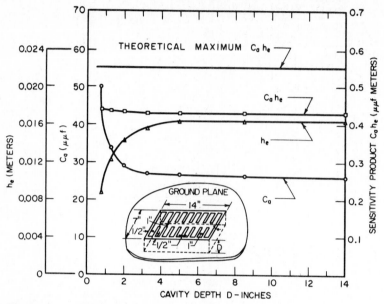

FIG. 14-5 Design data for a flush low-frequency antenna.

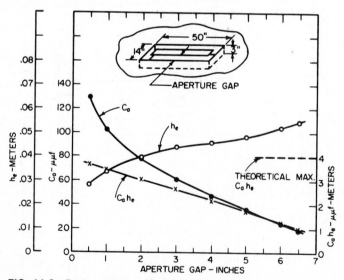

FIG. 14-6 Design data for a flush low-frequency antenna.

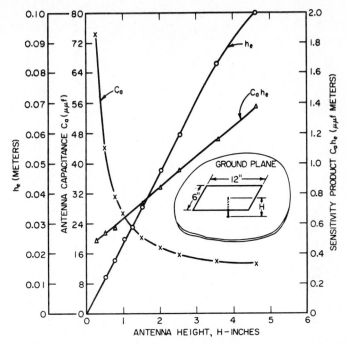

FIG. 14-7 Design data for a heavily top-loaded low-frequency antenna with flat-plate top loading.

A rule-of-thumb design limit for flush antennas may be derived on the basis of a quasi-static analysis of LF antenna performance. If any electric dipole antenna is short-circuited at its feed terminals and placed in a uniform electrostatic field with its equivalent-dipole axis aligned parallel to the field, charges of $+q$ and $-q$ will be induced on the two elements of the antenna. It may be shown that the product of the LF parameters h_e and C_a is related to the induced charge by the equation

$$h_e C_a = q/E \qquad\qquad (14\text{-}1)$$

where h_e and C_a are expressed in meters and picofarads respectively, q is expressed in picocoulombs, and the incident-field intensity E is expressed in volts per meter. The quantity q/E is readily calculated for a flush cavity-backed antenna of the type shown in Fig. 14-6, at least for the case in which the antenna element virtually fills the cutout in the ground plane. In this case, the shorted antenna element will cause practically no distortion of the normally incident field, and the number of field lines terminating on the element will be the same as the number which would terminate on this area if the ground plane were continuous. The value of q/E is hence equal to $\varepsilon_0 a$, where $\varepsilon_0 = 8.85 \times 10^{-12}$ F/m and a is the area of the antenna aperture in square meters. For a flush antenna with an element which nearly fills the antenna aperture we have

$$h_e C_a = \varepsilon_0 a \qquad\qquad (14\text{-}2)$$

The line labeled "theoretical maximum h_eC_a" in the curves of Figs. 14-5 and 14-6 was calculated from Eq. (14-2). With practical antenna designs, there will be regions of the aperture not covered by the antenna element, so that some of the incident-field lines will penetrate the aperture and terminate inside the cavity. As a result, the induced charge q will be smaller than that calculated above, and the product h_eC_a will be smaller than the value estimated from Eq. (14-2).

The helicopter presents a special antenna-design problem in the LF range as well as in other frequency ranges because of the shielding and modulation caused by the rotor blades. Measured curves of F_v obtained with an idealized helicopter model to demonstrate the effects of different antenna locations along the top of the simulated fuselage are shown in Fig. 14-8.

It is characteristic of rotor modulation of LF signals that relatively few modulation components of significant amplitude are produced above the fundamental rotor

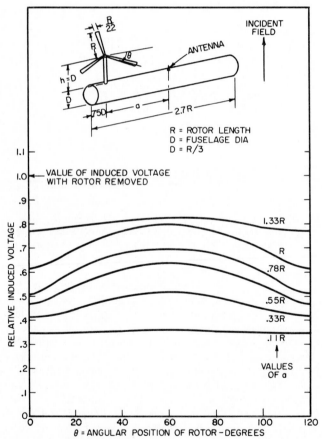

FIG. 14-8   Rotor modulation for a low-frequency receiving antenna on an idealized helicopter model.

modulation frequency, which is equal to the number of blade passages over the antenna per second. This fundamental frequency is of the order of 15 Hz for typical three-blade single-rotor helicopters. It should be noted that new sidebands will be generated about each signal sideband and about the carrier. It is hence not a simple matter to predict the effects of rotor modulation on a particular system unless tests have been made to determine the performance degradation of the airborne receiver due to these extra modulation components.

In the case of loop antennas, it is necessary to consider the distortion caused by the airframe in the magnetic field component of the incident wave. Unlike the electric field lines, the magnetic field lines distort in such a way that they avoid entering the conducting airframe. The field-line sketches in Fig. 14-9 illustrate the airframe effect for two cases in which the ground station is respectively to the side of the aircraft and

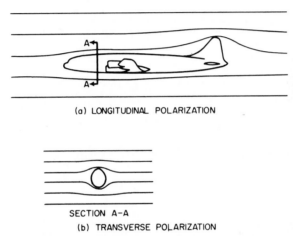

(a) LONGITUDINAL POLARIZATION

SECTION A–A

(b) TRANSVERSE POLARIZATION

FIG. 14-9 Magnetic field distortion caused by a conducting airframe for (*a*) longitudinal polarization and (*b*) transverse polarization.

ahead of the aircraft. For most locations near the top and bottom centerline, the local-field intensity is greater than the incident-field intensity in both cases. This field enhancement is important because it serves to increase the signal induced in a loop antenna and also affects the bearing accuracy of the direction-finder system. The ratio of local magnetic field intensity on the airframe surface to the incident magnetic field intensity is designated as a_{xx} for the case in which the incident field is transverse to the line of flight (the signal arrives from the front or rear of the aircraft) and as a_{yy} when the incident field is along the line of flight (the signal arrives from the side of the aircraft). These two coefficients give all the essential data for estimating the performance of a direction-finder loop antenna designed to take bearings on LF ground-wave signals, provided the loop is placed on the top or bottom centerline of the airframe. The amplitude and direction of the local field on the airframe surface may be calculated for any incident-field amplitude and direction (provided the latter is horizontal), once the coefficients a_{xx} and a_{yy} are known, by simply resolving the incident field into x and y components, multiplying these components by a_{xx} and a_{yy} respectively, and recombining the components.*

The ratio of local-field intensity to incident-field intensity, which in this case is a function of the angle of arrival of the wave, may be used as a curvature factor for estimating loop-antenna sensitivity in the same way that the factor F_v is used for monopole-antenna calculations.

The ADF system, which determines the direction of arrival of the signal by rotating its loop antenna about a vertical axis until a null is observed in the loop response (Fig. 14-10), is subject to bearing errors because of the difference in direction of the local incident magnetic fields. The relationship between the true and the apparent directions of the signal source is given by the equation

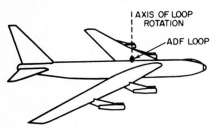

I AXIS OF LOOP ROTATION

ADF LOOP

$$\tan \phi_t = \frac{a_{xx}}{a_{yy}} \tan \phi_a \qquad (14\text{-}3)$$

FIG. 14-10   ADF loop antenna on aircraft.

where ϕ_t and ϕ_a are respectively the true and the apparent bearings of the signal source. A curve of the bearing error $(\phi_t - \phi_a)$ as a function of ϕ_t with the ratio $a_{xx}/a_{yy} = 2$ is shown in Fig. 14-11a. Figure 14-11b shows graphs of the maximum bearing error $|\phi_t - \phi_a|$ and the value of ϕ_t (in the first quadrant) at which the maximum bearing error occurs as functions of the ratio a_{xx}/a_{yy}.

Survey data showing the coefficient a_{xx} and the maximum bearing error for loop antennas along the top and bottom centerline of a DC-4 aircraft are shown in Fig. 14-12.

14-3   ADF ANTENNA DESIGN

The automatic-direction-finder loop antenna is normally supplied as a component of the ADF system, and the problem is to find a suitable airframe location. The two restrictions which govern the selection of a location are (1) that the cable between the loop antenna and receiver be of fixed length, since the loop and cable inductances form a part of the resonant circuit in the loop amplifier input, and (2) that the maximum bearing error which the system can compensate be limited to about 20°. As can be seen from the bearing-error data in Fig. 14-12, positions on the bottom of the fuselage forward of the wing meet the minimum bearing-error requirements on the aircraft type for which the data are applicable (and on other comparable aircraft). Such positions are consistent with cable-length restrictions in most cases since the radio-equipment racks are usually just aft of the flight deck.

---

*For locations off the centerline, an incident field polarized in one of the principal directions may cause local-field components in both principal directions (as well as in the z direction), and hence coefficients a_{xy}, a_{yx}, a_{zy}, and a_{zx} are needed in the general case to describe the local-field amplitude and direction in terms of the incident-field amplitude and direction.

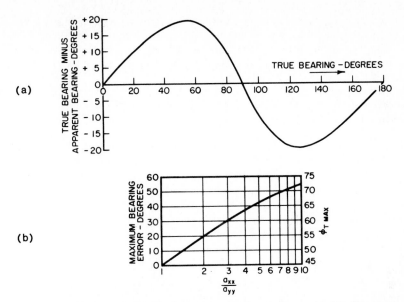

FIG. 14-11 (a) Bearing-error curve for $a_{zz}/a_{yy} = 2$. (b) The maximum bearing error and the true bearing at which the maximum bearing error occurs as functions of a_{zz}/a_{yy}.

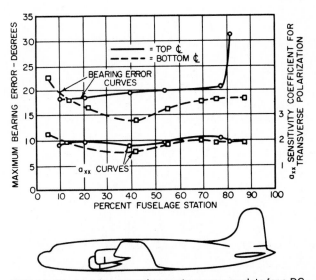

FIG. 14-12 Low-frequency loop-antenna survey data for a DC-4 aircraft showing maximum bearing error and sensitivity coefficient for transverse polarization for top and bottom centerline locations.

Bearing-error compensation is accomplished in most ADF systems by means of a mechanical compensating cam.[1,2] Standard flight-test procedures[1,3] are generally used in setting the compensating cams. In some loop designs[4] electrical compensation is achieved by modifying the immediate environment of the actual loop element in order to equalize the coefficients a_{xx} and a_{yy}.

Unlike the loop antenna, the sense antenna must be located within a limited region of the airframe for which the equivalent-dipole axis is essentially vertical in order to ensure accurate ADF performance as the aircraft passes over or near the ground station to which the receiver is tuned.[5] In Fig. 14-13 are shown the regions of confusion which exist above the ground station for an airborne ADF having a non-vertical sense-antenna pattern. The significance of such a confusion zone is that the ADF needle attempts to reverse and hence to indicate a bearing which is in error by 180° when the aircraft is within the zone. The intersection of the confusion zone with a surface of constant altitude is a circle; various types of needle behavior for different

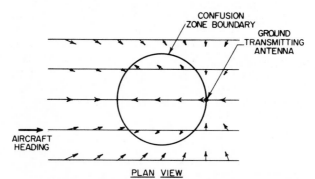

PLAN VIEW

NOTE: SMALL ARROWS SHOW DIRECTIONS WHICH ADF NEEDLE TRIES TO INDICATE AT VARIOUS POINTS ALONG A GROUP OF PARALLEL, COPLANAR FLIGHT TRACKS

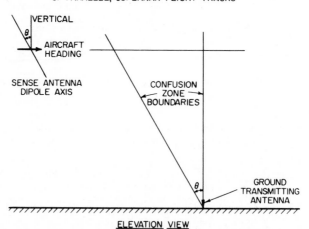

ELEVATION VIEW

NOTE: ANGLE BETWEEN CONFUSION ZONE BOUNDARIES IS EQUAL TO SENSE ANTENNA TILT ANGLE

FIG. 14-13 ADF overstation confusion zone.

amounts of course offset at the same altitude are illustrated in Fig. 14-13. Actually, the needle has a finite reversal time, so that the responses to the various reversal signals which it receives as the aircraft traverses a flight path near the station become superimposed to an extent which depends upon the aircraft altitude and speed as well as on the sense-antenna tilt angle. As a result, it is virtually impossible for the pilot to make an accurate determination of the time of station passage unless the confusion zone is made quite small.

The size of the zone is also dependent on the phase difference between the loop and sense signals at the point where they are mixed in the receiver. It can be shown[5] that, by introducing a controlled phase error of the proper sense at this point, the size of the confusion zone may be reduced very substantially, thereby relaxing the requirement for proper sense-antenna placement. Existing ADF receivers do not incorporate this feature, however, and hence a small sense-antenna tilt angle is usually a design requirement.

A typical flush ADF sense-antenna installation is illustrated by the C-141 aircraft.[6] The antenna is a cavity built into the upper fuselage fairing just aft of the wing crossover. It was decided during the design phase to use the Bendix ADF-73 automatic-direction-finder system, which to some extent determined the sense-antenna requirements. The basic antenna requirements were (1) antenna capacitance of 300 pF, (2) a minimum quality factor of 1.4, and (3) a radiation-pattern tilt angle between 0 and 10° downward and forward. Figure 14-14 shows the patterns of the dual sense antenna.

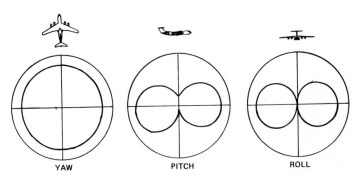

FIG. 14-14 Voltage radiation patterns for the C-141 ADF sense antenna.

The sensitivity of the ADF system varies directly with the sense-antenna quality factor, which is defined as $h_e C_a$, where h_e is the antenna effective height in meters and C_a is the antenna capacitance in picofarads. In actual practice the effective height is determined by a measurement of the antenna open-circuit voltage in microvolts divided by the received-signal field intensity in microvolts per meter.

The antenna effective height, for a top-loaded design, is directly proportional to the distance separating the loading element from the ground plane. Antenna capacitance is directly proportional to the element area and inversely proportional to the separation from the ground plane. Thus, as the separation is increased, h_e increases while C_a decreases. It can be shown that the quality factor will also increase for an external design. In the case of a flush-type cavity, the quality factor increases asymptotically with separation. With a limited separation, it is apparent that the desired quality factor may be obtained only by providing sufficient element area. Owing to aerodynamic and structural considerations, there is a limit to the area or separation which may be economically utilized, and a compromise is made.

The sense-antenna input of the ADF receiver is designed to receive an antenna circuit with a nominal capacitance of 3000 pF. The transmission line is composed of 60 ft (18.3 m) of RG-11/U. This represents 1230 pF, since RG-11/U has a capacitance of 20.5 pF per foot. The sense-antenna coupler (susceptiformer) transforms the remaining 1770 pF to the sense-antenna capacitance of 300 pF.

For the C-141, the electrical center of the aircraft (pattern tilt angle of zero) was initially located through radiation-pattern measurements by using a small model configuration; a tilt-angle change was noted when the feed point for the cavity ADF antenna was moved. A zero tilt angle was obtained, and it was found that an approximate 10-in (254-mm) movement of the feed point aft produced a $+10°$ tilt angle. Since a tilt angle of 0 to 10° is required, the measurements permitted selection of an optimum feed point ($+5°$ tilt). Figure 14-14 shows the yaw, pitch, and roll radiation patterns obtained with the optimum antenna feed point.

14-4   MOMENT-METHOD ANALYSIS OF LOW-FREQUENCY AND HIGH-FREQUENCY ANTENNA PATTERNS

The method of moments (MOM)[7] is a computer analysis method which can be employed to determine the radiation patterns of antennas on aircraft and missiles. A discussion of this technique is beyond the scope of this handbook, but user-oriented computer programs are available.[8,9]

The method of moments is based upon a solution of the electromagnetic-wave-scattering integral equations by enforcing boundary conditions at a number N of discrete points on the scattering surface. The resulting set of equations is then solved numerically to determine the scattered fields. This is typically accomplished by modeling the surface as a wire grid for which the self- and mutual impedances between the wires can be calculated. This technique, for a set of N wire elements which represents the aircraft such as shown in Fig. 14-15, results in an $N \times N$ impedance matrix. For a given antenna excitation, inversion of the $N \times N$ matrix yields the solution for the wire currents and hence the radiated fields of the antenna-aircraft system.

Computer programs which implement this technique for wire scatterers[7,10,11] and for surface patches[12] have been written. Some codes which employ special computer techniques are presently capable of handling matrices on the order of (200 $\times$ 200) to (1200 $\times$ 1200) elements. The number of wire or patch elements which must be employed to ensure a convergent numerical solution varies as a function of the particular code implementation. Typically, a wire element less than $\lambda/8$ in length must be used so that a $1\lambda^2$ surface patch requires a "square" wire-grid model containing 144 wires. Thus it is seen that as the surface area increases, the required matrix increases roughly as the square of the area which, in practice, restricts the MOM to electrically small- to moderate-sized geometries. Codes based on body-of-revolution approximations are also available[13] and are particularly applicable to helicopters and missiles.

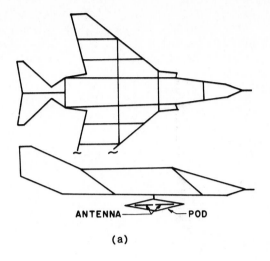

(a)

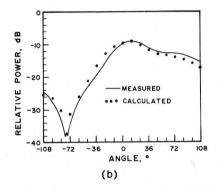

(b)

FIG. 14-15 MOM wire-grid analysis of low-frequency air-craft antenna patterns. (*a*) MOM wire-grid model of an RF-4C aircraft. (*b*) MOM calculated roll-plane pattern for a pod-mounted dipole array on the RF-4C aircraft. (*From Ref. 11,* © *1977 IEEE.*)

A wire-grid model which was used for the pattern analysis of a pod-mounted high-frequency (HF) folded-dipole array antenna on an RF-4C aircraft is shown in Fig. 14-15*a*. An MOM code employing piecewise sinusoidal basis functions[10,11] was used to calculate the roll-plane patterns for the array, and these results are compared with measured data in Fig. 14-15*b*.

A surface-patch method has also been employed to analyze the patterns of antennas on helicopters.[12] In this analysis, the helicopter surface was subdivided into a collection of curvilinear cells. A result of the analysis, which employed 94 surface-patch cells, is shown in Fig. 14-16 for the pitch and yaw planes. Generally, good results were obtained for this complex-shaped airframe. Also, we note that, by representing the helicopter rotor blades as wires, the rotor effects on the patterns can be computed.

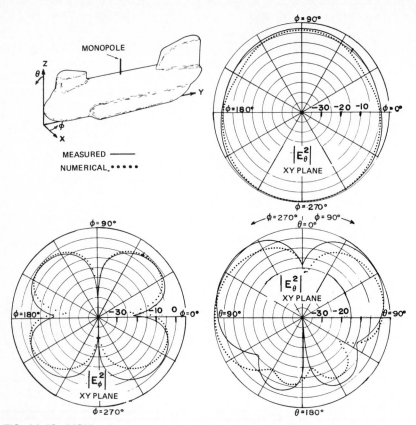

FIG. 14-16 MOM surface-patch-calculated patterns for the CH-47 helicopter. *(From Ref. 12, © 1972 IEEE.)*

For some LF applications, a simplified "stick model" of an aircraft may be sufficient,[14] and a model of an aircraft is shown in Fig. 14-17. The fuselage, wings, and empennage can be represented by either single or multiple thin wires if their maximum electrical width is approximately less than $\lambda/10$. Figure 14-17 shows a calculated result for a monopole mounted on a stick model, and the effects of the fuselage and wings can be seen by comparison with the unperturbed-monopole pattern.[15]

14-5 HIGH-FREQUENCY COMMUNICATIONS ANTENNAS

Aircraft antennas for use with communications systems in the 2- to 30-MHz range are required to yield radiation patterns which provide useful gain in directions significant to communications. Also, impedance and efficiency characteristics suitable for acceptable power-transfer efficiencies between the airborne equipment and the

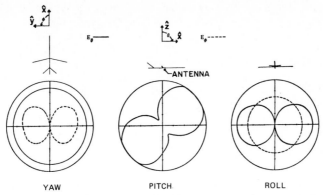

FIG. 14-17 MOM calculated voltage patterns for a blade antenna (tilted monopole) on a "stick model" aircraft.

radiated field are needed. A receiving antenna which meets these requirements will deliver to the input of a matched receiver atmospheric noise power under noise field conditions which is many times greater than the input-circuit noise in the communications receiver. When this is the case, no improvement in signal-to-noise ratio can be achieved by further refinement of the antenna design. Since the transmitting mode of operation poses the more stringent requirements, the remainder of the discussion of HF antennas is confined to the transmitting case. Sky-wave propagation is always an important factor at these frequencies, and because of the rotation of polarization which is characteristic of reflection from the ionosphere, polarization characteristics are usually unimportant; the effective antenna gain can be considered in terms of the total power density. At frequencies below 6 MHz, ionosphere (and ground) reflections make almost all the radiated power useful for communication at least some of the time, so that differences between radiation patterns are relatively unimportant in comparing alternative aircraft antennas for communication applications in the 2- to 6-MHz range. In this range, impedance matching and efficiency considerations dominate. For frequencies above 6 MHz, pattern comparisons are frequently made in terms of the average power gain in an angular sector bounded by cones 30° above and below the horizon.

In the 2- to 30-MHz range, most aircraft have major dimensions of the order of a wavelength, and currents flowing on the skin usually dominate the impedance and pattern behavior. Since the airframe is a good radiator in this range, HF antenna design is aimed at maximizing the electromagnetic coupling to the airframe. The airframe currents exhibit strong resonance phenomena that are important to the impedance of antennas which couple tightly to the airframe.

Wire Antennas

Wire antennas, supported between the vertical fin and an insulated mast or trailed out into the airstream from an insulated reel, are reasonably effective HF antennas on lower-speed aircraft. Aerodynamic considerations limit the angle between a fixed wire and the airstream to about 15°, so that fixed-wire antennas yield impedance characteristics similar to moderately lossy transmission lines, with resonances and antiresonances at frequencies at which the wire length is close to an integral multiple of $\lambda/4$. Figure 14-18 shows the input impedance of a 56-ft (17-m) fixed-wire antenna on a C-130 Hercules aircraft. Lumped reactances connected between the wire and the fin produce an effect exactly analogous to reactance-terminated lossy transmission lines.

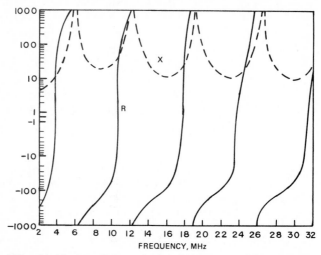

FIG. 14-18 Impedance of the C-130E right-hand long-wire antenna.

The average directive gain of these antennas in the sector $\pm 30°$ relative to the horizon (i.e., the fraction of the total radiated energy which goes into the sector bounded by the cones 30° above and below the horizontal plane) remains near 60 percent from 6 to 24 MHz. The efficiency of wire antennas is not high because of resistance loss in the wire itself and dielectric loss in the supporting insulators and masts. The resistance of commonly used wires is of the order of 0.05 Ω/ft at 4 MHz. The RF corona breakdown threshold of fixed-wire antennas is a function of the wire diameter and the design of the supporting fittings. To minimize precipitation static a wire coated with a relatively large-diameter sheath of polyethylene is frequently used, with special fittings designed to maximize the corona threshold. Even with such precautions, voltage breakdown poses a serious problem with fixed-wire antennas at high altitudes. Measurements indicate that standard antistatic strain insulators have an RF corona threshold of about 11 kV peak at an altitude of 50,000 ft (15,240 m)[16] and a frequency of 2 MHz. Such an insulator, placed between the vertical stabilizer and the aft end of the open-circuited antenna would go into corona at this altitude if the antenna were energized with a fully modulated AM carrier of about 150 W at 2 MHz.

Isolated-Cap Antennas

The airframe can be effectively excited as an HF antenna by electrically isolating a portion of the fin tip or a wing tip to provide antenna terminals. Figure 14-19a shows

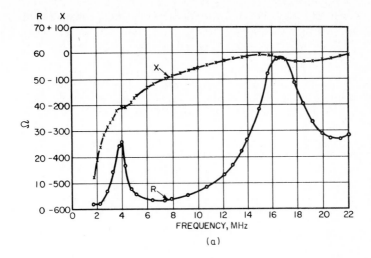

(a)

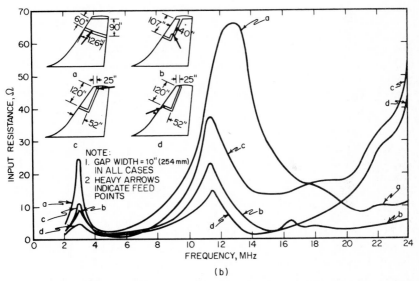

(b)

FIG. 14-19 Impedance characteristics of tail-cap antennas. (*a*) Input impedance of a 7-ft (2.1-m) tail-cap antenna on a DC-4 aircraft. (*b*) Effect of changing feed configurations on the input resistance of tail-cap antennas.

the impedance curves for a 7-ft (2.1-m) tail-cap antenna on an aircraft. The isolating gap was cut straight across the aircraft extremity, and the gap width was 6 in (152 mm). The sharp peak of input resistance at 4 MHz is due to coupling to the $\lambda/2$ resonance of the currents on the airframe, which at this frequency flow predominantly along a path extending from the fin tip to the fuselage, thence to the wing root, and along the wings to the tips. This resonance dominates the lower-frequency impedance characteristics of all cap-type antennas. The resistance peaks at higher frequencies are associated with current resonances on the wings and fuselage and, in the tail-cap cases, the empennage.

Probe Antennas

A variation of the isolated-cap antenna is the probe or bullet-probe antenna. The radiation-pattern and impedance-resonance effects of the empennage, fuselage, and wings also affect probe antennas. The marked advantage of the probe antenna over the tail-cap antenna lies in the relative ease of incorporation into the structure of the aircraft empennage. The C-141 bullet-probe antenna shown in Fig. 14-20[18] makes use of a portion of the structural intersection of the horizontal and vertical stabilizers in the T-tail configuration. The antenna is composed of the forward portion of the aerodynamic bullet with a probe extension. The plastic isolation gap in the bullet extends forward 11 in (279 mm), and the overall length of the antenna including the isolation band is 112 in (2845 mm). The maximum structural gap width was 13 in (330 mm), and the minimum gap width to withstand RF potentials was 3 to 4 in (76 to 102 mm). To obtain maximum airframe coupling, a maximum resistive component is required at the lowest expected resonant frequency. This resonance occurred at 2.4 MHz, which corresponds to a half-wavelength resonant condition caused by the wing-tip-to-horizontal-stabilizer-tip distance. Another parameter considered was the capacitive reactance of the antenna at 2.0 MHz. This reactance had to remain below 1000 Ω if existing couplers were to be used. This resulted in a gap width of 11 in for satisfactory performance. Figure 14-20 shows the impedance curves for the C-141 forward bullet-probe antenna.

A solid bulkhead was installed just aft of the isolation band for two reasons: (1) the metallic bulkhead acts as an electrostatic shield between the antenna and other

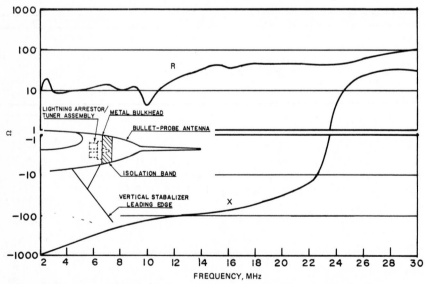

FIG. 14-20 Impedance curves for the C-141 aircraft forward bullet-probe antenna. *(From Ref 18.)*

metallic objects, and (2) the antenna tuner will mount on the bulkhead. Connection to the antenna is made through a lightning arrester which protrudes through the bulkhead into the gap area. The lightning arrester has a "birdcage" terminal which plugs into a flared tube tied electrically to the probe and metal portions of the forward bullet.

Shunt or Notch High-Frequency Antennas

Shunt- and notch-type HF antennas are becoming increasingly popular on modern high-speed aircraft because of drag reduction and higher reliability. The basic difference in the shunt and notch configurations is the ratio of length to width of the cutout (dielectric) portion of the aircraft. Figure 14-21 shows the basic configuration.[19]

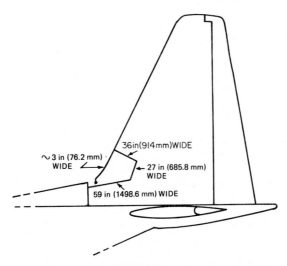

FIG. 14-21 Basic configuration of the C-130 notch high-frequency antenna. *(From Ref. 19.)*

The shunt or notch antenna is a kind of transmission line that uses the vertical stabilizer or wing to feed the remainder of the aircraft. It can be located at various points on the airframe, but since it functions as an inductive coupler, it should be located at a point of low impedance or high current. The shunt-fed loop current causes the airframe itself to act as the antenna. The particular portion of the airframe supporting the current path is a function of the frequency and physical lengths of parts of the airframe involved. For example, the wing-tip-to-horizontal-tip half-wavelength resonance on the C-5 aircraft occurs at approximately 2.5 MHz. The coupler feed point should always be located at the end of the shunt feed closest to the center of the aircraft.

The design of the antenna involves first finding a suitable location for the dielectric area and then arranging the dimensions of the shunt feed and dielectric area to provide an antenna that can be efficiently matched to 50 Ω with a relatively simple antenna coupler. Guidelines to observe are as follows: (1) The first choice for location on the airframe from a radiation-pattern and impedance standpoint is the root of the vertical stabilizer. An alternative is the wing root, but in this case the radiated vertically polarized energy is reduced. (2) The antenna coupler should be connected to the shunt feed near the stabilizer or wing root to provide maximum efficiency. (3) The dielectric material used to fill the opening should have low-loss characteristics. (4) The exact size and shape of the dielectric area is not a critical design parameter, but the dielectric area should be as deep as possible for maximum coupling to the airframe. Radiation efficiency is generally proportional to the dielectric area.

To avoid the need for a complicated antenna-tuning network, the antenna reactance at 2.0 MHz should be no less than $+j18$ Ω, and ideally parallel resonance should occur between 20 and 30 MHz. Series resonance should be completely avoided, and antenna reactance should be no lower than $-j100$ Ω at 30 MHz. To ensure good coupling to the airframe the parallel component of the antenna impedance should be less than 20,000 Ω.

Figure 14-21 shows the shunt-fed notch antenna for the C-130 aircraft. A portion of the dorsal and a portion of the leading-edge fairing are combined to yield a notch in the root of the vertical stabilizer. The total distance around the notch is approximately 160 in (4064 mm), or slightly smaller than the ideal periphery of approximately 200 in (5080 mm). This was necessary because of interference with the

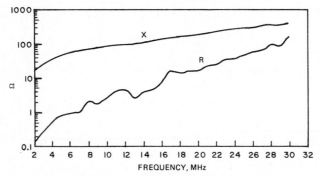

FIG. 14-22   Impedance of the shunt-fed HF notch antenna on the C-130 aircraft. *(From Ref. 19.)*

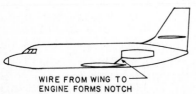

FIG. 14-23   Shunt-fed HF notch-antenna configuration on the Lockheed JetStar. *(From Ref. 20.)*

leading-edge deicing system. The impedance curve for this antenna is shown in Fig. 14-22.

A unique application of the shunt-fed HF antenna is found on the Lockheed JetStar aircraft.[20] Figure 14-23 is a sketch of the antenna. The antenna is fed from a coaxial feed on the upper surface of the wing, where a 32-in (813-mm) piece of phosphor-bronze stranded cable connects the center conductor to the leading lower edge of an aft-mounted engine nacelle. Excellent omnidirectional patterns are obtained, and the impedance is tuned by using an off-the-shelf coupler inside the pressurized area of the fuselage.

14-6   UNIDIRECTIONAL VERY-HIGH-FREQUENCY ANTENNAS

The marker-beacon, glide-slope, and radio-altimeter equipments require relatively narrowband antennas with simple patterns directed down or forward from the aircraft. This combination of circumstances makes the design of these antennas relatively simple. Both flush-mounted and external-mounted designs are available in several forms.

Marker-Beacon Antennas

The marker-beacon receiver operates on a fixed frequency of 75 MHz and requires a downward-looking pattern polarized parallel to the axis of the fuselage. A standard external installation employs a balanced-wire dipole supported by masts. The masts may be either insulated or conducting. Low-drag and flush-mounted designs are sketched in Fig. 14-24. The low-drag design is a simple vertical loop oriented in the

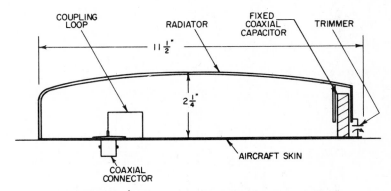

(a) SCHEMATIC OF THE COLLINS 37X-1 MARKER BEACON ANTENNA

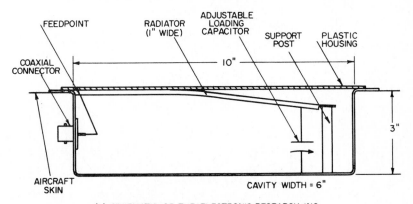

(b) SCHEMATIC OF THE ELECTRONIC RESEARCH INC. AT-134/ARN FLUSH MARKER BEACON ANTENNA

FIG. 14-24   Typical marker-beacon antennas.

longitudinal plane of the aircraft. The feed line is inductively coupled to the loop, which is resonated by a series capacitor, and the antenna elements are contained in a streamlined plastic housing. The flush design is electrically similar, but in this case the structure takes the form of a conductor set along the longitudinal axis of the open face of a cavity. To achieve the desired impedance level, the antenna conductor is series-resonated by a capacitor and the feed point tapped partway along the antenna element.

Glide-Slope Antennas

The glide-slope receiver covers the frequency range from 329 to 335 MHz and requires antenna coverage only in an angular sector 60° on either side of the nose and 20° above and below the horizon. This requirement can be met by horizontal loops or by vertical slots. Because of the narrow bandwidth, the antenna element need not be electrically large. Two variations on the loop arrangement are sketched in Fig. 14-25. Configuration *a* is a simple series-resonant half loop which can be externally mounted on the nose of the aircraft or within the nose radome. A variation of this antenna has two connectors for dual glide-slope systems. Configuration *b*, which is similar to the cavity-marker-beacon antenna of Fig. 14-24, is suitable for either external or flush mounting.

Altimeter Antennas

The radio altimeter, which operates at 4300 MHz, requires independent downward-looking antennas for transmission and reception. Proper operation requires a high

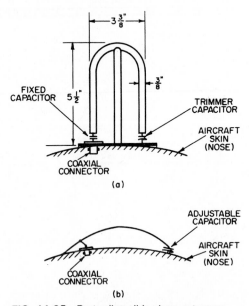

FIG. 14-25 Protruding glide-slope antennas.

degree of isolation between the transmitting and receiving elements, and horns are typically used, although microstrip antennas are becoming common.

14-7 OMNIDIRECTIONAL VERY-HIGH-FREQUENCY AND ULTRAHIGH-FREQUENCY ANTENNAS

In the frequency range in which the airframe is electrically large, the achievement of omnidirectional patterns, such as those used for short-range communications, is complicated by airframe effects. Since very-high-frequency (VHF) and ultrahigh-frequency (UHF) antennas of resonant size are structurally small, the required impedance characteristics can be achieved with fixed matching networks. Shadowing and reflection by the airframe result in major distortions of the primary pattern of the radiating element.

Airframe Effects on VHF and UHF Patterns

The tip of the vertical fin is a preferred location for omnidirectional antennas in the VHF and UHF range because antennas located there have a relatively unobstructed "view" of the surrounding sphere. Figure 14-26 shows the principal-plane patterns of a $\lambda/4$ stub on the fin tip of a B-50 aircraft at 1000 MHz. At this frequency, at which the mean chord of the fin is 10 wavelengths, the principal effect of the airframe on the radiation patterns is a sharply defined "optical" shadow region, indicated by the dashed lines in the figure. At the lower end of this frequency range and on smaller aircraft, the effect of the airframe on the patterns is more complicated. Deep nulls in the forward quadrants can occur because of the destructive interference between direct radiation and radiation reflected from the fuselage and wings. The latter contribution is more important for small aircraft since the ground plane formed by the surface of the fin tip is now sufficiently small to permit strong spillover of the primary pattern, which also can create lobing in the transverse-plane pattern. In this frequency range the null structure is strongly influenced by the position of the radiating element along the chord of the fin, and careful location of the antenna along the chord may result in improvements in the forward-horizon signal strength.

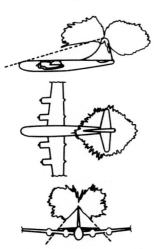

FIG. 14-26 Principal-plane patterns of a 1000-MHz stub antenna on the tip of the vertical stabilizer of a B-50 aircraft.

Most external antennas are located on the top or bottom centerline of the fuselage in order to maintain symmetry of the radiation patterns. Pattern coverage in such locations is limited by the airframe shadows. Figure 14-27 shows the patterns of a

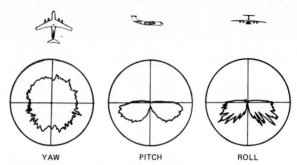

FIG. 14-27   Voltage radiation patterns of a 1000-MHz mon-opole on the bottom of a C-141 aircraft.

1000-MHz monopole on the bottom centerline of a C-141 aircraft. It is apparent that coverage is limited to the hemisphere below the aircraft. UHF antennas on the top of the fuselage yield patterns confined to the upper hemisphere, with a null aft owing to the shadow cast by the vertical fin. In many applications, as, for example, in sched-uled-airline operations, these pattern limitations are acceptable, and fuselage locations are frequently used.

The deep lobing in the roll-plane pattern of Fig. 14-27 is due to reflections from the strongly illuminated engine nacelles. In some locations similar difficulties are encountered because of reflection from the wing flaps when they are extended. Shad-ows and lobing due to the landing gear, when extended, are frequently troublesome for bottom-mounted antennas.

GTD Antenna Analysis

The MOM technique is limited to surfaces which are relatively small in terms of square wavelengths. However, as the frequency increases and the surface becomes large in terms of square wavelengths, the propagation of electromagnetic (EM) energy can be analyzed by using the techniques of geometrical optics. For example, if an incident field is reflected from a curved surface, ray optics and the Fresnel reflection coefficients yield the reflected field.

The geometrical theory of diffraction (GTD) uses the ray-optics representation of EM propagation and incorporates both *diffracted rays* and *surface rays* to account for the effects of edge and surface discontinuities and surface-wave propagation. In the case of edge diffraction, the diffracted rays lie on a cone whose half angle is equal to the half angle of the edge tangent and incidence vector. For a general curved sur-face, the incident field is launched as a surface wave at the tangent point, propagates along a geodesic path on the surface, and is diffracted at a tangent point toward the direction of the receiver. Each of these diffraction, launching, and surface propagation processes is described by appropriate complex diffraction and attenuation coefficients which are discussed in the literature.[21] In practice, these coefficients are implemented as computer algorithms which can be used as building blocks for a specific analysis.

In the GTD analysis, the received far- or near-zone field is composed of rays which are directly incident from the antenna, surface rays, reflected rays, and edge-diffracted rays. Also *higher-order effects* such as multiple reflections and diffractions can occur. Several computer models have been implemented to perform the required differential-geometry and diffraction calculations.[22,23,24] It is anticipated that these methods will become more user-oriented and powerful with the advent of increasingly sophisticated computer-graphics systems.

Figure 14-28 shows a model of a missile composed of a conically capped circular cylinder with a circumferential-slot antenna.[25] In this case, the radiated field is due to direct radiation from the slot, surface rays which encircle the cylinder, and diffracted rays from the cone-cylinder and cylinder-end junctions and from the cone tip. The computed results shown in Fig. 14-28 were obtained by employing the solution for a slot on an infinite circular cylinder to represent the direct- and surface-ray fields and by computing the diffracted fields from the junctions and tip as corrections to the infinite-cylinder result. The principal-plane pattern shown is in agreement with measured data, and comparable results have been obtained for the off-principal planes.[26]

An analysis due to Burnside et al.[22] approximates the fuselage as two joined spheroids, and the wings and empennage are modeled by arbitrarily shaped flat plates. In addition, the engines can be modeled by finite circular cylinders. Figure 14-29 shows a result obtained from this analysis of a $\lambda/4$ monopole on a KC-135 aircraft for the roll plane. Reference 23 presents a complete 4π-sr plot for a monopole on a Boeing 737 aircraft. The results, which compare well with measured data, illustrate the accuracy of the GTD analysis for VHF and UHF antennas.

It should be noted that a complete GTD model may not always be required for an engineering assessment of antenna performance. For example, if wing reflection and blockage effects are dominant, one can use a simpler model consisting of two

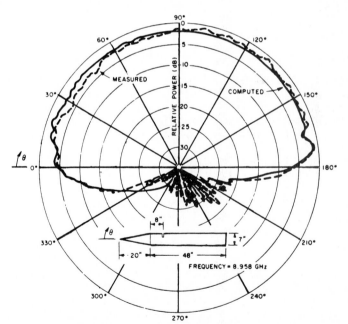

FIG. 14-28 Pattern of a circumferential slot on a conically capped circular cylinder. *(From Ref. 25, © 1972 IEEE.)*

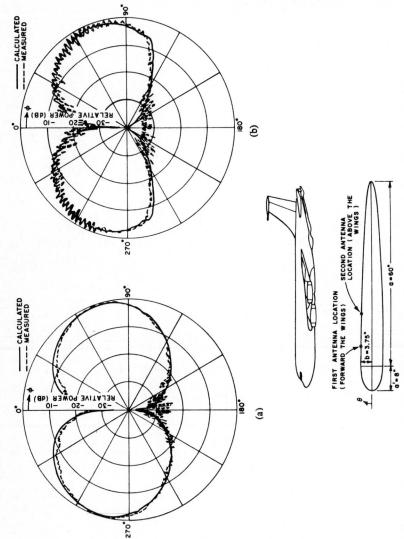

FIG. 14-29 (*a*) Roll-plane pattern (E_ϕ) for a 1:25-scale model of a KC-135 aircraft with a $\lambda/4$ monopole on the fuselage forward of the wings at a frequency of 34.92 GHz (model frequency). (*b*) Roll-plane pattern (E_ϕ) for a $\lambda/4$ monopole over the wings. (*From Ref. 22, © 1975 IEEE.*)

submodels: a submodel for an antenna on a finite (or perhaps an infinite) cylindrical fuselage and a flat-plate submodel for the wings.[24] In this case, the antenna-on-cylinder pattern results are used as the antenna illumination for calculation of the wing blockage, reflection, and diffraction.

Antennas for Vertical Polarization

The monopole and its variants are the most commonly used vertically polarized VHF and UHF aircraft antennas. The antenna typically has a tapered airfoil cross section to minimize drag. Simple shunt-stub matching networks are used to obtain a voltage standing-wave ratio (VSWR) below 2:1 from 116 to 156 MHz.

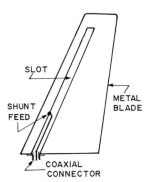

FIG. 14-30 Basic configuration of an all-metal VHF blade antenna.

The basic configuration of an all-metal VHF blade antenna is shown in Fig. 14-30. This antenna employs a shunt-fed slot to excite the blade. The advantage of this configuration is its all-metal construction, which provides lightning protection and eliminates p static. Figure 14-31 shows a low-aerodynamic-drag sleeve monopole for the 225- to 400-MHz UHF communications system. The cross section is diamond-shaped, with a thickness ratio of 5:1. A feature of this antenna[27] is the introduction into the impedance-compensating network of a shunt stub such that its inner conductor serves as a tension member to draw the two halves of the antenna together. In addition to providing mechanical strength, this inner conductor forms a direct-current path from the upper portion of the antenna to the aircraft skin, thereby providing lightning-strike protection.

A number of monopole designs have been developed for installation on a fin in which the top portion of the metal structure has been removed and replaced by a suitable dielectric housing. Various other forms of vertically polarized UHF and VHF radiators have been designed for tail-cap installation. The pickax antenna, consisting of a heavily top-loaded vertical element, has been designed to provide a VSWR of less than 3:1 from 110 to 115 MHz with an overall height of 15½ in (393.7 mm) and a length of 14 in (355.6 mm).

The basic flush-mounted vertically polarized element for fuselage mounting is the annular slot,[28] which can be visualized as the open end of a large-diameter, low-characteristic-impedance coaxial line. As seen from the impedance curve of Fig. 14-32, such a structure becomes an effective radiator only when the circumference of the slot approaches a wavelength. The radiation patterns have their maximum gain in the plane of the slot only for very small slot diameters and yield a horizon gain of zero for a slot diameter of 1.22.[29] This pattern variation is illustrated in Fig. 14-33. For these reasons and to minimize the structural difficulty of installing the antenna in an aircraft, the smallest possible diameter yielding the required bandwidth is desirable. For the 225- to 400-MHz band, the minimum practical diameter, considering construction tolerances and the effect of the airframe on the impedance, is about 24 in (610 mm).

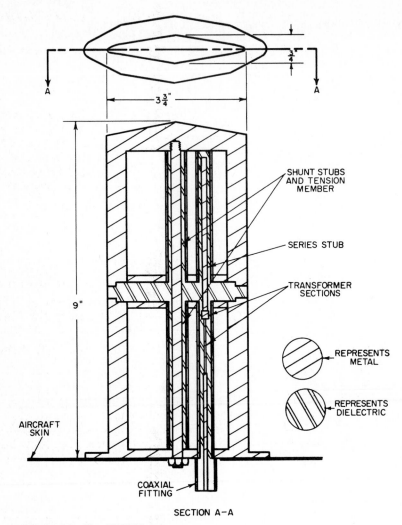

FIG. 14-31 The AT-256A antenna.

A VSWR under 2:1 can be obtained with this diameter and a cavity depth of 4.5 in (114.3 mm). Figure 14-34 shows a design by A. Dorne together with its approximate equivalent circuit. In the equivalent circuit the net aperture impedance of the driven annular slot and the inner parasitic annular slot is shown as a series resistance-capacitance (RC) circuit. The annular region 1, which is coupled to the radiating aperture through the mutual impedance between the two slots, and the annular region 2, which is part of the feed system, are so positioned and proportioned that they store primarily magnetic energy. The inductances associated with the energy storage in these regions are designated as L_1 and L_2 in the equivalent circuit. The parallel-tuned circuit in the equivalent circuit is formed by the shunt capacitance between vane 3 and horizontal

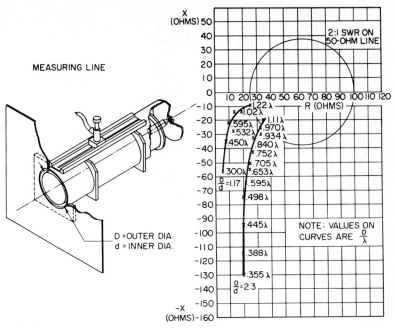

FIG. 14-32 Impedance of an annular-slot antenna.

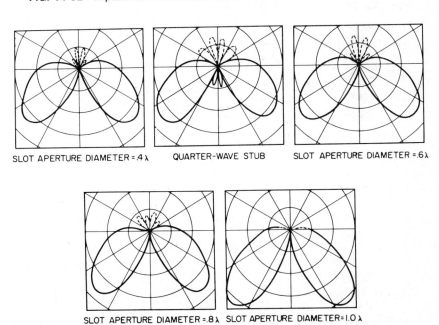

SLOT APERTURE DIAMETER = .4λ QUARTER-WAVE STUB SLOT APERTURE DIAMETER = .6λ

SLOT APERTURE DIAMETER = .8λ SLOT APERTURE DIAMETER = 1.0λ

FIG. 14-33 Radiation patterns of an annular-slot antenna and a quarter-wave stub on a 2½-wavelength-diameter circular ground plane.

disk 4, together with the shunt inductance provided by four conducting posts (5) equally spaced about the periphery of 4, which also serve to support 4 above 3. From this element inward to the coaxial line, the base plate is cambered upward to form a conical transmission-line region of low characteristic impedance. A short additional section of low-impedance line is added external to the cavity to complete the required impedance transformation.

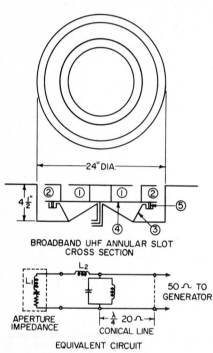

BROADBAND UHF ANNULAR SLOT
CROSS SECTION

EQUIVALENT CIRCUIT

FIG. 14-34 Annular-slot antenna for the 225- to 400-MHz band and its equivalent circuit.

Antennas for Horizontal Polarization

There are three basic antenna elements which yield omnidirectional horizontally polarized patterns: the loop, the turnstile, and the longitudinal slot in a vertical cylinder of small diameter. All three are used on aircraft, and all suffer from a basic defect. Because they must be mounted near a horizontal conducting surface of rather large extent (i.e., the top or bottom surface of the aircraft), their gain at angles near the horizon is low. The greater the spacing from the conducting surface, the higher the horizontal gain. For this reason, locations at or near the top of the vertical fin are popular for horizontally polarized applications, particularly for the VHF navigation system (VOR) which covers the 108- to 122-MHz range.

Figure 14-35 shows an E-fed cavity VOR antenna designed into the empennage tip of the L-1011 aircraft.[30] The antenna system consists of a flush-mounted dual E-slot antenna and a stripline feed network mounted on the forward bulkhead of the cavity between the two antenna halves. The antenna halves are mirror-image assemblies consisting of 0.020-in- (0.508-mm-) thick aluminum elements bonded to the inside surface of honeycomb-fiberglass windows. The feed network consists of a ring hybrid–power-divider device with four RF ports fabricated in stripline. The two input ports for connecting to the antenna elements are out-of-phase ports. The two output ports connect to separate VOR preamplifiers. The operating band of the antenna is 108 to 118 MHz with an input VSWR of less than 2:1 referred to 50 Ω. Radiation coverage is essentially omnidirectional in the horizontal plane. The principal polarization is horizontal with a cross-polarization component of more than 18 dB below horizontal polarization. The antenna gain is 3.6 dB below isotropic at band center. Figure 37-36 shows the principal-plane voltage radiation patterns.

The VOR navigation system is particularly vulnerable to the modulation effects of a helicopter rotor since, with this system, angular-position information is contained

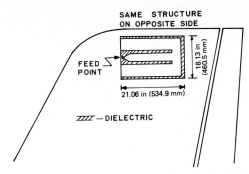

FIG. 14-35 *E*-slot VOR antenna for the L-1011 aircraft. *(From Ref. 30.)*

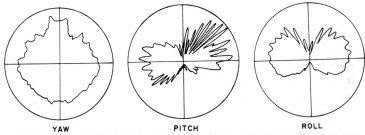

FIG. 14-36 Voltage radiation patterns for the *E*-slot VOR antenna on the L-1011 aircraft. *(From Ref. 30.)*

in a 30-Hz modulation tone which corresponds closely to the third harmonic of the fundamental blade-passage frequency on a typical helicopter. In Fig. 14-37 are shown two VOR antenna installations on an H-19 helicopter and the horizontal-plane radiation patterns of each. The fine structure on these patterns shows the peak-to-peak variation in signal amplitude due to passage of the rotor blades. The percentage of modulation of the signal received on the horizontal loop antenna is seen to be lower than that received on the ramshorn antenna. This is due partly to the shielding afforded the loop by the tail boom and partly to the fact that the loop antenna has inherently less response to scattered signals from the blades because its pattern has a null along its axis while the ramshorn antenna receives signals from directly above it very effectively.

Multiple-Antenna Systems

The limitations on omnidirectional coverage due to shadowing by the airframe can be overcome by the use of two or more antennas. A variety of diversity schemes are possible. If the pattern coverage of the two antennas is complementary, or at least approximately so, and if the separation between the two is a large number of wavelengths, the antennas can be connected directly together without the use of diversity techniques.[31] The resultant pattern is characterized by a large number of narrow lobes, with deep nulls only at those angles where the patterns of the two antennas have nearly equal amplitudes, as shown in Fig. 14-38. In view of the dynamic nature of the air-

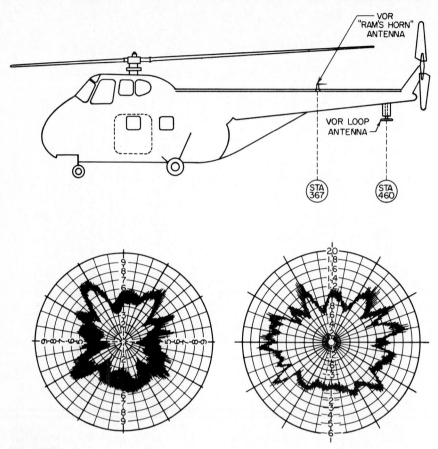

FIG. 14-37   Two VOR-antenna locations on the H-19 helicopter and horizontal-plane radiation patterns.

to-ground communications problem, it is easy to see that the time interval in which a given ground station is within one of the nulls is small, especially if the two patterns overlap in the directions broadside to the aircraft.

14-8   HOMING ANTENNAS

Airborne homing systems permit a pilot to fly directly toward a signal source. Although the navigational data supplied by a homing system are rudimentary, in the sense that only the direction and not the amount of course correction required are provided, the system does not require special cooperative ground equipment. Because of the symmetry of airframes, satisfactory homing-antenna performance can be obtained in frequency ranges in which direction-finder antennas are unusable because of airframe effects.

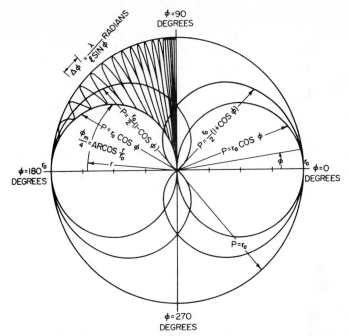

FIG. 14-38 Resultant of two cardioid patterns connected in parallel; r_0 = maximum range, ℓ = spacing between antennas, ϕ_m = total angle for which the range $\geq r$.

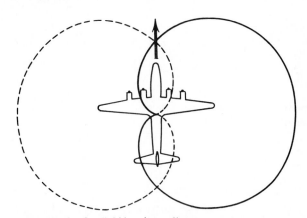

FIG. 14-39 Cardioid homing patterns.

The principle of operation used in airborne homing systems is illustrated in Fig. 14-39. Two patterns, which are symmetrical with respect to the line of flight and which ideally are cardioids, are generated alternately in time either by switching between separate antennas or by alternately feeding a symmetrical antenna array in two modes. The homing system compares the signals received under these two conditions and presents an indication that the pilot is flying a homing course or, if not, in which direction to turn to come onto the homing course. The equisignal condition which leads to the on-course indication can also arise for a reciprocal course leading directly away from the signal source. The pilot can resolve this ambiguity by making an intentional turn after obtaining the on-course indication and noting whether the direction of the required course correction shown by his or her instrument is opposite the direction of the intentional turn (in which case the pilot is operating about the correct equisignal heading) or in the same direction (in which case the pilot is operating about the reciprocal heading).

One of the basic problems in the design of a homing-antenna system is the proper compromise between directivity patterns and system sensitivity. Consider the idealized homing array shown in Fig. 14-40, which is assumed to be in free space. With mutual

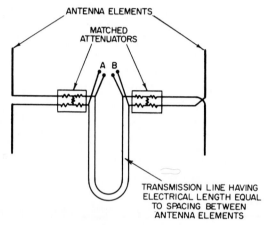

FIG. 14-40   A homing array.

impedance between the two elements neglected, it is readily shown that, by arranging the feed system as indicated, the array will have a null in one direction along the line joining the midpoints of the two elements and that the null direction reverses as the feed is switched between the two sets of feed terminals. Figure 14-41 shows the horizontal-plane patterns for such an array with various element spacings. For small spacings, the pattern becomes a cardioid. The pattern quality, as determined by the pattern slope at the on-course heading (a steep slope is desirable since this will lead to a clear indication of course error due to a small deviation from the homing course) and by the difference in response between the left-hand and right-hand patterns for courses other than the homing course and the reciprocal course, becomes poorer as the spacing is increased. The relative system sensitivity for the homing-course direction, on the other hand, increases as the element spacing is increased, being proportional to the sin $(\pi s/\lambda)$, where s/λ is the spacing between elements in wavelengths. Since most homing systems are designed for relatively wide frequency bands, a compromise element spac-

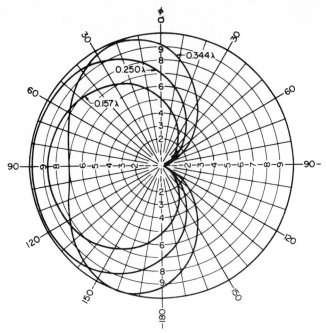

FIG. 14-41   The effect of dipole spacing on the pattern of a homing array.

ing which will avoid severe pattern deterioration at the high-frequency end while retaining sufficient system sensitivity at the low-frequency end is chosen. The sensitivity of the system is increased by increasing the directive gain of the individual elements, and hence broadband elements of resonant size are used wherever possible. At the lower frequencies, when resonant-size elements are structurally too large, the largest elements practicable are used.

Since it is difficult to predict accurately the airframe effects on homing-antenna patterns, the design is usually a step-by-step process in which experimental-model measurement data are used to supplement the free-space antenna-design concepts.

At sufficiently low frequencies, it is possible to use a symmetrical pair of balanced vertical elements as illustrated in Fig. 14-42 to achieve essentially free-space antenna characteristics. The elements are decoupled from currents and charges

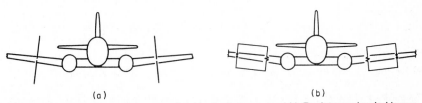

(a)   (b)

FIG. 14-42   Aircraft homing array. (*a*) Vertical elements. (*b*) Resistance-loaded loops.

induced in the wings by virtue of the balanced construction of the elements and from fuselage effects by virtue of their spacing from the fuselage. This technique is limited at low frequencies by the increasing difficulty of maintaining the degree of symmetry between the upper and lower halves of each element necessary to isolate the antenna from airframe resonance effects[32] and at high frequencies because of scattering from the fuselage and empennage. For frequencies as high as 10 MHz, balanced dipoles of this type have been found to provide patterns which are remarkably close to the free-space dipole pattern.

The use of separate antennas to generate the right- and left-hand cardioids is a sound design procedure for frequencies that are above the range for which the homing-array technique is applicable. A homing system using two resistance-loaded loop antennas is illustrated in Fig. 14-42. The balanced configuration is used to isolate the antennas from airframe resonance effects, and the elements may be placed well outboard from the fuselage. Since the low-signal portions of the cardioid patterns are oriented toward the scattering sources on the airframe, this technique can provide clean homing patterns.

REFERENCES

1   *Instruction Book for Type NA-1 Aircraft Navigation System,* Bendix Radio Div., Bendix Aviation Corp., Baltimore, Md.

2   L. R. Mullen, "The Marconi AD 7092 Series of ADF Receivers," *IRE Trans. Aeronaut. Navig. Electron.,* vol. AN-2, no. 4, December 1955.

3   J. T. Bolljahn and R. F. Reese, "Electrically Small Antennas and the Low-Frequency Aircraft Antenna Problem," *IRE Trans. Antennas Propagat.,* vol. AP-1, no. 2, October 1953, pp. 46–54.

4   A. A. Hemphill, "A Magnetic Radio Compass Antenna Having Zero Drag," *IRE Trans. Aeronaut. Navig. Electron.,* vol. AN-2, no. 4, December 1955.

5   H. H. Ward, 3d, "Analysis of the Over-Station Behavior of Aircraft Low-Frequency ADF Systems," *IRE Trans. Aeronaut. Navig. Electron.,* vol. AN-2, no. 4, December 1955, pp. 31–41.

6   P. M. Burdell, "C-141A ADF Sense Antenna Development," Eng. Rep. 5832, Lockheed-Georgia Company, May 28, 1963.

7   R. F. Harrington, *Field Computation by Moment Methods,* The Macmillan Company, New York, 1968.

8   G. L. Burke and A. J. Poggio, "Numerical Electromagnetic Code (NEC)—Method of Moments, Part 1: Program Description—Theory," Tech. Doc. 116, Naval Electronic Systems Command (ELEX 3041), July 18, 1977.

9   Computer Program Librarian, Ohio State University ElectroScience Laboratory, 1320 Kinnear Road, Columbus, Ohio 43212.

10   J. H. Richmond, "A Wire-Grid Model for Scattering by Conducting Bodies," *IEEE Trans. Antennas Propagat.,* vol. AP-14, November 1976, pp. 782–786.

11   J. J. Wang and C. E. Ryan, Jr., "Application of Wire-Grid Modelling to the Design of a Low-Profile Aircraft Antenna," *IEEE Antennas Propagat. Int. Symp. Dig.,* Stanford University, June 20–22, 1977, pp. 222–225.

12   D. L. Knepp and J. Goldhirsh, "Numerical Analysis of Electromagnetic Radiation Properties of Smooth Conducting Bodies of Arbitrary Shape," *IEEE Trans. Antennas Propagat.,* vol. AP-20, May 1972, pp. 383–388.

13   J. R. Mantz and R. F. Harrington, "Radiation and Scattering from Bodies of Revolution," *J. App. Sci. Res.,* vol. 20, 1969, p. 405.

14 Edmund K. Miller and Jeremy A. Landt, "Direct Time Domain Techniques for Transient Radiation and Scattering from Wires," *IEEE Proc.,* vol. 68, no. 11, November 1980, pp. 1396–1423.

15 V. K. Tripp, Engineering Experiment Station, Georgia Institute of Technology, private communication, March 1982.

16 R. L. Tanner, "High Voltage Problems in Flush and External Aircraft HF Antennas," *IRE Trans. Aeronaut. Navig. Electron.,* vol. AN-1, no. 4, December 1954, pp. 16–19.

17 O. C. Boileau, Jr., "An Evaluation of High-Frequency Antennas for a Large Jet Airplane," *IRE Trans. Aeronaut. Navig. Electron.,* vol. AN-3, no. 1, March 1956, pp. 28–32.

18 B. S. Zieg and W. P. Allen, "C-141A HF Antenna Development," Eng. Rep. ER 5101, Lockheed-Georgia Company, July 1963.

19 P. M. Burdell, "C-130 HF Notch Antenna Design and Development," Eng. Rep. LG 82 ER 0036, Lockheed-Georgia Company, March 1982.

20 B. S. Zieg, "JetStar HF Antenna," *Tenth Ann. USAF Antenna R&D Symp.,* University of Illinois, Monticello, Oct. 3–7, 1960.

21 R. C. Hansen (ed.), *Geometric Theory of Diffraction,* selected reprint ser., IEEE Press, Institute of Electrical and Electronics Engineers, New York, 1981.

22 W. D. Burnside, M. C. Gilreath, R. J. Marhefka, and C. L. Yu, "A Study of KC-135 Aircraft Antenna Patterns," *IEEE Trans. Antennas Propagat.,* vol. AP-23, May 1975, pp. 309–316.

23 C. L. Yu, W. D. Burnside, and M. C. Gilreath, "Volumetric Pattern Analysis of Airborne Antennas," *IEEE Trans. Antennas Propagat.,* vol. AP-26, September 1978, pp. 636–641.

24 W. P. Cooke and C. E. Ryan, Jr., "A GTD Computer Algorithm for Computing the Radiation Patterns of Aircraft-Mounted Antennas," *IEEE Antennas Propagat. Int. Symp. Prog. Dig.,* Laval University, Quebec, June 2–6, 1980, pp. 631–634.

25 C. E. Ryan, Jr., "Analysis of Antennas on Finite Circular Cylinders with Conical or Disk End Caps," *IEEE Trans. Antennas Propagat.,* vol. AP-20, July 1972, pp. 474–476.

26 C. E. Ryan, Jr., and R. Luebbers, "Volumetric Patterns of a Circumferential Slot Antenna on a Conically-Capped Finite Circular Cylinder," ElectroScience Lab., Ohio State Univ. Res. Found. Rep. 2805-1, Cont. DAAA21-69-C-0535, 1970.

27 H. Jasik, U.S. Patent 2,700,112.

28 A. Doring, U.S. Patent 2,644,090.

29 A. A. Pistolkors, "Theory of the Circular Diffraction Antenna," *IRE Proc.,* vol. 36, no. 1, January 1948, p. 56.

30 N. R. Ray, "L-1011 VOR Antenna System Design and Development," Eng. Rep. ER-10820, Lockheed-Georgia Company, Aug. 13, 1970.

31 A. G. Kandoian, "The Aircraft Omnidirectional Antenna Problem for UHF Navigational Systems," *Aeronaut. Eng. Rev.,* vol. 12, May 1953, pp. 75–80.

32 P. S. Carter, Jr., "Study of the Feasibility of Airborne HF Direction Finding Antenna Systems," *IRE Trans. Aeronaut. Navig. Electron.,* vol. AN-4, no. 1, March 1957, pp. 19–23.

BIBLIOGRAPHY

Bennett, F. D., P. D. Coleman, and A. S. Meier: "The Design of Broadband Aircraft Antenna Systems," *IRE Proc.,* vol. 33, October 1945, pp. 671–700.

Bolljahn, J. T., and J. V. N. Granger: "The Use of Complementary Slots in Aircraft Antenna Impedance Measurements," *IRE Proc.,* vol. 39, no. 11, November 1951, pp. 1445–1448.

Granger, J. V. N.: "Shunt Excited Flat Plate Antennas with Applications to Aircraft Structures," *IRE Proc.,* vol. 38, no. 3, March 1950, pp. 280–287.

————: "Design Limitations on Aircraft Antenna Systems," *Aeronaut. Eng. Rev.,* vol. 11, no. 5, May 1952, pp. 82–87.

————: "Designing Flush Antennas for High-Speed Aircraft," *Electronics,* vol. 11, March 1954.

———— and T. Morita, "Radio-Frequency Current Distributions on Aircraft Structures," *IRE Proc.,* vol. 39, no. 8, August 1951, pp. 932–938.

Haller, G. L.: "Aircraft Antennas," *IRE Proc.,* vol. 30, no. 8, August 1942, pp. 357–362.

Hurley, H. C., S. R. Anderson, and H. F. Keary: "The Civil Aeronautics Administration VHF Omnirange, *IRE Proc.,* vol. 39, no. 12, December 1951, pp. 1506–1520.

Kees, H., and F. Gehres: "Cavity Aircraft Antennas," *Electronics,* vol. 20, January 1947, pp. 78–79.

Lee, K. S. H., T. K. Liu, and L. Marin: "EMP Response of Aircraft Antennas," *IEEE Trans. Antennas Propagat.,* vol. AP-26, no. 1, January 1978, pp. 94–99.

Moore, E. J.: "Factor of Merit for Aircraft Antenna Systems in the Frequency Range 3–30 Mc," *IRE Trans. Antennas Propagat.,* vol. AP-3, August 1952, pp. 67–73.

Raburn, L. E.: "A VHF-UHF Tail-Cap Antenna," *IRE Proc.,* vol. 39, no. 6, June 1951, pp. 656–659.

————: "Faired-In ADF Antennas," *IRE Conv. Rec.,* vol. 1, part 1, March 1953, pp. 31–38.

Sinclair, G., E. C. Jordan, and E. W. Vaughan: "Measurement of Aircraft Antenna Patterns Using Models," *IRE Proc.,* vol. 35, no. 12, December 1947, pp. 1451–1462.

Tanner, R. L.: "Shunt-Notch-Fed HF Aircraft Antennas," *IRE Trans. Antennas Propagat.,* vol. AP-6, no. 1, January 1958, pp. 35–43.

Chapter 15

Seeker Antennas

James M. Schuchardt

Millimeter Wave Technology, Inc.

Dennis J. Kozakoff

Millimeter Wave Technology, Inc.

Maurice M. Hallum III

U.S. Army Missile Command
Redstone Arsenal

15-1 INTRODUCTION

This chapter will focus on antennas forward-mounted in a missile functioning in the role of seeker of target emissions. Such a seeker antenna is a critical part of the entire airborne guidance system, which includes the missile radome, seeker antenna, radio-frequency (RF) receiver, antenna gimbal, autopilot, and airframe. The seeker has several functions: to receive and track target emissions so as to measure line of sight and/or line-of-sight angular rate, to measure closing velocity, and to provide steering commands to the missile autopilot and subsequently to the control surfaces.[1-6] The signals received by the missile-mounted seeker antenna or antennas are thus utilized in a closed-loop servocontrol system to guide the missile to the target.

The seeker RF elements, radome and antenna, initially discriminate in angle through the seeker antenna's pencil beam. This beam can be steered mechanically (the whole antenna moves), electromechanically (an antenna element such as a subreflector moves), or electronically (there is a phased-array movement). In some situations, the antenna is stationary and only forward-looking; thus beam motion occurs only if the entire missile airframe rotates. Fixed seeker-antenna beams are used when a pursuit navigation (or a variant) guidance algorithm is used. The use of movable seeker-antenna beams occurs when a proportional navigation (or a variant) guidance algorithm is used.

15-2 SEEKER-ANTENNA RADIO-FREQUENCY CONSIDERATIONS

Seeker antennas are defined as forward-mounted antennas in a missile or a projectile which function in the role of reception of target emissions. The target emissions result from either target-reflected radar signals, antiradiation homing (ARH) in the case of an emitting target such as ground or airborne radar, or passive (radiometric) emissions which occur because of natural background radiation in accordance with the Planck equation.

Table 15-1 presents a summary of commonly employed seeker-antenna types; detailed performance and design criteria for many of the basic antenna radiator types are found in the appropriate chapters in this handbook. General information on many of these antenna elements relative to various seeker categories is presented below.

Many system factors must also be considered in the antenna-design process. Antenna beamwidth determines whether spatial resolution of multiple targets occurs.[7] The earlier that resolution takes place in a flight path, the more time the missile will have to correct errors induced by tracking the multiple-target centroid. Low sidelobes are important because they decrease electronic-jamming effects. This is particularly important if the missile gets significantly close to the jammer during the engagement. Noise injected in this manner can increase the final miss distance.

The body-fixed antenna configuration confronts designers with unique problems. The beam may be very broad for a large field of view (FOV), or it may be rapidly steered to form a tracking beam.[4] The broad-FOV approach requires that the airframe and the autopilot be more restricted in their responsiveness. The steered-beam approach requires rapid beam forming and signal processing to isolate the missile-body rotational motion properly from the actual target motions.

TABLE 15-1 Missile-Borne Seeker-Antenna Summary

Basic radiator	Body-fixed				Gimballed				Comments
	Monopulse	Sequential lobing	Conscan*	Multimode	Monopulse	Sequential lobing	Conscan	Multimode	
Stripline	X	X	X		X	X			Narrowband, intermediate power, dual-polarization capability
Microstrip	X	X	X	X	X	X		X	Narrowband, intermediate power, dual-polarization capability
Spiral				X	X	X		X	Broadband, circular polarization, low power
Ring array					X	X			Narrowband, high power, linear polarization
Paraboloid			X		X	X	X		High power, dual-polarization capability
Lens			X		X	X	X		High power, dual-polarization capability
Slots	X	X	X	X	X	X			Narrowband, linear polarization
Monopoles	X	X			X	X			Narrowband, linear polarization

*Achieved with a rolling airframe and squinted beams.

It is often necessary to be able to integrate and test key receiver front-end elements and the RF elements of the transmitter as part of the seeker-antenna assembly. For example, it is generally possible to integrate in one package the feed antenna, monopulse comparator, mixer or mixers, local oscillator, and transmitter oscillator.[8]

15-3   IMPACT OF AIRFRAME ON ANTENNA DESIGN

The missile-borne guidance antenna enters the picture as depicted in Fig. 15-1. The figure shows the major functional features present in missiles and projectiles with on-board guidance. For a body-fixed on-board sensor antenna, the body isolation and beam steering are not always present. The implementations of these functions vary from system to system. Factors such as operational altitude and range, targets to be engaged, missile speed, and terrain over which missions are to be performed also influence the design.[9,10]

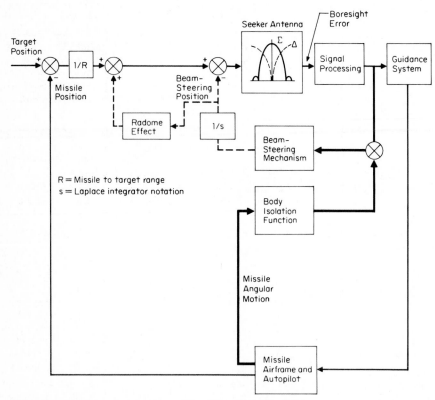

FIG. 15-1 Seeker antenna and overall guidance functional diagram.

The airframe and the autopilot steer the missile according to the guidance-computer commands to minimize the final miss distance at impact. The dynamic behavior of the missile in response to commands consists of attitude changes that cause the antenna to receive through an ever-changing portion of the radome. This, coupled with the natural geometry changes that occur as an engagement takes place, makes the antenna appear to have time-varying directivity and gain properties. The response characteristics of the airframe and the autopilot generate a pseudo-noise influence of the radome-antenna combination.

15-4 SEEKER-ANTENNA MECHANICAL CONSIDERATIONS

Generally, the seeker antenna is of the maximum size possible within the limits of the missile-body diameter in order to maximize antenna gain and minimize the antenna beamwidth. Alternately conformal- and flush-mounted antennas (most often forward-mounted) utilize the entire conical forward area to achieve maximum gain.[11]

Gimballed antenna systems often must mechanically steer not only the antenna structure but also the attendant beam-forming network and critical transmitter and receiver elements. As the mechanical-steering rates become excessive, the conformal-mounted antenna with electronically steered beams must be used.[10-16] When all the seeker-antenna and associated hardware are gimballed together, the use of lightweight materials may achieve the desired results. Lightweight techniques include:[8,17]

1 The use of lightweight honeycomb materials (including low-density dielectric foams) for both filler and structural-load-bearing surfaces. Foam reflectors and waveguide elements including feed horns can be machined and then metallized by vacuum deposition, plating, or similar techniques.

2 The use of stripline techniques. Using multilayer techniques, one can have a layer with microstrip radiating elements, a layer with beam-forming networks, and a layer with beam-switching and signal-control elements such as a diode attenuator and phase shifters. The use of plated-through holes and/or pins to make RF connections can eliminate cables and connectors. Excess substrate can be removed and printed-circuit-board edges plated in lieu of the use of mode-suppression screws.

3 The use of solid dielectric lenses, above Ku band, with no significant weight penalty. Lenses can also be zoned to remove excess material.[18]

15-5 APERTURE TECHNIQUES FOR SEEKER ANTENNAS

Precise control of both amplitude and phase to permit aperture illumination tailoring is necessary for achieving a well-behaved antenna pattern.[19-25] Aperture tapering of rotationally symmetric illuminations starts with the feed itself and continues by varying the energy across the aperture by several methods: corporate-power splitting, aperture-element proportional size or spacing, surface control of the reflector or lens surface, and element thinning. Several of these aspects are discussed below.

Reflector or Lens Antennas

A consideration in the design of monopulse reflector antennas is the four-horn-feed design and the effects of aperture blockage. One criterion posed for an optimum monopulse feed configuration is maximizing the product of the sum times the derivative of the difference error signal.[26] The monopulse horn size for optimum monopulse sensitivity is shown in Fig. 15-2. The impact of optimizing monopulse feed design on antenna performance is illustrated in Fig. 15-3, in which a 5.5-in- (139.7-mm-) diameter aperture is assumed in the calculations. These data illustrate that at millimeter wavelengths (above 30 GHz) the effects of aperture blockage on antenna performance are small for this size of antenna even when a reflector is used.

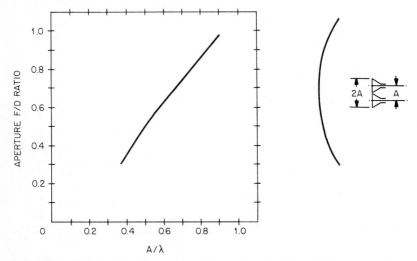

FIG. 15-2 Monopulse antenna horn size or spacing for optimum monopulse sensitivity.

Other criteria, such as the requirement for equal E- and H-plane antenna beamwidths, may enter into the details of monopulse feed design. An example of a monopulse-feed antenna operating near 95 GHz that meets equal-beamwidth requirements is shown in Fig. 15-4.[27] Here, the subaperture dimensions are very small [0.080 by 0.100 in (2.032 by 2.54 mm)], and an electroformed process was required in the feed fabrication.

A wide-angle-scan capability can be achieved by using a twist-reflector concept with a rotatable planar mirror.[28,29] In this configuration, the forward nose of the missile can be approximately paraboloidal-shaped, or the paraboloid can be located inside a higher-fineness-ratio radome. This paraboloid is composed of horizontal metal strips. The movable planar twist reflector uses 45°-oriented strips ¼λ above the planar metallic reflector. (Grids may also be used to combine widely separated frequencies to permit dual-band operation.[30])

Beam steering is obtained by moving the planar mirror. The steering is enhanced because for every 1° that the mirror moves the beam moves 2°. By using a parallel-

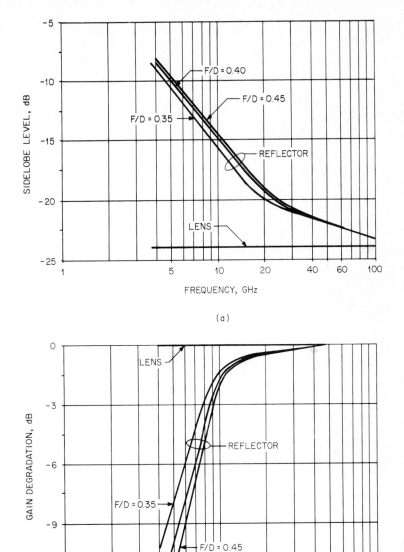

FIG. 15-3 Performance of optimized four-horn monopulse-fed 5.5-in- (139.7-mm-) diameter apertures. (*a*) Sidelobe level. (*b*) Gain degradation.

ogram gimbal to move the planar mirror, beam accelerations of $20,000°/s^2$ have been demonstrated.[29] In this situation a 42° beam scan was achieved with 21° of mirror motion, and less than 2 dB of sidelobe degradation of the sum pattern occurred.

By using a corrugated horn with a rotationally symmetric feed pattern providing a -17-dB edge taper (capable of achieving a -30-dB sidelobe level), one can further taper by varying the inner- and outer-surface contour of a collimating lens to achieve a circular Taylor amplitude distribution ($\bar{n} = 7$) with a -40-dB sidelobe level.[18] Practically, one can make the lens out of a material having an $\varepsilon_r = 6.45$ (titania-loaded polystyrene) and follow up with a surface-matching layer made out of a material having an $\varepsilon_r = 2.54$ (Rexolite-polystyrene).

Planar-Array Antennas

Array techniques can be used to provide a nearly planar (flat-plate) aperture. Commonly used are arrays that provide symmetric patterns and offer a maximum use of the available circular aperture with reduced grating-lobe potential. In such an array, higher aperture efficiency with low sidelobes can be achieved

FIG. 15-4 Monopulse-feed antenna (95-GHz operation). *(After Ref. 27.)*

because of reduced aperture blockage and spillover. In monopulse applications, it is noted that the ring array contains no central element and exhibits quadrantal symmetry, as indicated in Fig. 15-5.

The choices in a ring-array design include the number of rings and the number of elements in each ring and the feed network.[31,32] Constraints include:

1 Minimum distance of the outer elements to the antenna edge

2 Minimum spacing of the elements as impacted by excess mutual coupling (between elements and quadrants and feed geometry)

3 Desired radiation pattern (as impacted by the density tapering)

Optimization of small ring arrays (having diameters $<10\lambda$) is often carried out heuristically or nonlinearly. Examples of array geometries that have been analytically determined to have -17- to -20-dB sidelobes are given in Table 15-2 for the array geometry shown in Fig. 15-5a.

Array elements can be of many types.[33-40] A two-dimensional array of waveguide slots (Fig. 15-5b) excited by a network of parallel waveguides forming the antenna structural supports has been used.[39] Waveguide slot arrays have demonstrated low cross-polarized response as well. The bandwidth of a waveguide slot array is inversely proportional to the size of the array. A variety of outer contours can be utilized to conform to the space available.

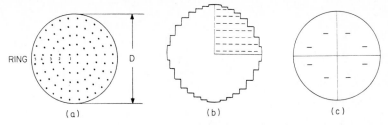

FIG. 15-5 Monopulse planar-array geometries. (*a*) Five-ring monopulse array. (*b*) Waveguide slot array. (*c*) Thinned-slot (image) ring array.

TABLE 15-2 Data on Five-Ring Monopulse Uniformly Excited Array*

Ring number	Case A† Directivity = 27.6 dB Maximum sidelobe level = −20 dB Radius, λ	Case B† Directivity = 28.1 dB Maximum sidelobe level = −17 db Radius, λ
r_1	0.45	0.50
r_2	1.25	1.33
r_3	2.08	2.16
r_4	3.06	2.98
r_5	3.72	3.72

*After Ref. 31.

†The maximum outer diameter = 8.16λ for both cases.

Techniques for thinning these types of arrays reduce complexity.[41] By using an image-element approach, a reduction of the number of elements by over 90 percent is possible (Fig. 15-5c). This method also readily allows for integration of the monopulse comparator.

Multimode and Single-Mode Spirals

The multimode spiral antenna is a very broadband, broad-beamed, circularly polarized antenna which is amenable to ARH applications.[42] Such an antenna can also be synthesized by using a circular array.[43] The circuit to resolve monopulse sum-and-difference patterns is shown in Fig. 15-6; typical angular coverage is from ±30 to 40°. The rather wide instantaneous FOV makes these antennas attractive for body-fixed applications.

The printed-circuit construction of four-arm spirals limits their use to low-power applications. Thus, this type of antenna is almost always used for passive (nontransmitting) applications. Loading techniques can be employed to reduce size, permitting operation at low frequencies.[44] Current fabrication technology permits high-frequency operation to above 40 GHz. Single-mode spirals are also very useful as elements in small arrays.

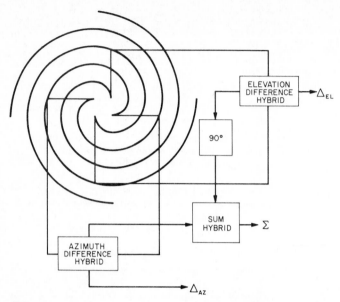

FIG. 15-6   Monopulse circuit for a four-arm spiral antenna.

15-6   SMALL ARRAYS FOR SEEKER ANTENNAS

Arrays of two, three, four, and five elements can be used for seeker antennas in small interferometers. These types of arrays are usually fixed-mounted (nongimballed) and used when pursuit navigation algorithms are suitable. With the addition of a gyroscope to sense missile angular motion, line-of-sight data can be derived.[45]

Good sum-and-difference or phase monopulse angle-tracking performance can be achieved when the array elements, the array beamwidth, and the array element spacing are properly chosen. Broadband performance is often achievable by using antenna elements such as spirals or log-periodic antennas.

The receiving arrays as shown in Fig. 15-7 are either two-element (rolling missile) or four-element (stabilized missile). The third or fifth element can be used for transmission. The equations for the receiving voltage pattern are

$$\Sigma_4 = \left[1 + \cos \left(\frac{\pi d}{\lambda} \sin \theta \right) \right] f(\theta)$$

$$\Sigma_2 = \left[\cos \left(\frac{\pi d}{\lambda} \sin \theta \right) \right] f(\theta)$$

$$\Delta_2 = \left[\sin \left(\frac{\pi d}{\lambda} \sin \theta \right) \right] f(\theta)$$

where $f(\theta)$ = element pattern [$\cos^m(\theta)$ is often used.]
   d = center-to-center element spacing
   λ = operating wavelength

These equations can be used to determine either the principal-plane (0°) or the intercardinal-plane (45°) pattern. Figure 15-8 indicates the appropriate array dimen-

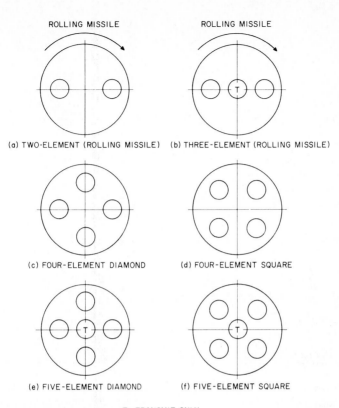

ROLLING MISSILE

ROLLING MISSILE

(a) TWO-ELEMENT (ROLLING MISSILE)   (b) THREE-ELEMENT (ROLLING MISSILE)

(c) FOUR-ELEMENT DIAMOND

(d) FOUR-ELEMENT SQUARE

(e) FIVE-ELEMENT DIAMOND

(f) FIVE-ELEMENT SQUARE

T = TRANSMIT ONLY.

FIG. 15-7   Small-array geometries.

sions and equations to use for each case. Also shown are the two comparator networks used for either diamond or square arrays. Small arrays can be used by themselves, or they can illuminate apertures such as reflectors or lenses. Control of the array element pattern is important in reducing energy spillover beyond the difference-pattern lobes. Often one can achieve this control by using end-fire techniques with the elements. For example, dielectric rods can be used for waveguide, printed-circuit, and spiral elements. An example of this approach is shown in Fig. 15-9.

A four-element array can serve as an RF seeker by utilizing a phase processor. The effects of system errors for such an array on antenna patterns, S curves, and angular sensitivity as ascertained analytically can be used to form a catalog of results that can be employed to diagnose fabrication errors or tolerances based on measured data.[46]

15-7   RADOME EFFECTS

The introduction of a protective radome over a seeker antenna results in an apparent boresight error (BSE) because of the wavefront phase modification over the seeker-

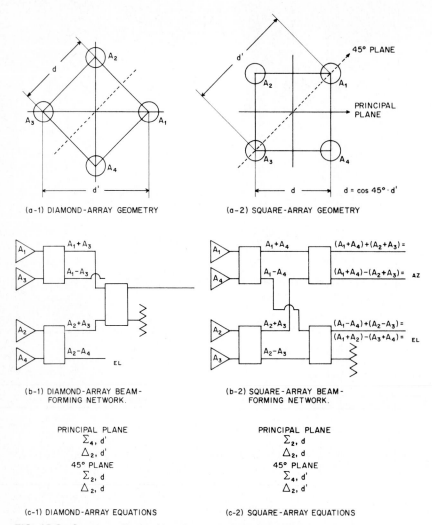

(a-1) DIAMOND-ARRAY GEOMETRY

(a-2) SQUARE-ARRAY GEOMETRY

$d = \cos 45° \cdot d'$

(b-1) DIAMOND-ARRAY BEAM-
FORMING NETWORK.

(b-2) SQUARE-ARRAY BEAM-
FORMING NETWORK.

(c-1) DIAMOND-ARRAY EQUATIONS

PRINCIPAL PLANE
Σ_4, d'
Δ_2, d'
45° PLANE
Σ_2, d
Δ_2, d

(c-2) SQUARE-ARRAY EQUATIONS

PRINCIPAL PLANE
Σ_2, d
Δ_2, d
45° PLANE
Σ_4, d'
Δ_2, d'

FIG. 15-8 Square or diamond interferometer array parameters.

antenna aperture. These effects need to be carefully considered in the selection of a radome design since they impact seeker-system performance.[9,47–49]

In the case of pursuit guidance, the miss distance is proportional to the BSE. Conversely, in the case of proportional guidance, it is the boresight-error slope (BSES) that more directly impacts miss distance. The relationship between BSE or BSES and miss distance is not straightforward but generally is quantified only by a hardware-in-the-loop (HWIL)[50,51] or computer simulation that takes into account a variety of scenarios and flight conditions. A BSE requirement of better than 10 mrad (or a BSES

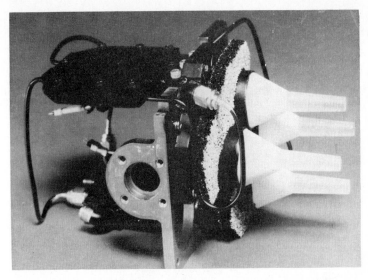

FIG. 15-9 Four-element seeker array using dielectric rods and spiral antennas. *(Photo courtesy Eaton Corporation, AIL Division.)*

requirement better than 0.05 deg/deg and maximum transmission loss of 1.0 dB are typical for many radome types.

The method most often employed for the evaluation of radome effects on seeker-antenna performance utilizes open-loop testing in a suitable anechoic-chamber facility.[52] This requires a precalibration of the monopulse error-channel sensitivity in volts per degree without the radome and subsequently measuring the error-channel outputs with the radome over the seeker antenna.

Broadband ARH antennas generally require the use of multilayer radome walls to obtain a required radome bandwidth commensurate with antenna performance. Figure 15-10 illustrates hemispherical-radome performance for half-wave, full-wave, *A* and *B* sandwich walls. Half-wave radomes generally have a useful bandwidth on the order of 5 to 20 percent, depending on dielectric constant and radome-fineness ratio. Higher-order wall radomes have considerably narrower useful bandwidths and thus are limited to narrowband-antenna applications. Figure 15-10 also demonstrates transmission loss versus frequency for a half-wave wall radome when radome-fineness ratio is taken as a parameter.

15-8   EVALUATION OF SEEKER ANTENNAS

Seeker-antenna testing proceeds from conventional methods utilizing anechoic-chamber techniques with the antenna alone.[53-57] Next, open-loop testing including critical elements of the seeker electronics (S curves) is most often performed. Ultimately a variety of simulations are used to ascertain the seeker-antenna performance in both benign and complex environments.

The many nonlinear elements in the guided missile make complete closed-form

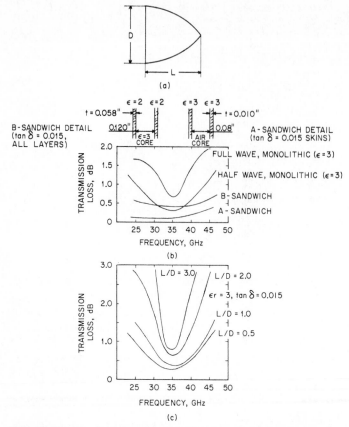

FIG. 15-10   Missile-radome transmission data. (*a*) Radome geometry. (*b*) Hemispherical-radome (*L/D* = 0.5) performance for a low-velocity (<Mach 1) missile. (*c*) Monolithic half-wave radome performance for various fineness (*L/D*) ratios.

analytical simulations unsuitable for final analyses. The precise simulation of these elements is also difficult. The final analyses are best done with an HWIL simulation.[49-51] As many hardware items as possible should be inserted, leaving only the propulsion, aerodynamics, and instrument feedback to be simulated analytically.

There are several concepts for implementing a simulation with seeker hardware. Placing the seeker in an anechoic chamber with the target of appropriate characteristics moved about according to the engagement geometry is one method. Flying a target with the seeker on an orientation table is another.

The HWIL simulation serves as a tool not only for statistical performance analysis but for provision of validity information for an improved all-analytical model of the system. The HWIL method provides the only valid present means of evaluating the interaction of the radome-antenna combination in a missile-guidance-seeker environment.

REFERENCES

1   J. F. Gulick, "Overview of Missile Guidance," *IEEE Eascon Rec.*, September 1978, pp. 194–198.
2   W. Kelley, "Homing Missile Guidance—A Survey of Classical and Modern Techniques," *Southcon Tech. Prog.*, January 1981.
3   C. F. Price et al., "Performance Evaluation of Homing Guidance Laws for Tactical Missiles," final rep. on N00014-69-C-0391 (AD761626), January 1973.
4   P. G. Savage, "A Strapdown Phased Array Radar Tracker Loop Concept for a Radar Homing Missile," *AIAA Guid. Cont. Flight Conf.*, August 1969, pp. 1–8.
5   G. M. Siouris, "Noise Processor in a Homing Radar Seeker," *NTZ J.*, July 1973, pp. 321–323.
6   G. L. Slater and W. R. Wells, "Optimal Evasive Tactics against a Proportional Navigation Missile with Time Delay," *J. Spacecr. Rockets,* May 1973, pp. 309–313.
7   H. G. Oltman and M. E. Beebe, "Millimeter Wave Seeker Technology," *AIAA Guid. Cont. Conf.*, August 1978, pp. 148–158.
8   C. R. Seashore et al., "MM-Wave Radar and Radiometer Sensors for Guidance Systems," *Microwave J.*, August 1979, pp. 41–59.
9   P. Garnell and D. J. East, *Guided Weapons Control Systems,* Pergamon Press, London, 1977.
10   R. A. Chervolk, "Coherent Active Seeker Guidance Concepts for Tactical Missiles," *IEEE Eascon Rec.,* September 1978, pp. 199–202.
11   R. C. Hansen (ed.), *Conformal Antenna Array Design Handbook,* AIR310E, U.S. Navy, Washington, September 1981.
12   P. C. Bargeliotes and A. F. Seaton, "Conformal Phased Array Breadboard," final rep. on N0019-76-C-0495, ADA038-350.
13   T. W. Bazire et al., "A Printed Antenna/Radome Assembly (RADANT) for Airborne Doppler Navigation Radar," *Fourth Europ. Microwave Conf. Proc.,* Montreux, September 1974, pp. 494–498.
14   H. S. Jones, Jr., "Some Novel Design Techniques for Conformal Antennas," *IEE Conf. Antennas Propagat.,* London, Nov. 28–30, 1978, pp. 448–452.
15   Naval Air Systems Command, "Conformal Antennas Research Program Review and Workshop," AD-A015-630, April 1975.
16   A. T. Villeneuve et al., "Wide-Angle Scanning of Linear Arrays Located on Cones," *IEEE Trans. Antennas Propagat.,* vol. AP-22, January 1974, pp. 97–103.
17   J. S. Yee and W. J. Furlong, "An Extremely Lightweight Electronically Steerable Microstrip Phased Array Antenna," *IEEE Antennas Propagat. Int. Symp. Dig.,* May 1978, pp. 170–173.
18   D. K. Waineo, "Lens Design for Arbitrary Aperture Illumination," *IEEE Antennas Propagat. Int. Symp. Dig.,* October 1976, pp. 476–479.
19   E. T. Bayliss, "Design of Monopulse Antenna Difference Patterns with Low Sidelobes," *Bell Syst. Tech. J.,* May–June 1968, pp. 623–650.
20   L. J. Du and D. J. Scheer, "Microwave Lens Design for a Conical Horn Antenna," *Microwave J.,* September 1976, pp. 49–52.
21   P. W. Hannan, "Optimum Feeds for All Three Modes of a Monopulse Antenna (Parts I and II)," *IRE Trans. Antennas Propagat.,* vol. AP-9, September 1961, pp. 444–464.
22   R. W. Kreutel, "Off-Axis Characteristics of the Hyperboloidal Lens Antenna," *IEEE Antennas Propagat. Int. Symp. Dig.,* May 1978, pp. 231–234.
23   S. Pizette and J. Toth, "Monopulse Networks for a Multielement Feed with Independent Control of the Three Monopulse Modes," *IEEE Microwave Theory Tech. Int. Symp. Dig.,* April–May 1979, pp. 456–458.
24   R. L. Sak and A. Sarremejean, "The Performance of the Square and Conical Horns as Monopulse Feeds in the Millimetric Band," *IEE Ninth Europ. Microwave Conf. Proc.,* September 1979, pp. 191–195.

25 P. A. Watson and S. I. Ghobrial, "Off-Axis Polarization Characteristics of Cassegrainian and Front-Fed Antennas," *IEEE Trans. Antennas Propagat.*, vol. AP-20, November 1972, pp. 691–698.

26 D. R. Rhodes, *Introduction to Monopulse,* McGraw-Hill Book Company, New York, 1959.

27 D. J. Kozakoff and P. P. Britt, "A 94.5 GHz Variable Beamwidth Zoned Lens Monopulse Antenna," *IEEE Southeastcon Proc.*, April 1980, pp. 65–68.

28 E. O. Houseman, Jr., "A Millimeter Wave Polarization Twist Antenna," *IEEE Antennas Propagat. Int. Symp. Dig.*, May 1978, pp. 51–53.

29 D. K. Waineo and J. F. Koneczny, "Millimeter Wave Monopulse Antenna with Rapid Scan Capability," *IEEE Antennas Propagat. Int. Symp. Dig.*, June 1979, pp. 477–480.

30 L. Goldstone, "Dual Frequency Antenna Locates MM Wave Transmitters," *Microwave Syst. N.*, October 1981, pp. 69–74.

31 D. A. Huebner, "Design and Optimization of Small Concentric Ring Arrays," *IEEE Antennas Propagat. Int. Symp. Dig.*, May 1978, pp. 455–458.

32 A. R. Lopez, "Monopulse Networks for Series-Feeding an Array Antenna," *IEEE Trans. Antennas Propagat.*, vol. AP-16, July 1968, pp. 68–71.

33 S. W. Bartley and D. A. Huebner, "A Dual Beam Low Sidelobe Microstrip Array," *IEEE Antennas Propagat. Int. Symp. Dig.*, June 1979, pp. 130–133.

34 C. W. Garven et al., "Missile Base Mounted Microstrip Antennas," *IEEE Trans. Antennas Propagat.*, vol. AP-25, September 1977, pp. 604, 616.

35 C. S. Malagisi, "Microstrip Disc Element Reflect Array," *IEEE Eascon Rec.*, September 1978, pp. 186–192.

36 S. Nishimura et al., "Franklin-Type Microstrip Line Antenna," *IEEE Antennas Propagat. Int. Symp. Dig.*, June 1979, pp. 134–137.

37 P. K. Park and R. S. Elliott, "Design of Collinear Longitudinal Slot Arrays Fed by Boxed Stripline," *IEEE Trans. Antennas Propagat.*, vol. AP-29, January 1981, pp. 135–140.

38 *Flat Plate Antennas,* Rantec Div., Emerson Electric Co., June 1969.

39 G. G. Sanford and L. Klein, "Increasing the Beamwidth of a Microstrip Radiating Element," *IEEE Antennas Propagat. Int. Symp. Dig.*, June 1979, pp. 126–129.

40 D. H. Schaubert et al., "Microstrip Antennas with Frequency Agility and Polarization Diversity," *IEEE Trans. Antennas Propagat.*, vol. AP-29, January 1981, pp. 118–123.

41 W. H. Sasser, "A Highly Thinned Array Using the Image Element," *IEEE Antennas Propagat. Int. Symp. Dig.*, June 1980, pp. 150–153.

42 J. D. Dyson, "Multimode Logarithmic Spiral Antennas," *Nat. Electron. Conf.*, vol. 17, October 1961, pp. 206–213.

43 J. M. Schuchardt and W. O. Purcell, "A Broadband Direction Finding Receiving System," *Martin Marietta Interdiv. Antenna Symp.*, August 1967, pp. 1–14.

44 T. E. Morgan, "Reduced Size Spiral Antenna," *IEE Ninth Europ. Microwave Conf. Proc.*, September 1979, pp. 181–185.

45 E. R. Feagler, "The Interferometer as a Sensor for Missile Guidance," *IEEE Eascon Rec.*, September 1978, pp. 203–210.

46 L. L. Webb, "Analysis of Field-of-View versus Accuracy for a Microwave Monopulse," *IEEE Southeascon Proc.*, April 1973, pp. 63–66.

47 G. Marales, "Simulation of Electrical Design of Streamlined Radomes," *AIAA Summer Computer Simulation Conf.*, Toronto, July 1979, pp. 353–354.

48 A. Ossin et al., "Millimeter Wavelength Radomes," Rep. AFML-TR-79-4076, Wright-Patterson Air Force Base, Dayton, Ohio, July 1979.

49 K. Siwak et al., "Boresight Errors Induced by Missile Radomes," *IEEE Trans. Antennas Propagat.*, vol. AP-27, November 1979, pp. 832–841.

50 R. F. Russell and S. Massey, "Radio Frequency System Simulator," *AIAA Guidance Cont. Conf.*, August 1972, pp. 72–861.

51 D. W. Sutherlin and C. L. Phillips, "Hardware-in-the-Loop Simulation of Antiradiation Missiles," *IEEE Southeastcon Proc.*, Clemson, S.C., April 1976, pp. 43–45.

52 J. M. Schuchardt et al., "Automated Radome Performance Evaluation in the RFSS Facility at MICOM," *Proc. 15th EM Windows Symp.*, Atlanta, June 1980.

53 R. C. Hansen, "Effect of Field Amplitude Taper on Measured Antenna Gain and Side-lobes," *Electron. Lett.,* April 1981, pp. 12–13.

54 R. C. Hansen et al., "Sidewall Induced Boresight Error in an Anechoic Chamber," *IEEE Trans. Aerosp. Electron. Syst.,* vol. AES-7, November 1971, pp. 1211–1213.

55 D. J. Kaplan et al., "Rapid Planar Near Field Measurements," *Microwave J.,* January 1979, pp. 75–77.

56 A. S. Thompson, "Boresight Shift in Phase Sensing Monopulse Antennas Due to Reflected Signals," *Microwave J.,* May 1966, pp. 47–48.

57 R. D. Monroe and P. C. Gregory, "Missile Radar Guidance Laboratory," *Range Instrumentation–Weapons System Testing and Related Techniques,* AGARD-AG-219, vol. 219, 1976.

Chapter 16

Direction-Finding Antennas and Systems

Hugh D. Kennedy
William Wharton

Technology for Communications International

Radio Direction-Finding Systems

A radio direction-finding (DF) system is basically an antenna-receiver combination arranged to determine the azimuth of a distant emitter. In practice, however, the objective of most DF systems is to determine the location of the emitter.[1] Virtually all DF systems derive emitter location from an initial determination of the arrival angle of the received signal. Figure 16-1*a* shows how the location of the emitter is found if azimuth angles are measured at two DF stations connected by a communication link and separated by a distance that is comparable with the distance to the emitter. The determination of emitter location by using azimuth angles measured at two or more D𝖥 stations is known as *horizontal or azimuth triangulation*.

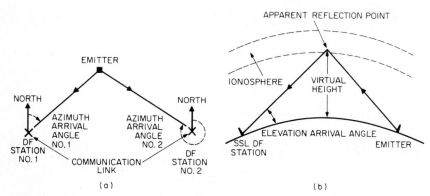

FIG. 16-1   Emitter location by triangulation. (*a*) Horizontal (azimuth) triangulation. (*b*) Vertical triangulation (single-station location).

In the high-frequency (HF) band (approximately 2 to 30 MHz), emitter location by using only a single DF station is possible. Such an arrangement is known as a *single-station-location (SSL) system*.[2] A DF-SSL system can only be used when the signal arrives at the DF station after refraction through the ionosphere (i.e., via sky-wave propagation), as illustrated in Fig. 16-1*b*. The DF-SSL system must be able to measure both the azimuth and the elevation angles[3] of the signal arriving at the DF station. Furthermore, the height of the ionosphere must be either known or determined. The measured elevation angle, in conjunction with knowledge of the height of the ionosphere, enables the distance to the emitter to be established. This process is known as *vertical triangulation*. Emitter location is then calculated by using azimuth and distance.

Applications of Direction-Finding Systems

There are three principal applications for DF systems which influence the details of their design:

1   DF systems that are designed to determine the unknown location of an emitter. Such systems may be fixed or movable.

2 Navigation systems designed to determine the location of the DF system itself with respect to emitters of known location. The DF system is on a moving vehicle (ship or aircraft), and azimuth angles are measured to two or more emitters at the same time or at successive times on the same emitter. The determination of the position of the vehicle is the complement of the horizontal triangulation process illustrated in Fig. 16-1*a*.

3 Homing systems that are designed to guide a vehicle carrying a DF system toward an emitter which may be either a beacon of known location or an emitter of unknown location.

System Approach to Direction Finding

The essential components of any DF system are shown in Fig. 16-2. They comprise:

1 An antenna system to collect energy from the arriving signal

2 A receiving system to measure the response of the antenna system to the arriving signal

3 A processor to derive the required DF information (for example, azimuth and elevation angles and emitter location) from the output of the receiver

4 An output device to present the required DF information in a form convenient to the user

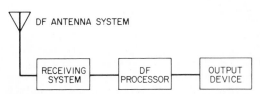

FIG. 16-2 Essential components of a DF system.

In the past, each component of a DF system was often regarded as a separate design problem. Attention was focused mainly on the antenna because in general it imposed the major constraints on performance. Some form of symmetry was usually required, and the choice of antenna type was usually quite limited. However, the advent of small high-speed digital computers has meant that DF-system designers now have more freedom in their choice of DF antennas and that full use can now be made of antenna-system responses. That is, most older systems made use of either amplitude response or phase comparison but not both. Furthermore, most older systems were restricted to the use of only one linear polarization. The employment of real-time signal-processing techniques now makes it possible to use both phase and amplitude responses, to use antennas of any polarization, and to respond to signals of any polarization. It is therefore desirable that a DF system for all but the simplest applications be designed as a complete integrated system to take full advantage of these possibilities. This chapter will discuss antenna-design principles in the context of complete DF systems.

16-2   RADIO-WAVE PROPAGATION

Propagation Characteristics

The performance of a DF system depends on the nature of the signals arriving at the DF antenna. In turn, the nature of these signals depends on the mode of propagation, which will fall into one of two broad categories:

1   Signals that arrive directly at the DF antenna from the emitter with near-zero elevation angle. Included in this category are low-frequency (LF), medium-frequency (MF), and HF surface waves which will be vertically polarized and very-high-frequency (VHF), ultrahigh-frequency (UHF), and microwave direct-wave signals which may be of any polarization.

2   HF signals that are refracted from the ionosphere and arrive at the DF antenna with an elevation angle that may vary from a few degrees to nearly 90°. Irrespective of their transmitted polarization, these signals will be elliptically polarized upon arrival at the DF antenna system. That is, the two polarization components will vary independently with time.[4,5]

Whether the signals arriving at the DF antenna fall into category 1 or category 2 or include both categories depends primarily on the frequency band and secondarily on the daily and seasonal variations of the ionosphere.

Frequencies in the Low- and Medium-Frequency Bands

In these bands, during the day only a vertically polarized surface wave is propagated since sky waves are absorbed in the D layer of the ionosphere. However, at night the D layer disappears, and signals are refracted from the ionosphere so that sky-wave signals can arrive at the DF antenna.

The surface wave is attenuated with distance so that, depending on time of day, transmitter power, ground conductivity, and distance of the DF site from the emitter, three types of received signal may exist:

1   Near the emitter the vertically polarized surface wave will predominate because the sky wave will always be small with respect to it.

2   With increasing distance, the surface wave becomes attenuated. At night there will be a zone (known as the *fading zone*) in which sky wave and surface wave are comparable in magnitude. Fading occurs owing to interaction between surface and sky waves.

3   At distances beyond the fading zone the surface wave becomes highly attenuated so that the only receivable signal at such distances will be the night-time sky wave.

Magnitude of the surface wave (which primarily determines the distance of the fading zone from the emitter) increases with increasing ground conductivity and decreasing frequency. Thus, in the LF and lower MF bands the signal available at the DF site is, in most cases, predominantly a surface wave arriving at zero elevation angle. But in the upper MF band the signal may be a combination of surface and sky waves or, alternatively, a sky wave alone, especially at night.

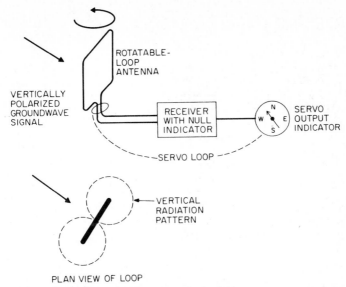

PLAN VIEW OF LOOP

FIG. 16-3   Sample of a system using a rotatable loop.

A simple rotating-loop arrangement, as shown in Fig. 16-3, has an amplitude response to vertically polarized signals shaped like a figure of eight. During daylight hours, when only a vertically polarized field is present, the two nulls will give an accurate determination of bearing if means are provided to remove the 180° ambiguity. However, the amplitude response of the loop to horizontally polarized signals is also a figure of eight, but rotated 90° in space in relation to the vertically polarized response. At night, in the presence of a sky wave, which will have both vertically and horizontally polarized field components, there will be no nulls or maxima with a fixed relationship to the loop orientation. Consequently, an accurate measurement of bearing is impossible. This phenomenon is referred to as *night effect* or *polarization error*.[6] It illustrates a fundamental difficulty that arises when sky waves are received by DF systems that depend only on the amplitude of the signal and when the antenna system is responsive to horizontal polarization.

Frequencies in the High-Frequency Band

In this band the attenuation of the surface wave over land is high, as is illustrated in Fig. 16-4. In many cases only a sky wave arrives at the DF site. However, over seawater the surface and sky waves will be comparable in magnitude at much greater distances.

At distances up to about 2000 km the sky-wave signal may arrive via one refraction from the ionosphere (a *one-hop mode*). But at any except the very shortest distances multihop signals, in addition to the one-hop signal, can occur. The arriving signal will then consist of a number of components of different amplitudes, of random phase, and with different elevation angles. As these components vary with time in both amplitude and phase, the arriving wavefront will be distorted and the azimuth and

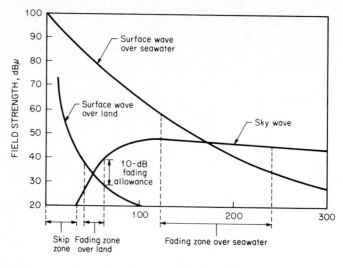

FIG. 16-4   Approximate variation of HF field strength with distance for 1-kW emitter power.

elevation angles of the signal will vary.[7] When DF measurements on such complex signals are necessary, simple amplitude-only DF systems are inadequate. However, systems employing digital computers and using suitable analytical programs can provide accurate measurements of bearing, elevation angle, and emitter location under these conditions (see Sec. 16-4).

Frequencies in the Very-High-Frequency and Ultrahigh-Frequency Bands

At frequencies above about 50 MHz the ionosphere becomes transparent to radio transmission. Propagation is by direct wave, and polarization is determined by the transmitting antenna. It may be vertical or horizontal or may contain both vertical and horizontal components. Furthermore, range is limited because the signal is rapidly attenuated beyond the line of sight.

16-3   DIRECTION-FINDING-SYSTEM PLANNING

Planning a DF system (which may comprise a single station or a network of stations) must take into account a number of factors of major importance:

- Geographic area in which the emitters are located in relation to the DF stations.
- Frequencies of the emitters.
- Number of DF stations necessary to ensure adequate accuracy of measurement.

- Response time (the minimum time in which a measurement must or can be made).
- Physical limitations of the system. These may include the space available for antennas, the need (or not) for a movable system, and limitations on equipment size, weight, and complexity.

Geographic Area and Frequency Coverage

The geographic area in which emitters are located in relation to the available locations of the DF stations will determine the maximum and minimum distances over which signals must be received. These maximum and minimum distances, together with the frequency range in which the emitters operate, enable the most likely propagation modes to be identified so that the optimum DF system can be chosen.

Emitter-Location Systems   If the area in which the emitters are situated is known and will remain fixed and if all the emitters are within propagation range of the proposed DF sites, a fixed-site emitter-location DF system will be chosen. Fixed-site DF systems have the advantage that it is possible to use electrically large antennas and antenna arrays, which are faster and more accurate than small-aperture systems. Furthermore, electrically large antenna elements have low noise figures, enabling DF measurements to be made on relatively low-field-strength signals unless the received signal-to-noise ratio is limited by external noise.

On the other hand, the area of the emitters may vary or be located so that it is not possible to receive adequate signals from the emitters at fixed DF sites. In this case it may be necessary to use a movable DF system with electrically small antenna elements. (But movable DF systems can employ large-aperture arrays. See Sec. 16-5.)

Navigation and Homing Systems   If the objective is aid to navigation or homing, the DF system will be mounted on a moving vehicle. This will impose a serious limitation on the antenna system. As an offset to the antenna disadvantage, such systems normally have the advantage of operating over short ranges in a fixed and narrow frequency band using direct or surface waves. The need for electrically large antennas is not so great as would be the case if sky waves were involved.

Number and Location of Stations in a Direction-Finding Network

The number and location of stations in a DF network have a direct effect on the accuracy with which an emitter can be located. The limitations of radio propagation as well as the usual geometric limitations on the accuracy of triangulation must be considered.

Horizontal Triangulation (Azimuth)   Under optimum conditions, measurements of bearing from two separated DF sites are sufficient to determine location. In practice, location of the DF sites may not be optimum so that three sites are the practical minimum. By using just two sites, minimum location error will occur when the two lines of bearing intersect at $90°$. With reference to Fig. 16-5, if radius R is the standard deviation of error for two lines of bearing intersecting at right angles, then R' will be the major-axis radius of an ellipse of error when the two lines intersect at less than $90°$.

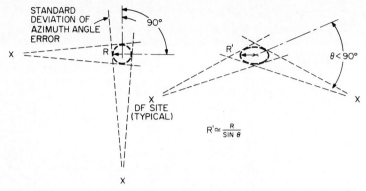

FIG. 16-5   Horizontal triangulation, using two DF sites.

If the two sites are not optimally sited or if one site is unable to determine a line of bearing for any reason, a third site will prove valuable. A third site is likely to increase location accuracy in any case, but only a small further increase in accuracy will likely result from more than three sites. (See Fig. 16-6.)

The minimum spacing between DF sites should depend on the expected distance to the emitter. For a given site spacing, there is a maximum range at which an emitter can be located with any reasonable accuracy. This means that, quite apart from the limitation of DF range imposed by propagation conditions, a limitation is also imposed by the spacing between the DF sites.

Ideally, the sites of a DF system would be placed uniformly round the edge of the area containing the emitters, as illustrated in Fig. 16-7a, but geographic features, national boundaries, and communication problems between the DF sites sometimes make this impossible. In such cases the best arrangement is to locate the sites of a DF network as close as possible to the area containing the emitters, keeping the sites as far apart as possible.

The following general rules, which are illustrated in Fig. 16-7b, c, and d, give a useful guide for DF siting:

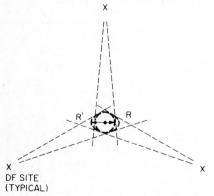

FIG. 16-6   Horizontal triangulation, using three DF sites.

- As shown in Fig. 16-7b, DF sites should be evenly spaced, and the distance between adjacent sites should be at least equal to the distance across the area containing the emitters.

- If none of the DF sites can be located close to the area containing the emitters, the distance between sites should be approximately equal to the distance between the centroid of the area and the centroid of the DF sites, as shown in Fig. 16-7c.

If one of the DF sites is so close as to be in the sky-wave skip zone of one of the emitters (so that no signal is available), the foregoing spacing criteria will maximize the probability that one or more of the other sites will receive a signal. This assumes that the emitter is transmitting a signal to a location in the area containing the other emitters.

In addition to the foregoing DF-station-siting criteria, the various DF sites should be located with a nonparallel baseline, as illustrated in Fig. 16-7d. Arranging the sites in a triangle or a square will minimize azimuth triangulation error.

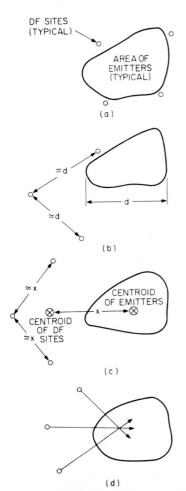

Vertical Triangulation (Single-Station Location) Single-station location (SSL) against sky-wave signals by definition requires only one site. Azimuth measurement is used to determine direction to the emitter, and elevation measurement, in conjunction with the height of the ionosphere, is used to determine distance to the emitter.[8] Location error of the emitter will depend on the accuracies with which the DF system can determine direction and distance. These two parameters are subject to different causes of error:

- The accuracy with which direction can be determined will be identical to that of a single conventional DF system.

- The accuracy with which distance can be determined will depend on the errors inherent in measurement of elevation angle and on the accuracy with which the height of the ionosphere can be calculated or measured. (See Fig. 16-8.)

The accuracy of a DF-SSL system will be higher for azimuth than for distance. Although single-station location is generally less accurate than multistation triangulation, SSL can offer significant advantages of reliability, simplicity, and speed of operation since interstation communications and correlation of separate measurements are not necessary.

A number of DF-SSL sites can be used in a network for improved emitter-

FIG. 16-7 Distribution of DF sites for minimum horizontal triangulation error. (a) Ideal distribution of DF sites. (b) Approach to ideal distribution of DF sites. (c) Optimum distribution if no DF site can be near the target area. (d) DF site location with respect to other fixed sites. Sites should be located on nonparallel baselines and as nearly in a triangle or a square as possible.

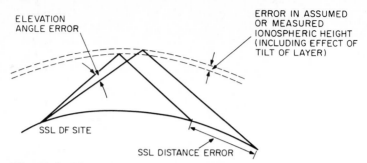

FIG. 16-8 Effect of angular and ionospheric errors on SSL distance measurement.

location accuracy. The form of the location process will be different from the conventional azimuth triangulation process. (See Sec. 16-6.)

Measurement Accuracy and Speed

The measurement accuracy of all DF and emitter-location systems is affected to a first order by six main parameters:

1 Aperture of the antenna system

2 Instrumental accuracy of the measuring system,[9] which includes the effects of errors in antenna performance, site error due to local scattering and ground irregularity, equipment and processing error, and operator error in the case of manually operated DF systems

3 Ionospheric behavior in the case of systems operating in the upper part of the MF band and in the HF band

4 Received signal-to-noise ratio

5 Integration time, that is, the time over which the signal is averaged in the measurement process

6 The distance of the emitter from the DF site or sites

Because the effects of noise and of some forms of ionospheric error are decreased with integration time, measurement speed (inverse of the minimum time that is required for a measurement) and accuracy (the accuracy with which an emitter can be located) are linked together. There is a speed-times-accuracy product that is relatively stable for a given DF system (see Sec. 16-7). Large-aperture systems tend to be fast and accurate, whereas small-aperture systems tend to be slow and less accurate.

For this reason, it is desirable to use the largest practicable aperture for skywave HF DF antenna systems. The advantages of large antenna aperture are:

1 The effect of noise and interference can be reduced (i.e., the signal-to-noise ratio available at the antenna output for a given field strength can be increased) if the aperture of the array is used to form beams. Such arrays are able to discriminate against noise and cochannel interference arriving from directions other than that of the emitter.

2 The effect of sky-wave error is reduced. Sky-wave error occurs when more than one sky-wave mode is present and results in a standing-wave pattern which changes shape slowly with time (in terms of seconds or minutes). This is often referred to as wave interference.[10] See Fig. 16-9.

3 As a result of improved signal-to-noise ratio and reduction of sky-wave error, the integration time required for a given measurement accuracy is reduced.

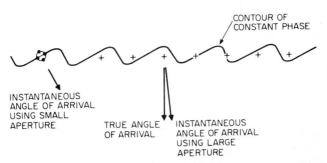

FIG. 16-9 Instantaneous angle-of-arrival errors of small and large aperture under wave-interference conditions. • = sampling ports of a small-aperture system (such as an Adcock). + = sampling ports of a large-aperture system.

4 Errors due to reradiation from obstructions in the vicinity of the DF site are reduced by the directivity of formed beams in the same way that the effect of external noise is reduced. This is true even if beams are not formed since errors contributed by scattered signals at individual elements of the array tend to be random and tend to cancel. Thus, large well-filled apertures are less sensitive to reradiation than are small or sparsely filled apertures.

The use of a large-antenna aperture therefore increases speed, accuracy, and resolution[11] (the ability of the system to distinguish between emitters that are in relatively close proximity).

Physical Limitations

Available Antenna Space   The size of the antenna array of a DF system will be determined by its application (whether fixed or movable), by the frequency band of the emitters, and by the required accuracy and speed of measurement. The largest antenna arrays are required in the HF and upper-MF bands where sky-wave signals are received and variations of phase and amplitude of the arriving wavefront are at a maximum (see Fig. 16-9). In the LF and lower-MF bands, surface-wave signals which have stable phase and amplitude characteristics are received, so that relatively small-aperture antenna arrays provide adequate speed and accuracy. In the VHF and UHF bands, antennas having a large electrical aperture require relatively little physical space.

Thus, antenna arrays may be small when high accuracy and high speed are not primary operating requirements and when sky-wave signals are not likely to be received.

System Mobility   If a DF system is to be fully mobile (meaning that the whole operational system including the antennas is vehicle-mounted), the size of the antenna system will be limited by the size of the vehicle. This restriction is most demanding for land-based systems, less so for ships and aircraft. The restriction is acceptable in the LF and MF bands, permits a reasonable electrical aperture in the VHF and UHF bands, but is often unacceptable in the HF band. Land-based mobile HF DF systems will have relatively low accuracy, low speed, and low resolution. Their use will be limited to applications in which the DF vehicles can be sited near the target emitters.

If the DF system is to be transportable, as opposed to fully mobile, the restrictions on antenna-system size are far less severe. If the land area available at each DF location is sufficient, it is possible to form a large antenna array by spacing out a number of small loop or whip antennas (see below, Fig. 16-17). The accuracy, speed, and resolution of such a system will be greater than is possible with a fully mobile system. However, transportable antenna elements will be electrically small and consequently less sensitive than electrically large elements, so that low-level signals may not be detectable.

Size, Weight, and Complexity of Direction-Finding Equipment   The advent of solid-state electronics permits the design of sophisticated measuring, processing, and output-display equipment that is light and portable. Thus, the size and weight of such equipment is not usually a limiting system-design consideration. It is customary to use modular design so that units can be added to the basic DF system to satisfy more complex operational requirements such as networking, signal acquisition, signal monitoring, and signal processing.

16-4   DIRECTION-FINDING-SYSTEM DESIGN

Types of Direction-Finding Systems

DF systems can be grouped into two broad classifications in accordance with the way in which their antenna systems obtain information from the arriving signals of interest and in which this information is subsequently processed. Scalar DF systems obtain and use only scalar numbers about the signal of interest. Vector or phasor systems obtain vector numbers about these signals. Scalar systems work with either amplitude or phase, while vector systems work with both amplitude and phase.

Scalar Systems   The simplest scalar system is the rotary loop illustrated in Fig. 16-3, which depends on the figure-of-eight symmetry of its vertically polarized amplitude response. Most scalar systems employing amplitude response depend upon some form of symmetry. Examples are the Adcock and Watson-Watt systems (see Sec. 16-5) and any circularly disposed system using a rotating goniometer.

Scalar systems employing phase response are typified by interferometers and Doppler systems (see Sec. 16-5). These systems employ multiport antenna systems that provide phase differences.

Scalar systems are capable of measuring either azimuth or elevation angles of arrival, or both. However, scalar systems of reasonable size and complexity are not well suited for the resolution of the individual angles of arrival of multimode signals (such as signals received in the fading zone).

Vector or Phasor Systems   Vector systems have the ability to obtain and use vector or phasor information from the arriving signals of interest. That is, they make use of both amplitude and phase. Vector systems require the use of multiport antennas and at least two amplitude- and phase-coherent receivers. Wavefront-analysis systems, described in the following paragraphs, are a special class of vector system especially intended for the resolution of multicomponent wave fields.[12,13,14]

Wavefront-Analysis Direction-Finding Systems   Wavefront-analysis (WFA) systems employ multiport antenna arrays and a suitable receiver-processor-output system so that the defining parameters of the incident wave or waves can be determined. That is, a single incident wave (or a single arriving ray; either wave or ray theory may be assumed) is defined by four parameters: arriving azimuth and elevation angles and relative amplitude and phase (i.e., polarization) of the electric vector.[15] Similarly, a two-component wave (i.e., of two modes) is defined by 10 parameters. Thus, a WFA system must have a sufficient number of antenna ports and sufficient measuring and processing capability so that the desired number of unknown parameters can be resolved.

In practical systems, the received signals are usually processed in digital form, fully preserving relative amplitude and phase, so that the unknown wave parameters can be extracted from the set of antenna-port responses. Given the contemporary state of minicomputer capacity, two wave (or two ray) fields can be resolved.

WFA systems require that the phase and amplitude response to signals of any arriving angle and any polarization be known for every antenna port. Such responses can be determined with considerable precision for any arbitrary antenna element or array of elements.[16] Thus, WFA systems have the important advantage that they may employ virtually any arbitrary array of antenna elements provided only that:

1   The aperture of the array is consistent with the desired speed and accuracy of the system.

2   The number of elements is large enough so that the unknown wave parameters can be determined.

3   The antenna angular patterns, polarization responses, and sensitivity are consistent with the desired application.

Consequently WFA systems can employ antenna arrays of virtually any shape or form, symmetrical or not, including circular or randomly placed land-based arrays or arrays mounted on irregular vehicles (such as ships or aircraft).

Receivers

Receivers comprise the measuring equipment of a DF system. They are used as RF voltmeters to measure antenna responses and to provide responses to the DF processor. In systems in which the azimuth angle is determined by observation of a null or a beam maximum, a single receiver can be used, since the processor will require only information concerning the amplitude of the response pattern of the antenna system.

However, in any system in which measurement is based on an amplitude and/or phase comparison, a number of alternative receiver arrangements are possible (see Fig. 39-10):

1 Single-channel receiver with switch to compare the outputs of two or more antenna ports sequentially (Fig. 16-10a)
2 Dual-channel receiver without switch to compare the outputs of two antenna ports simultaneously (Fig. 16-10b)

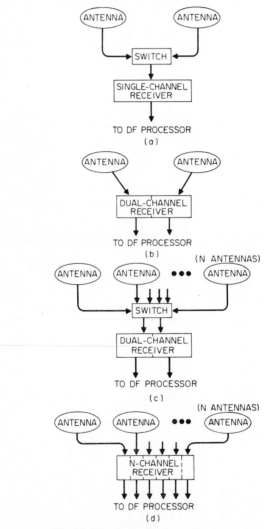

FIG. 16-10 Alternative receiver arrangements.

3   Dual-channel receiver with switch to compare the outputs of three or more antenna ports sequentially and allow any two to be compared simultaneously (Fig. 16-10*c*)

4   *N*-channel receiver without switch to compare the outputs of *N* antenna ports simultaneously (Fig. 16-10*d*)

Dual-channel and multichannel receivers are usually arranged to operate from a common frequency-synthesizer source, so that the phases of the output signals will be in the same relationship as the phases of the incoming signals. Dual-channel and multichannel receivers may also be matched in gain and/or phase. Alternatively, they may be made gain- and phase-stable over their measurement bandwidth and their relative gain and phase normalized for each measurement.

DF receivers are usually provided with selectable-measurement bandwidths. Narrowing the bandwidth can reduce the effect of adjacent channel interference but usually requires an increase in measurement time.

Direction-Finding Processor

The function of the processor is to calculate the required DF and emitter location on the basis of the signal voltages at the output of the receiver system. The complexity of the processor varies with the nature of the calculations required to deduce emitter location. At one extreme, the processor may be as simple as an angle scale on a rotating loop. At the other extreme, a digital processor such as in a multiport WFA system may be required.[17]

Processors for WFA systems usually consist of small digital computers having sufficient speed and memory for their applications. The software for such systems is usually modular, so that a variety of antenna systems can be accommodated and optional capabilities such as networking, signal acquisition, and signal monitoring can be provided.

Figure 16-10*c* and *d* shows two antenna-receiver arrangements suitable for use with a WFA system. The output ports of the antenna elements are connected to the switch for routing to the receivers. Alternatively, the antenna-element responses may be combined into groups to form directive beams.

The wavefront-analysis technique[18] is essentially the inverse of the antenna-port-response computation as previously described.[16] That is, the WFA system, under the control of its processor, measures the responses of the antenna ports to the incoming wave field. Since the current distribution on the antenna array and the resulting antenna-port responses are uniquely determined by the incoming wave field and since the antenna-port responses have previously been computed with precision, the processor is able to solve for the wave-field parameters by inverting the matrix consisting of the measured and the computed antenna-port responses.

The computed wave-field parameters consist of the arrival angles (azimuth and elevation) and the instantaneous sense of polarization of up to two waves (or rays). Each computation also results in a complex correlation coefficient which indicates the quality of that particular computation. Thus, a means for eliminating results seriously contaminated with noise or interference is provided.

The DF processor may also be provided with software so that the measured elevation angle, in conjunction with either stored or measured ionospheric data, can be used to compute a distance to the target emitter, resulting in SSL capability. The software and hardware associated with real-time ionospheric-data systems can become complex and extensive if the ultimate in SSL accuracy is desired.

The WFA technique is fully automatic (human intervention is not required), and it can be very fast. Depending on the configuration of the particular WFA system, individual measurements (or *cuts*) can be made in as little as 0.1 s.

Output-Display Arrangements

The purpose of output-display equipment is to provide the required DF information in a form suitable to the user. The simplest form of display is the scale on the control of a rotating-loop antenna. Scalar DF systems usually employ some form of analog display such as a cathode-ray tube (CRT) or graphic plotter. Some scalar systems are equipped with automatic bearing processors so that numerical values of bearing angle can be used remotely.

Vector (such as WFA) systems often provide a variety of local and remote displays. Visual displays in graphic and tabular form can be furnished, as can hard copies of both. Interactive displays can also be provided so that operators can edit and enhance the accuracy and usefulness of results. Data can also be easily transmitted to remote locations.

16-5   DIRECTION-FINDING ANTENNA ELEMENTS AND ARRAYS

In general, a DF antenna system is an array of individual antenna elements arranged to provide the responses required by the particular system. The antenna elements and arrays are of standard types that are used in other radio-communications applications, and their basic theory of operation is explained elsewhere. In this section, therefore, only those properties of the elements and arrays that are of importance in DF applications are discussed in detail.

Antenna Elements

An antenna element for a DF system is essentially a single-port (two-terminal) arrangement of conductors for the interception and collection of radio-frequency energy. A wide variety of element types are used in DF systems. These include monopoles, dipoles, loops, log-periodic antennas, and current sensors. The choice of a particular type of element is dictated by the DF application and the frequency band in which the target emitters operate. Those scalar HF DF systems which measure only the amplitude of the signal incident on the antenna must use antenna elements that are sensitive only to vertically polarized signals (monopoles, vertically polarized dipoles, and vertically polarized log-periodic antennas). Otherwise, polarization error will occur in the presence of sky waves.

Monopoles   Monopoles are vertically polarized elements operated over a ground plane. Since they economize in height (they are shorter than a vertical dipole resonating at the same frequency), they are suitable for applications in the LF, MF, and HF bands.

DF systems using arrays of monopole elements are sensitive only to the vertically polarized component of the signal and are thus suitable for scalar (amplitude-only) systems. However, all monopoles have a pattern null at zenith and are not well suited for short-range, high-angle signals. Sleeve monopoles (sometimes referred to as *elevated-feed monopoles*)[19] provide a high sensitivity to low-elevation-angle signals over a wide bandwidth because there is no reverse loop of current when the monopole length exceeds a half wavelength.

Dipoles   Dipoles are used at frequencies in the HF, VHF, and UHF bands at which their length can be accommodated. Vertically polarized dipoles have most of the same advantages and disadvantages of vertical monopoles. Horizontally polarized dipoles can be arranged to be sensitive to signals arriving at high elevation angles. Horizontal dipoles typically have low sensitivity to signals arriving at low elevation angles so that they minimize the effect of local human-made interference, which is usually propagated by surface wave.

Loops   Broadband (untuned) single-turn loop antennas are particularly suitable for wideband DF applications in the HF band. Loop sizes from about 2 m² to 4 m² represent a good compromise between sensitivity (a larger size is better) and freedom from reradiation when used in arrays and ease of handling when used in transportable systems (a smaller size is better).

Loops are responsive to both vertical and horizontal polarization. Whereas this is a serious disadvantage with amplitude-only systems when sky waves are present, it is a distinct advantage when used with WFA systems. A pair of crossed loops, suitably phased, is responsive to radiation of at least one sense of polarization from virtually all angles of arrival in half space. That is, such loops are sensitive to ground waves and to sky waves from all but very low elevation angles.

Loops, being electrically small and broadband, have a high noise factor (a substantial amount of internal noise is generated by conductor, mismatch, and earth losses[20]). Thus, when the external noise level is low (at the higher frequencies in the HF band and at quiet locations), a DF system with loop antennas may become internally noise-limited. With weak signals longer integration time may be required to provide acceptable measurement accuracy.

Current Sensors   In shipboard and airborne DF-SSL systems, the incoming wave induces RF currents over the entire hull or aircraft body. If the vehicle has dimensions of the order of a wavelength or more, the vehicle itself becomes part of the antenna system, and its effect cannot be ignored. However, for any incoming wave (or set of waves) the current distribution on the vehicle will be unique and will therefore define the incoming wave. Thus, the whole hull and superstructure of the ship or the aircraft body can be used as the DF antenna if the current distribution is sampled by a wavefront-analysis DF system. A convenient form of current sensor can be constructed from a small ferrite core with a coil winding mounted on a backing plate that can be secured to the hull. An essential step in designing such a system is to determine the response of every sensor, wherever located on the vehicle, to both amplitude and phase of waves arriving from any direction.

Arrays of Direction-Finding Antenna Elements

Factors to be considered in the selection of antenna arrays are:

● Coverage (the range and azimuth sector over which the target emitters are located). This will determine the form of the array.

● Expected propagation modes of the arriving signals (surface-wave, direct-wave, sky-wave, or multimode). These will determine the required elevation response of the antenna array and the type of antenna elements to be used.

● Combination of measurement speed and measurement accuracy. These will determine the required aperture of the antenna array.

● Physical requirements of the DF system (fixed or movable and the related limitation on space available for the array). These will affect the physical form of the elements and the size of the array.

DF antenna arrays can be conveniently divided into three categories:

1   Antenna arrays with a single or multiport output that are rotated either mechanically or electrically so that azimuth can be determined by a scalar system based on knowledge of the amplitude of the directional responses

2   Antenna arrays with a multiport output in which the azimuth and elevation angle are determined by measuring the phase difference of the signals at the output ports

3   Antenna arrays with a multiport output from which azimuth angles, elevation angles, and polarizations of the components of a multimode signal can be determined by a vector system

Single-Port or Multiport Antennas with Rotation of a Directional Pattern
This type of antenna system requires a scalar system to measure the amplitude of the signal. The antenna radiation pattern will be symmetrical about some axis, forming in general either a directive beam or a sharp null. In the DF operation the antenna array may be rotated mechanically (if it is sufficiently small), in which case it will have a single port. Alternatively, the antenna array may be rotated electrically by means of a goniometer (an electromechanical device that takes a weighted sum of the outputs of a number of fixed antenna elements and produces a response pattern that can be rotated). This type of antenna will inevitably have a multiport output, and the azimuth of the emitter will be determined by the maximum response (if the beam maximum is used) or the minimum response (if a null is used). It is often convenient to use the beam maximum when searching for the emitter and the null to establish the precise azimuth.

An example of a DF system using an antenna array with a symmetrical beam of known response is the Adcock, which is shown in two-port fixed form in Fig. 39-11. A basic form of the Adcock antenna array[21,22] comprises four vertical monopoles connected so that a figure-of-eight response pattern is produced. The pattern is rotated by a goniometer until the null is directed toward the emitter. The Adcock array differs from the rotating loop of Fig. 16-3 in that there are no unshielded horizontal conductors, so that polarization error due to the horizontally polarized component of sky waves is eliminated. Since there are two nulls in the figure-of-eight response pattern, there will be an ambiguity of 180° in the determination of azimuth. It is necessary to switch the monopoles to provide a cardioid radiation pattern to determine the correct azimuth. Thus, the measurement is made in two stages, the first to determine two precise azimuths by means of a null output and the second to determine which of these azimuths is the correct one.

The Adcock antenna array is typical of a narrow-aperture DF system and suffers from the disadvantages of low speed and/or accuracy, as discussed in Sec. 16-3. There are many variations of the basic Adcock design, most of which employ more than four elements with the objective of increasing aperture without increasing spacing error.[23] But all these must be considered narrow-aperture systems.

An alternative to the conventional Adcock system is the Adcock Watson-Watt arrangement,[24] one form of which is illustrated in Fig. 16-12. In this arrangement, the two antenna ports are connected (by means of two receivers) to the X and Y plates of a CRT so that the display is a line indicating the azimuth of the emitter on a surrounding circular scale calibrated in degrees. The advantage of this arrangement over the Adcock with a goniometer is that an instantaneous reading of azimuth rather than the intermittent reading provided by the goniometer is obtained. The Watson-Watt display is subject to jitter and variation caused by modulation, fading, and noise so that considerable integration is required for accurate measurement.

HORIZONTAL CONDUCTORS AND SHIELDED CABLES INSTALLED UNDER GROUND SCREEN TO MINIMIZE PICKUP OF HORIZONTALLY POLARIZED FIELDS

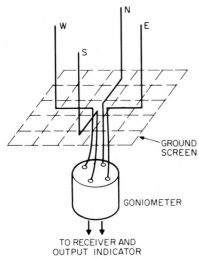

FIG. 16-11 Two-port, four-element narrow-aperture Adcock DF antenna.

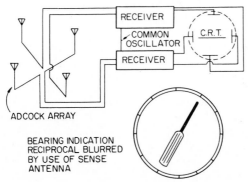

FIG. 16-12 Adcock antenna with a Watson-Watt instantaneous cathode-ray-tube display.

Some DF systems with a rotating directional pattern are capable of high speed and high sensitivity. These are wide-aperture multiport antennas of many elements disposed in a circle. Rotation of the response pattern is obtained by a rotating capacity-coupled goniometer. The goniometer usually combines responses from adjacent antenna elements so as to form highly directive sum-and-difference beams. An example of a wide-aperture DF system using a circular array of elements, together with goniometer pattern rotation, is the Wullenweber,[25] which is illustrated in Fig. 16-13. As in the case of the Adcock, the Wullenweber system with a goniometer measures only the amplitude (and hence only the azimuth angle) of the signal. To minimize polarization error the elements are monopoles.

The monopoles are arranged symmetrically around a cylindrical screen, and each is connected to a stator segment of the goniometer. The rotor segments span about 100° of arc and are connected to the switch by delay lines D_1, D_2, and D_3, whose lengths are equal to the free-space delays of the signal, as shown in Fig. 16-13. The signals from the antenna elements in operation at any moment combine in phase at the receiver and produce a sharp beam, since the arrangement functions as a broadside array. As shown in Fig. 16-13, the delay lines can be optionally split into two groups which are connected in antiphase, thus producing a rotating null as opposed to a rotating beam. Either the sum or the difference output can be connected to the receiver input. The output of the receiver is connected to a CRT display with a synchronized rotating time base so that the response pattern of the antenna appears as a polar display centered on the direction of the emitter. When searching for an emitter, the sum mode is normally used, but when the emitter has been identified, the difference mode is used so that the sharp null in the response pattern can display azimuth angle with maximum accuracy.

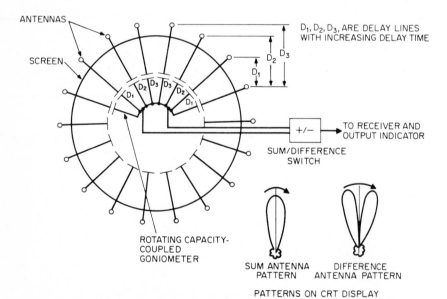

FIG. 16-13 Wullenweber wide-aperture antenna.

Multiport Antennas Using Phase Difference   Measurements of phase differences between the ports of a multielement antenna enable both azimuth and elevation angle of the arriving signal to be determined. One system of this type uses the Doppler technique.[26] In principle, an antenna element could be moved in a circular path so that the instantaneous frequency of the received signal would be modified. In practice, it is usually inconvenient to rotate an antenna element physically so the quasi-Doppler antenna arrangement shown in Fig. 16-14a is used instead. A rotating commutator is used to couple a receiver in rapid sequence to the elements of the array, thereby introducing a frequency shift on the received signal which is extracted by a frequency discriminator. As illustrated in Fig. 16-14b, the frequency shift is proportional to sin ($\theta - \theta_0$) cos ϕ_0, where θ is the angular position of the rotor, θ_0 is the azimuth angle of the received signal, and ϕ_0 is the elevation angle of the received signal. By using this expression, the azimuth angle is given by the angular position of the rotor at which zero instantaneous frequency shift occurs. Ambiguity can be removed by taking account of the angles at which maximum positive and negative frequency shifts occur. Also by using this expression, the peak-to-peak amplitude of the instantaneous fre-

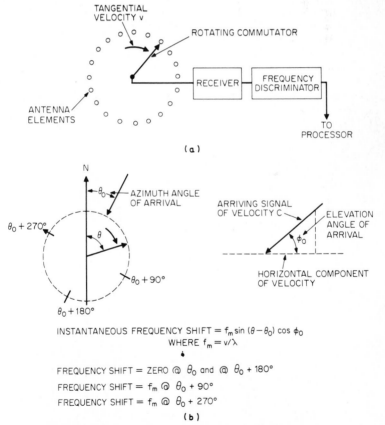

FIG. 16-14   Quasi-Doppler DF system. (a) General arrangement. (b) Relationship between frequency shift and angles of arrival.

quency shift is proportional to the cosine of the elevation angle so that the latter can be determined.

In practice, it is found that the quasi-Doppler arrangement will give accurate measurements of azimuth angle with short integration times. However, accurate measurement of elevation angle with multimode HF signals requires that one of the modes be dominant. The required integration time will depend on the relative levels of the minor modes. Furthermore, azimuth-angle accuracy drops off at high elevation angles as the amplitude of the instantaneous frequency shift approaches zero.

Another DF system making use of phase differences between the signals at the ports of a multiport antenna is the interferometer,[27] one form of which is shown in Fig. 16-15. This arrangement comprises five identical antenna elements, which can be loops, crossed loops, or monopoles. Since the distance between elements A and B is large in terms of a wavelength, the phase difference at their output ports will be large and will be very sensitive to angle of arrival relative to baseline AB. But just because AB is greater than $\lambda/2$, different arrival angles will produce the same phase difference (see Fig. 16-15). This ambiguity can be resolved by the less accurate measurement of phase difference between elements A and E, provided their spacing is less than $\lambda/2$ at the highest frequency of operation. The same principles apply to the line of elements A, D, C.

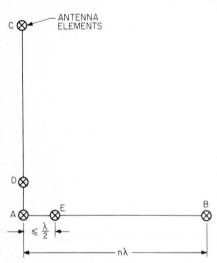

FIG. 16-15 Interferometer DF system. Example: If $n = 3$, arrival angles of 43°, 66.5°, and 86° relative to line AB will all result in a phase difference of 70° between A and B. The phase difference between A and E will resolve this ambiguity.

The loci of the angles of arrival with respect to the two baselines will be cones in half space, with the baselines as their axes. The line of intersection of the half cones about the two baselines will be the direction of arrival of the signal. The processor will compute the line of intersection, which will indicate both azimuth and elevation angles of arrival.

Scalar systems such as interferometers are not generally able to resolve the individual components of a multimode signal such as will occur under conditions of wave interference.

Multiport Antennas for Wavefront-Analysis Direction-Finding Systems

When wide-aperture systems are required, the Wullenweber type of antenna shown in Fig. 16-13 can be used, but the land area required is large. This is due to the fact that a ground screen extending well beyond the antenna elements is required for control of the near-field environment so as to minimize site errors. A more efficient wide-aperture system which achieves equal or better performance over a greater bandwidth and has a much smaller radius is shown in Fig. 16-16. This system comprises a symmetrical ring of quasi-log-periodic antennas (LPAs) arrayed so that their main lobes point inward (toward the center of the ring), thus conserving space. The array can

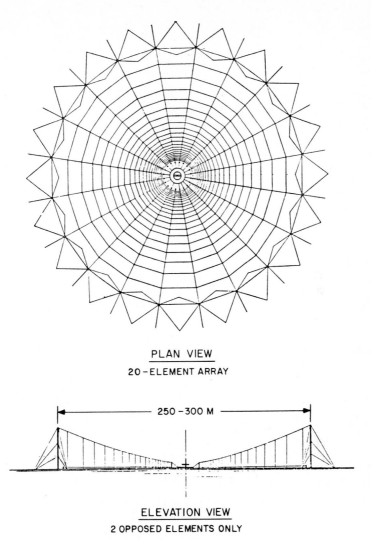

PLAN VIEW
20-ELEMENT ARRAY

ELEVATION VIEW
2 OPPOSED ELEMENTS ONLY

FIG. 16-16 Circular array of quasi-log-periodic antennas, suitable for HF DF-SSL and monitoring, horizontally and/or vertically polarized.

comprise vertically and/or horizontally polarized LPAs, as illustrated in Fig. 16-16. The ground screen need not extend beyond the outer antenna elements because each LPA faces inward toward a controlled and symmetrical environment in its near field.

Loops are also used for medium-aperture WFA arrays suitable for fixed or movable use. Loops are preferred over monopoles because they do not have a vertical pattern null and are therefore sensitive to high-angle sky waves. If space for the array is limited and if the number of individual elements is to be minimized, loops or crossed loops, as illustrated in Fig. 16-17, can be used. With this type of array, beams usually are not formed. The crossed-loop responses can be combined, or they may be used individually.

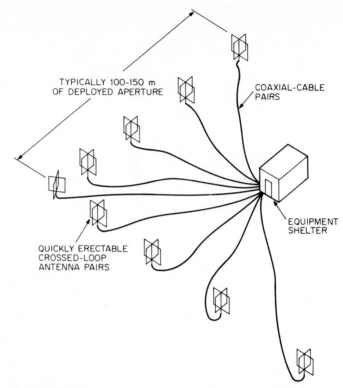

TYPICALLY 100-150 m
OF DEPLOYED APERTURE

COAXIAL-CABLE
PAIRS

EQUIPMENT
SHELTER

QUICKLY ERECTABLE
CROSSED-LOOP
ANTENNA PAIRS

FIG. 16-17   Typical deployment of one unit of a transportable HF DF-SSL system.

16-6   EMITTER-LOCATION ACCURACY

Emitter-location accuracy depends on two different kinds of error. The first, angular-measurement error, affects the accuracy of emitter location when azimuth triangulation is used and when SSL techniques are used. The second, SSL distance error, affects the accuracy of emitter location only when SSL techniques are used.

Angular-measurement error is the fundamental performance parameter of all DF systems.[28] SSL distance error is applicable only to sky-wave propagation.

Emitter-location error (or accuracy) can be affected by both angular-measurement error and (if used) SSL distance error. Figure 16-18 shows the hierarchy of emitter-location errors.

Angular-Measurement Error

Angular-measurement accuracy is usually taken to mean the ability of an ideal DF system to measure *direction to the target emitter*. In actuality, DF systems measure *direction of arrival* of the signal radiated by the emitter. These two directions may be sensibly identical for ground-wave and direct-wave signals, but for sky-wave signals

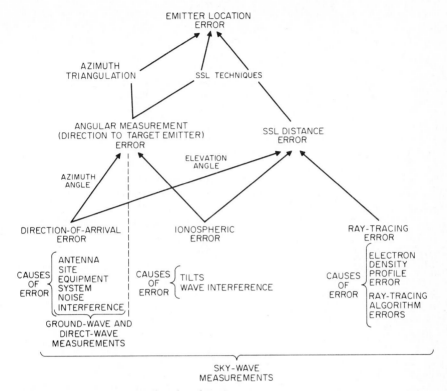

FIG. 16-18 Hierarchy of emitter-location error.

they will generally be different, often by a significant amount because of ionospheric error.

Thus, angular-measurement error must account for (1) direction-of-arrival errors such as those contributed by antenna, equipment, site, noise, and interference and (2) ionospheric errors.

Direction-of-Arrival Error   This type of error can be caused by antenna siting or construction errors, unequal cable lengths, unequal gain or phase match, antenna-pattern errors, computational or interpolation errors, reradiation and scattering from nearby objects, noise errors, and cochannel interference errors. For well-designed large aperture systems installed at good sites, the sum of all the errors caused by system design, construction, and siting should not exceed 0.1 to 0.5°.

Noise errors can be reduced to insignificance with sufficient integration time. Cochannel interference errors can be reduced by employing filtering in the domains of frequency, time, or space or some combination thereof. In view of the effect of operational choice upon the reduction of noise and interference errors, it is not practical to give a numerical allowance for them.

Ionospheric Error   This error results from wave interference, from large-scale ionization gradients (tilts) caused by hourly and seasonal solar variation, and from traveling ionospheric disturbances. The magnitude of ionospheric error varies with geographic location, with azimuth direction of signal propagation, and with distance to the emitter. Some of these errors are systematic, while others are random. The component of angular-measurement error due to ionospheric error can be reduced to insignificance with very long integration times (hours or days or seasons), usually much longer than is practical. Short-term statistical allowances for ionospheric errors have been suggested,[29,30] but these are of little operational use. It is possible to apply a tilt correction either by measuring tilt in near real time or by forecasting tilt on the basis of large-scale stored ionospheric data.

Sky-wave DF systems present a special problem in angular terminology because they do not measure azimuth angles directly (in the conventional meaning of terrestrial geometry) except for signals arriving at 0° elevation angle. In fact, signals arriving from zenith (90° elevation angle) convey no information at all about azimuth. For these reasons, it is convenient to use the concept of *great-circle degrees* when discussing direction-of-arrival accuracy (but not when discussing azimuth angular-measurement accuracy). Figure 16-19 defines this concept.

The ionospheric-signal path from the true position of the emitter to the DF site can be idealized as two straight-line segments joined at their apparent reflection point in the ionosphere. If the ionosphere is spherical and geocentric, these two line segments would define a great-circle plane. But if the ionosphere is tilted, the two line segments from *true emitter position* to DF site will not define a great-circle plane, although a great-circle plane would be defined by the two line segments from *apparent emitter position* to DF site. The direction-of-arrival error can be defined as the difference between the actual and the ideal straight-line segments from their reflection points to the DF site. For small error angles, great-circle error angles can be converted to azimuth error angles by dividing by the cosine of the elevation angle.

The great-circle-degree concept of angular error is necessary so that azimuth error results of signals arriving from different elevation angles can be compared. For example, if the horizontal great-circle angle-of-arrival error of three signals is, say, 1.0° (great circle) and if their elevation angles happen to be 0°, 30°, and 60°, their errors expressed in azimuth degrees will be 1.0°, 1.15°, and 2.0°, respectively.

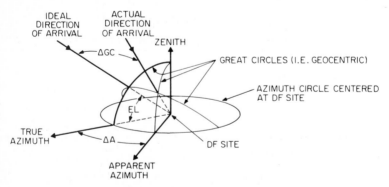

ΔGC = ERROR EXPRESSED IN GREAT-CIRCLE DEGREES

ΔA = ERROR EXPRESSED IN AZIMUTH DEGREES

$\Delta A = \dfrac{\Delta GC}{\cos EL}$ (FOR SMALL-ERROR ANGLES)

FIG. 16-19 Definition of direction-of-arrival error, expressed in great-circle degrees.

Lateral Error   *Lateral error* is defined as the location error, perpendicular to the measured azimuth, at the true distance (see Fig. 16-20). This is the component of location error caused by angular-measurement error.

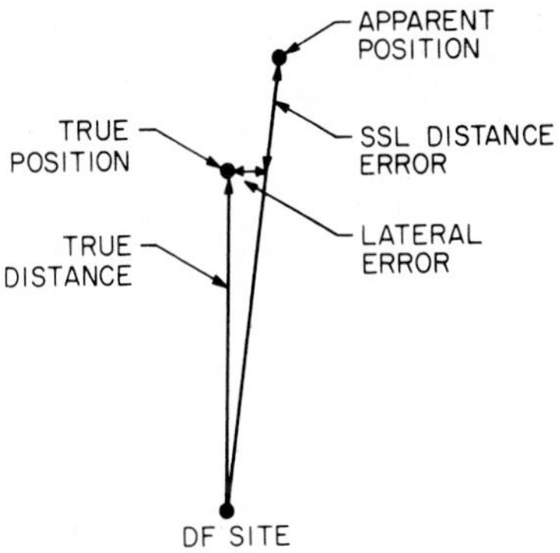

FIG. 16-20   Geometry of emitter-location error.

For surface waves and direct waves, in which the signal arrives at zero elevation angle, lateral error is directly proportional to angular-measurement error. If, for example, total angular-measurement error were 1°, lateral error would be as in Table 16-1.

The result of angular-measurement error as a function of true distance is very different for sky waves, in which the signal arrives at an elevation angle determined by the true distance and the height of the ionosphere. Figure 16-21 shows percent lateral and SSL distance errors as a function of true distance, assuming 1.6° (great-circle) direction-of-arrival error allowance and an ideal spherical-mirror ionosphere of 300-km height. The lateral error shown in Fig. 16-21 is presented in Table 16-2.

TABLE 16-1

True distance, km	Lateral error, km	Lateral error, % of true distance
50	0.9	1.8
100	1.8	1.8
500	9	1.8

SSL Distance Error

SSL distance error will be a function of elevation angle error and of the ability to trace the direction of the arriving ray backward through its zone of ionospheric refraction to the location of the emitter. Of these two error components, elevation angle error will be the lesser. Well-designed DF systems are able to measure elevation angle as accurately as they are able to measure azimuth angle except at very low elevation angles, at which the projected vertical aperture of most DF antenna systems approaches zero.

The ability of a DF-SSL system to trace the direction of the arriving ray backward to the location of the emitter will depend on the accuracy of the available ionospheric data, the quality of the ray-tracing algorithm, and the degree of uncorrected ionospheric tilt.

In their simplest form, ionospheric data may consist of stored hourly median

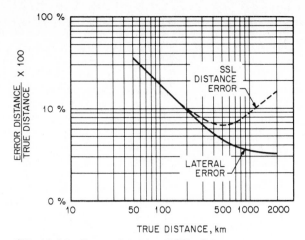

FIG. 16-21 Percent lateral and SSL distance errors as a function of true distance for sky-wave signals reflected from an ideal spherical-mirror ionosphere. Assumptions: (1) Perfect-mirror ionosphere at 300-km height. (2) 1.6° (great-circle) direction-of-arrival error.

TABLE 16-2

True distance, km	Lateral error, km	Lateral error, % of true distance
50	17.5	35
100	18	18
500	22	4.4
1000	34	3.4
2000	65	3.25

values published by numerous government agencies.[31] Such data can be used with or without tilt correction. In their most complex form ionospheric data may consist of near-real-time electron-density profiles obtained by a network of ionospheric sounders throughout the geographic area of interest and combined into a map of electron-density contours. Such a map will include tilt information.

Ray-tracing algorithms of varying complexity are available to convert elevation angles into distances. Most of these algorithms provide a one-hop distance based on a single measured value of elevation angle. More complex ray-tracing algorithms combined with WFA systems capable of resolving multiple angles of arrival show promise of an ability to provide multiple-hop SSL distances.

By referring again to Fig. 16-21, it is clear that SSL distance error is equal to or greater than lateral error for reasons of geometry. That is, for distances greater than about 200 km (for a 300-km-ionosphere) distance error is more sensitive to a given arrival-angle error than is lateral error. Furthermore, Fig. 16-21 does not account for ray-tracing error, which will further degrade SSL distance error. The SSL distance error shown in Fig. 16-21 is presented in Table 16-3.

Figure 16-22 shows the data from Fig. 16-21 and Tables 16-2 and 16-3 plotted out to 1000 km.

TABLE 16-3

True distance, km	SSL distance error, km	SSL distance error, % of true distance
50	17.5	35
100	18	18
500	33.5	6.7
1000	91	9.1
2000	310	15.5

Emitter-Location Error

As can be seen from Fig. 16-18, emitter-location error depends upon how emitter location is to be accomplished. Azimuth triangulation using a network of DF sites and SSL are the two techniques to be considered.

Azimuth Triangulation with a Network   With azimuth triangulation, location error will depend on the lateral error of the individual lines of bearing, on the number of sites reporting, and on locations of the sites with respect both to the emitter and to each other.[7] (See Sec. 16-3.)

If two DF sites are optimally situated with respect to the emitter, that is, if their lines of bearing intersect at 90°, then emitter-location error will approximately equal lateral error. If additional well-situated DF sites are used and if the individual lateral errors are random, then an increase in the number of DF sites will reduce emitter-location error.

Use of SSL Techniques   SSL techniques are clearly of greatest utility if network operation is not possible for some reason. But if the individual DF sites of a working network do provide SSL results, overall emitter-location accuracy can be improved. This can be accomplished at the network control center where the lines of bearing and SSL

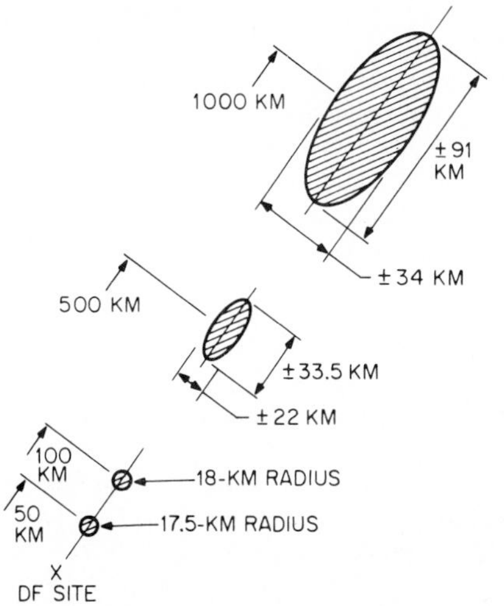

FIG. 16-22   Uncertainty of emitter location when SSL technique is used, employing data from Fig. 16-21.

results from all the sites are processed to give a best-point estimate (BPE). The most rational procedure is to require each DF-SSL site to furnish a quality rating for each line of bearing and for each SSL result. The BPE algorithm should be designed to test the available results for convergence, giving weight to the various quality ratings. This procedure, making best use of whatever results are available, will, on the average, produce the most reliable and accurate emitter-location results.

16-7   PERFORMANCE OF DIRECTION-FINDING SYSTEMS

Specifying and testing the performance of DF systems are more complex than might be supposed, especially for systems operating against sky waves. This section suggests how performance can be specified and what quality of results may be expected with several kinds of systems.

Factors Affecting Performance

Emitter-location accuracy of DF-SSL systems operating against sky waves is affected by at least the six kinds of parameters first given in Sec. 16-3. That is, the quality of results will be influenced to a first order by all of the following:

1 Antenna aperture
2 Instrument or system accuracy
3 Ionospheric behavior
4 Received S/N ratio
5 Integration time
6 Distance to emitter

The first two of these parameters are under the control of the system designer. For any particular system and for any frequency or band of frequencies these parameters may be regarded as fixed. Instrument or system accuracy is the parameter most likely to be specified by most system suppliers, even though it will seldom be the determining parameter with respect to emitter-location accuracy. Furthermore, instrument or system accuracy is difficult to measure accurately, especially with sky-wave DF-SSL systems.

The third parameter, ionospheric behavior, varies continuously. Corrections can be applied by the use of statistical ionospheric data or by the use of near-real-time ionospheric data. The last three parameters, S/N ratio, integration time, and distance, can obviously vary widely and are primarily under the control of the operator of the emitter.

It follows that a performance specification should take account of all six parameters. Furthermore, a testing program should permit separation of the different contributions to emitter-location error so the system can be rationally evaluated.

Performance Comparison of Direction-Finding Systems

It is useful to have a method for comparing the accuracy of different DF systems or of the same system under different conditions. But in view of the six parameters having a first-order effect on DF accuracy, fully rigorous comparisons are likely to be too complex to be useful.

Figure 16-23 suggests a practical method of azimuth accuracy comparison for different kinds of systems operating under widely different conditions. It is not meant to be rigorous, but it is useful for comparative purposes. It shows the nature of the relations between accuracy and integration time and between accuracy and aperture.

Figure 16-23 accounts for the first two of the six parameters, namely, aperture and instrument or system accuracy, by means of the four kinds of systems, A to D. Regarding the third parameter, ionospheric behavior, Fig. 16-23 implies that there is some amount of ionospheric error which is unlikely to be eliminated within the integration times of interest, but the figure does not account for large-scale solar-induced tilts.

The fourth parameter, the received S/N ratio, is assumed to be at least high enough to yield plausible results at the shortest integration times shown for each kind of system. The slope of measurement error versus integration time reflects the fact that each doubling of integration time is equivalent to increasing the S/N ratio by 3 dB and that a 6-dB S/N improvement reduces the effect of random errors by a factor of approximately 2. The fifth parameter, time, is the independent variable. The figure is approximately independent of the sixth parameter, distance, because the independent variable (measurement-error angle) is expressed in great-circle degrees.

Figure 16-23 is useful in visualizing the following ideas:

1 Large-aperture systems are faster and more accurate than small-aperture systems.

2 Manual systems (wherein a human operator makes the bearing estimate) probably require 5- to 10-s minimum integration to obtain any plausible result.

3 Fully automatic systems, combined with wide-aperture antennas, are between one and two orders of magnitude faster than manual systems.

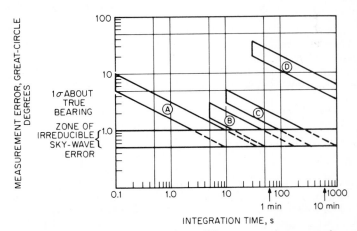

FIG. 16-23 Comparison of sky-wave DF-system performance. A = wide-aperture fully automatic WFA systems. B = medium-aperture goniometric systems such as Wullenwebers and wide-aperture interferometers. C = Adcock and Watson-Watt systems (narrow aperture). D = vehicular-mounted systems (zero aperture).

REFERENCES

1 F. E. Terman, *Radio Engineering,* McGraw-Hill Book Company, New York, 1947, p. 817.

2 G. S. Sundaram, "Ground-Based Radio Direction-Finding System," *Int. Def. Rev.,* January 1981.

3 P. J. D. Gething, J. G. Morris, E. G. Shepherd, and D. V. Tibble, "Measurement of Elevation Angles of H.F. Waves," *IEE Proc.,* vol. 116, 1969, pp. 185–193.

4 J. Ames, "Spatial Properties of the Amplitude Fading of Continuous HF Radio Waves," *Radio Sci.,* vol. 68D, 1964, pp. 1309–1318.

5 B. H. Briggs and G. J. Phillips, "A Study of the Horizontal Irregularities of the Ionosphere," *Phys. Soc. Proc.,* vol. B63, 1950, pp. 907–923.

6 R. L. Smith-Rose and R. H. Barfield, "The Cause and Elimination of Night Errors in Radio Direction Finding," *J. IEE,* vol. 64, 1926, pp. 831–838.

7 P. J. D. Gething, *Radio Direction-Finding and the Resolution of Multicomponent Wave-Fields,* Peter Peregrinus Ltd., Stevenage, England, 1978, chap. 14.

8 R. F. Treharne, "Vertical Triangulation Using Skywaves," *Proc. Inst. Radio Electron. Eng.,* vol. 28, 1967, pp. 419–423.

9 P. J. D. Gething, "High-Frequency Direction Finding," *IEE. Proc.,* vol. 113, no. 1, January 1966, pp. 55–56.

10 Gething, 1978, p. 35.

11 Gething, 1978, chaps. 8 and 9.

12 D. Cawsey, "Numerical Methods for Wavefront Analysis," *IEE. Proc.,* vol. 119, 1972, pp. 1237–1242.

13 P. J. D. Gething, "Analysis of Multicomponent Wave-Fields," *IEE. Proc.,* vol. 118, 1971, pp. 1333–1338.

14 J. M. Kelso, "Measuring the Vertical Angles of Arrival of HF Skywave Signals with Multiple Modes," *Radio Sci.,* vol. 7, 1972, pp. 245–250.

15 Gething, 1978, p. 175.

16 R. L. Tanner and M. G. Andreasen, "Numerical Solution of Electromagnetic Problems," *IEEE Spectrum,* vol. 4, no. 9, September 1967, pp. 53–61.

17 H. V. Cottony, "Processing of Information Available at the Terminals of a Multiport Antenna," *Antennas Propagat. Int. Symp.,* Sendai, Japan, 1971, pp. 9–10.

18 R. L. Tanner, "A New Computer-Controlled High Frequency Direction-Finding and Transmitter Locating System," paper presented to NATO-AGARD Symposium of the Electromagnetic Wave Propagation Panel, Lisbon, May–June 1979.

19 R. W. P. King, *Theory of Linear Antennas,* Harvard University Press, Cambridge, Mass., 1956, p. 407.

20 H. A. Wheeler, "Fundamental Limitation of Small Antennas," *IRE Proc.,* September 1947.

21 F. Adcock, "Improvements in Means for Determining the Direction of a Distant Source of Electro-Magnetic Radiation," British Patent 130,490, 1919.

22 F. Adcock, "Radio Direction Finding in Three Dimensions," *Proc. Inst. Radio Eng. (Australia),* vol. 20, 1959, pp. 7–11.

23 Gething, 1966, p. 51.

24 R. A. Watson-Watt and J. F. Herd, "An Instantaneous Direct-Reading Goniometer," *J. IEE (London),* vol. 64, 1926, p. 11.

25 H. Rindfleisch, "The Wullenweber Wide Aperture Direction Finder," *Nachrichtenech. Z.,* vol. 9, 1956, pp. 119–123.

26 C. W. Earp and R. M. Godfrey, "Radio Direction Finding by Measurement of the Cyclical Difference of Phase," *J. IEE (London),* part IIIA, vol. 94, March 1947, p. 705.

27 W. Ross, E. N. Bradley, and G. E. Ashwell, "A Phase Comparison Method of Measuring the Direction of Arrival of Ionospheric Radio Waves," *IEE Proc.,* vol. 98, 1951, pp. 294–302.

28 Gething, 1966, pp. 58, 59.

29 T. B. Jones and J. S. B. Reynolds, "Ionospheric Perturbations and Their Effect on the Accuracy of HF Direction Finders," *Radio Electron. Eng.,* vol. 45, no. 1–2, January–February 1975.

30 A. D. Morgan, "A Qualitative Examination of the Effect of Systematic Tilts in the Ionosphere on HF Bearing Measurements," *J. Atm. Ter. P.,* vol. 36, 1974, pp. 1675–1681.

31 *Ionospheric Communications Analysis and Prediction Program (IONCAP),* version 78.03, Institute for Telecommunication Sciences, Boulder, Colo., March 1978.

Chapter 17

ECM and ESM Antennas

Vernon C. Sundberg

GTE Systems

Daniel F. Yaw

Westinghouse Defense and Electronic Systems

Electronic-countermeasures (ECM) and electronic-support-measures (ESM) systems place a great variety of requirements on the antenna or antennas used, depending on the specific function being performed by the system, the characteristics of the target, and the environment within which the system is functioning. The breadth of these requirements is such that virtually any type of antenna, from a simple monopole to the most sophisticated phased array, may eventually be used in an ECM or ESM system. This chapter describes a few "typical" antenna subsystems used in ECM or ESM systems.

The functions to be performed by an antenna in an ESM system can generally be divided into two broad categories: (1) to monitor the environment for signals and (2) to determine the direction of arrival of an incoming signal (direction finding or tracking). The primary antennas on an ECM system are used to direct a significant amount of jammer power toward the threat emitter. ECM systems also use passive antennas for direction finding.

17-2   SURVEILLANCE ANTENNAS

The first and primary function of an ESM system is that of monitoring the environment for signals or, more particularly, for certain specific signals or types of signals. The knowledge that certain signals are being radiated can be of great importance to the battlefield commander or the combatant himself. A few typical examples of surveillance antennas are described in the following paragraphs.

Biconicals

One of the most useful omnidirectional antennas is the biconical horn,[1,2] particularly since the advent of meander-line-array[3] aperture polarizers. The most popular method of feeding the biconical for broadband operation is with coaxial line. This method produces a vertically polarized antenna. Cylindrical-aperture polarizers can be used to convert the polarization to either circular or slant-45° linear polarization.

By using meander-line-array polarizers to produce circular polarization, axial ratios can typically be held to less than 3 dB. For broadband operation, a flat disk of resistance card in the biconical is needed to absorb the cross-polarized reflections off the polarizer, however small these may be. If these low-level reflections were not absorbed, they would be re-reflected by the biconical and would radiate as the undesired sense of circular polarization and rapidly degrade the axial-ratio performance. At the present state of development, meander-line-array polarizers are limited to two octaves or less, which becomes the controlling factor for a circularly polarized biconical.

Slant-45° linear polarizers can provide broader bandwidths of operation. Bandwidths of 12:1 have been achieved on biconicals by using this type of polarizer. The slant-45° linear polarizer consists of a series of cylindrical printed-circuit-board (PC-board) layers, each with a grid of linear parallel lines. The PC boards are spaced in the radial dimension by approximately a quarter wavelength, with the innermost layer having its lines oriented orthogonally to the polarization of the antenna. On each succeeding sheet, the grid is tilted at an increased angle relative to the innermost board until the last sheet has its lines at 45° to the innermost sheet and orthogonally to the desired slant-linear-polarization orientation. Usually this is accomplished with four or five layers.

Another useful circularly polarized, omnidirectional antenna is the Lindenblad,[4] which covers octave bandwidths with broader elevation patterns.

Radial-Mode Horns

Radial-mode horns[5] can provide increased gain relative to an omnidirectional antenna when the azimuthal angular sector to be monitored is less than 360°. Radial-mode horns (Fig. 17-1) consist of a wave-launching section and a radial section in which a cylindrical wave moves out and radiates from the cylindrical aperture. The azimuthal (or E-plane) pattern is relatively flat-topped with a beamwidth of 0.95 θ, independent of frequency, provided that the wave-launcher slot width w is less than 0.5λ. Useful angles of θ have been found to vary from approximately 90° up to at least 200°. The radius R, measured from the point at which the back walls would intersect, must be greater than some minimum length (which is a function of θ) to maintain the flat-topped-pattern shape. The minimum radius R required increases rapidly as the angle θ decreases to 90°.

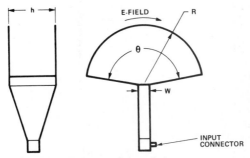

FIG. 17-1 E-plane radial-mode horn.

The elevation (or H-plane) pattern of this antenna varies with frequency in the normal manner of an aperture with a cosine amplitude distribution. Bandwidths of up to 2.25:1 have been achieved with this antenna.

The radial-mode horn itself generates a polarization that is perpendicular to the cylindrical axis. This linear polarization can be converted to either sense of circular polarization by using a cylindrical section of meander-line-array polarizer or to slant-45° linear polarization by the polarizer previously described. The polarization has also been converted to the orthogonal linear polarization very successfully by using essentially a double thickness of meander-line-array polarizer.

Figure 17-2 is an azimuthal (E-plane) pattern of an E-plane radial-mode horn with a meander-line polarizer taken with a linearly polarized transmit antenna rotating in polarization.

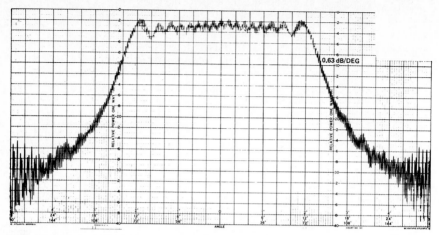

FIG. 17-2 Azimuthal pattern of an *E*-plane radial mode horn ($R = 8.9\lambda$, $\theta = 179°$).

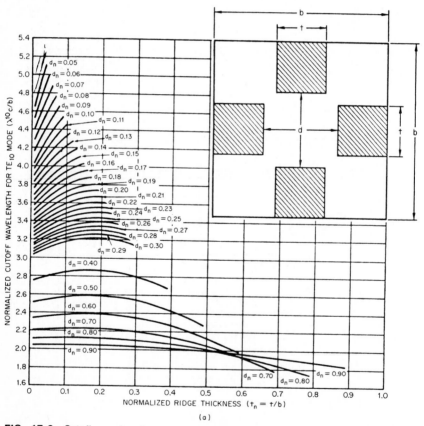

FIG. 17-3 Cutoff wavelength versus ridge thickness for various gap widths. (*a*) TE$_{10}$ mode. (*b*) TE$_{30}$ mode.

Ridge-Loaded Horns

Among other broadband antennas having increased gain that could be used for monitoring are the frequency-independent log-periodic dipole array and spiral antennas and ridge-loaded horns. These are used either directly or as feeds for reflectors. The frequency-independent log-periodic dipole array and spiral antennas have been discussed elsewhere[24] and will not be discussed further here.

Ridges are introduced into a pyramidal horn for the same reason that they are used in ridged waveguide, that is, to expand the frequency range of operation. Ridges on both broad walls of a rectangular waveguide (doubly ridged horns) lower the cutoff frequency of the TE_{10} mode of the waveguide. The ridges, if properly selected, will also raise the cutoff frequency of the TE_{30} mode slightly. (If the waveguide is straight and the input probe is centered, asymmetric modes will not be excited.) For the proper waveguide and ridge configuration, the ratio of TE_{30} to TE_{10} cutoff frequencies can be over 12:1 for doubly ridged horns. Curves for the design of doubly ridged horns can be found in Ref. 6.

Lenses are often employed to correct the aperture phase errors of horns operating over bandwidths of more than 5:1 to maximize their gain at the higher frequencies.

Quadruply ridged horns,[7] which provide dual orthogonal linear polarizations, can be operated over bandwidths as large as 9:1. Figure 17-3 shows the TE_{10} and TE_{30} cutoff wavelengths versus ridge thickness for various gap widths, as published in Ref. 7.

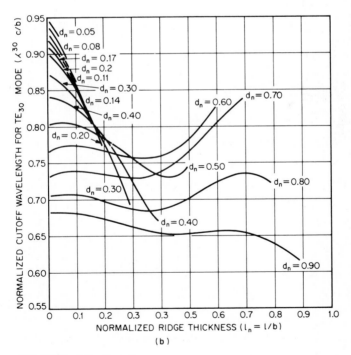

FIG. 17-3 (*Continued*).

17-3 DIRECTION-FINDING ANTENNAS

Direction-finding (DF; determination of the azimuth bearing to a given emitter) techniques applicable to ESM systems would include a rotating antenna sweeping a beam around in azimuth to determine the direction of highest signal intensity and amplitude- and phase-comparison techniques. If the DF system is searching for a target emitter that is relatively stable as a function of time and if sufficient time will be available for acquiring, identifying, and performing the DF function, a simple rotating-antenna beam may be adequate. On the other hand, if the emitter-signal strength varies greatly as a function of time, more sophisticated techniques, such as phase or amplitude comparison with simultaneous reception on all beams, may be required. The DF accuracy of any of the three techniques will be directly proportional to the beamwidth of the antenna in the azimuthal plane.[8-12] These techniques are discussed in detail in Chap. 16.

Rotating Antennas

An example of a rotating-beam-antenna system is shown in Fig. 17-4. The antennas can be spun at up to 300 r/min or pointed for surveillance purposes. In this system a pair of conical spirals is used to cover the 0.25- to 2-GHz range. The spirals are fed through a 180° hybrid to provide both sum and difference patterns in the azimuthal plane. A coaxial switch allows selection of sum or difference patterns.

Four compound sectoral horns provide fan beams for DF or monitoring purposes over the frequency bands of 2 to 4, 4 to 8, 8 to 12, and 12 to 18 GHz. Dielectric lenses

FIG. 17-4 Rotating direction-finding system.

are used in the horns for phase correction in the azimuth plane. Meander-line polarizers convert the vertical linear polarization in the horns to right-hand circular polarization in the radiated field. A horizontal absorber fence reduces the effects of the metal surfaces of the horns on the conical spiral patterns. DF accuracy of this system against emitters with high signal strength and amplitude stability varied from 5° at the low frequency to 0.4° at the high frequency of each horn.

This system used a five-channel rotary joint to bring the five RF bands off the rotating platform.

Shaped-Beam Antennas

Another form of rotating-beam antenna uses a specially shaped reflector to control the elevation pattern coverage for ESM purposes. The doubly curved reflector is used to shape the main-beam radiation pattern in the elevation plane for specific system requirements while focusing the beam in the azimuth plane for maximum directivity. For instance, for an air or surface search system, a constant-power return from the target as a function of elevation or depression angle is optimum. The amplitude pattern for such an optimization is nominally proportional to $\csc^2 \theta$. Another application of the shaped-reflector system for shipborne ESM systems is provided when the system is designed to maintain a fixed elevation-plane beamwidth across the frequency range. This type of system is often installed on unstabilized shipborne platforms. In this case the beam must be sufficiently broad to compensate for the pitch and roll of the platform and maintain a constant illumination of the potential threats.

The doubly curved reflector has no exact solution, and any solution obtained is not unique. The design methods outlined in the literature[13,14,15] are based on an iterative geometrical-optics process that tends to converge rapidly. The geometrical-optics method transforms the primary radiation pattern of the feed into the desired secondary radiation pattern in the elevation plane. Although this method of surface determination was originally developed for narrowband systems, it is useful for broadband ESM applications provided that the illuminating primary pattern is relatively constant over the frequency range of interest. Therefore, the shaped-beam technique works well with primary feeds such as log-periodics, spirals, and, to some extent, scalar electromagnetic horns, whose beamwidths are a slowly varying function over at least a waveguide bandwidth.

Measured radiation patterns of doubly curved reflectors closely approximate the desired theoretical radiation-pattern shape and hence validate the geometrical-optics design process. As one might expect, the larger the reflector in terms of wavelengths, the better the conformance to the theoretical pattern shape. However, even for small reflectors (0.6- to 0.75-m diameter) typical of some ESM applications, pattern-shaping techniques are extremely useful. Frequency ranges of 18:1 are entirely feasible by using this technique. Such an antenna (0.75-m diameter) is shown in Fig. 17-5. The antenna is linearly polarized at 45° and provides a minimum elevation beamwidth of 18° to 18 GHz. As the frequency of this antenna decreases, a crossover point is reached where the physical aperture size dominates and the elevation-plane beamwidth becomes inversely proportional to frequency. At the low end of the frequency range, the beamwidth increases to 24°. Some aperture gain is sacrificed even at the low end of the frequency range because of the phase distortion associated with shaping the radiation pattern for the high end of the frequency range.

FIG. 17-5 Shaped-beam antenna (1 to 18 GHz). *(Courtesy of Condor Systems, Inc.)*

Amplitude Comparison

Amplitude-comparison DF requires two or more overlapping antenna patterns whose peaks are separated by some angular extent in the azimuth angle. By measuring the relative amplitude on the two patterns and comparing the measured with a priori data of the pattern shapes, the angle between the antenna boresight and the emitter's bearing can be determined. Narrower beamwidths provide better DF accuracy, but only over narrower azimuthal sectors. Hence, a trade-off must be made between gain and DF accuracy against sector coverage or number and volume of antennas required.

DF accuracy is reduced by the polarization sensitivity of the pattern. Because the polarization of the incident signal is normally unknown, any variation of the pattern shape as a function of polarization will introduce error in the angle determination. Pattern changes that are a function of frequency can be accommodated by storing data in the computer as a function of frequency, but polarization effects cannot be similarly accommodated unless the incoming-signal polarization is measured.

Pairs of phase- and amplitude-matched conical spirals, arrayed in a frequency-independent manner, can be used in amplitude-comparison DF systems. In a circular array of spiral pairs, the summed outputs of each pair can be compared to obtain a DF bearing. DF is often accomplished by using a circular array of single spirals (either conical or flat cavity-backed), but the pairs provide a narrower azimuthal beamwidth and hence better accuracy. Corrugated horns are also useful in amplitude-comparison systems. A circular array of almost any type of directive antenna elements can be used for amplitude comparison; however, to obtain similar DF accuracy over a broad bandwidth, the pattern performance of the elements must be relatively independent of frequency.

17-4 TRACKING ANTENNAS

A tracking antenna is designed to provide the information required by the electronic system to change periodically the pointing of the antenna in both azimuth and elevation so that the antenna boresight is repositioned to the target emitter's location. When this is accomplished sufficiently often, the antenna will track the emitter in a smooth motion. Three main techniques are used: conical scan, sequential lobing, and monopulse.[16] These techniques have been described in Chap. 11. Examples of how these techniques are applied in broadband systems will be described in the following paragraphs.

Conical Scan

Conical scan is achieved by mechanically moving the feed in a circle about the focal point in the focal plane of a paraboloidal reflector. Moving the phase center of the feed antenna off the focal axis causes the secondary beam to squint off the focal axis in the opposite direction; moving the feed in a circle then causes the peak of the secondary beam to move on a conical surface. If the feed maintains a fixed plane of polarization as the feed rotates, the feed is said to *nutate* as compared with a *rotating* feed, in which the plane of the polarization rotates with the feed. For ESM applications, for which linear polarization is required, the antenna must have a nutating feed because rotating the feed would modulate the signal at twice the scan rate, destroying the tracking information.

A conical-scan tracking antenna,[17] which operates over the 1- to 11-GHz band, is shown in Fig. 17-6a. The feed is a linearly polarized, ridge-loaded horn. A dielectric lens is used in the aperture to correct phase error. The decrease in secondary beamwidth with frequency is limited by the fact that above approximately 4 GHz the feed horn illuminates only a portion of the reflector. As the frequency increases above 4 GHz, the area illuminated is reduced sufficiently as a function of frequency to maintain essentially a constant secondary beamwidth (Fig. 17-6b). This allows the feed to be nutated at a constant radius about the focal point, with the crossover level being maintained within 4 dB of the beam peaks. Gain and boresight shift as measured on this antenna are also shown in Fig. 17-6b. While the constant-beamwidth technique results in a system whose gain does not increase with frequency, it does result in a system with a constant acquisition angle, which minimizes acquisition time.

FIG. 17-6a 1- to 11-GHz conical-scan tracking antenna.

Sequential-Lobing Antennas

Sequential lobing is very similar to conical scan, the difference being that whereas in conical scan the beam is moved continuously in a conical fashion, in sequential lobing the beam is switched between four positions. The beam is stepped from beam right to beam up, to beam left, to beam down. Signal strength is sampled at each position and compared with the opposite position to obtain tracking information.

Figure 17-7a shows a four-horn sequential-lobing tracking antenna. The horns are doubly ridged and hence provide linear polarization which can be rotated by remote control. Dielectric lenses correct any phase error in the aperture.

The beam-forming network used with this antenna system is shown in Fig. 17-7b. In the normal tracking mode, all four horns are combined with the proper phase to form the beam in one of the desired beam positions, the phase being introduced by the comparator quadrature hybrids. The comparator hybrids inherently provide a crossover level of approximately 3 dB on boresight independently of frequency. Diode switches select the proper path for each horn to achieve or change the beam position. Because the horns are separated by a fixed distance, grating lobes which can provide a multiple-tracking axis develop at the higher frequencies. To ensure that the true boresight is achieved, an acquisition mode is provided. Each horn is physically pointed 5° off the boresight direction. By combining only the horns on the right side, a beam squinted to the right is obtained. The width of the beam is the same as the individual horn beamwidth and hence is sufficiently wide to cover all grating lobes. Coaxial switches bypass the comparator hybrids for this mode. This mode provides a sufficient

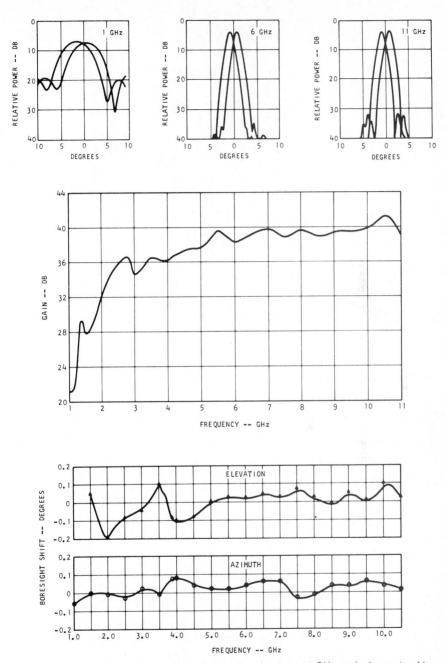

FIG. 17-6b Patterns, gains, and boresight shifts of a 1- to 11-GHz conical-scan tracking antenna.

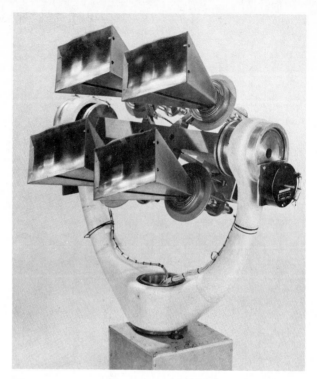

FIG. 17-7a Four-horn sequential-lobing tracking antenna (1 to 5 GHz).

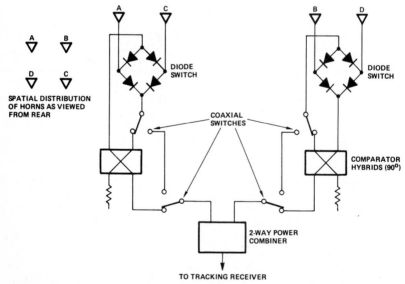

FIG. 17-7b Beam-forming network of the four-horn sequential-lobing tracking antenna (1 to 5 GHz).

error gradient to ensure that the system will lock on at the true boresight when it is switched to the normal tracking mode for better tracking accuracy.

Two-Channel Monopulse

Figure 17-8 shows a two-channel monopulse feed consisting of eight log-periodic monopole arrays on an octagonal cone. The outputs from the eight arrays are combined in a beam-forming network to form circularly polarized sum and difference patterns. For two-channel monopulse, both patterns are figures of revolution, with the difference pattern having a point null. The phase of the sum pattern changes from 0 to 360° around the boresight axis; for the difference pattern, the phase changes from 0 to 720°. Thus, the phase difference between the patterns defines the angle ϕ about the boresight axis (relative to an established reference plane) from which the target signal is arriving (Fig. 17-9), while the sum-to-difference-amplitude ratio defines the angle θ off the axis.

FIG. 17-8 Two-channel monopulse feed.

Two-channel monopulse systems lend themselves to broadband (36:1 bandwidth) tracking systems because beam-forming-network errors result in boresight shift rather than null fill-in, as is the case for three-channel systems. Null fill-in degrades tracking accuracy owing to degradation of error slope, while boresight shift in two-channel systems produces bias errors which in principle can be eliminated by calibration techniques.

Two-channel monopulse patterns are also inherent in multimode spiral antennas[18] in addition to the multiple log-periodic monopole array configuration. However, since spiral patterns turn about the boresight axis as a function of frequency, the reference plane at which the sum and difference channels are in phase also turns

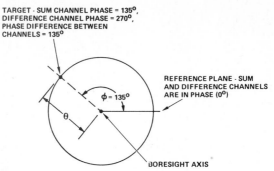

FIG. 17-9   Target location using two-channel monopulse techniques.

about the axis as a function of frequency, whereas in the log-periodic array configuration this plane is independent of frequency. Phase-compensation networks can be employed to stabilize the reference plane when spirals are used.

It should also be noted that, given a priori knowledge of the two-channel monopulse system's patterns and the orientation of the reference plane, this system can also be used for direction finding by measuring the amplitude ratio and phase difference between the sum and difference channels.

17-5   ECM TRANSMIT ANTENNAS

The antenna has been considered to be a transformer used to couple efficiently signals to or from free space. In that context, the problem for ECM-antenna designers has been to impedance-match the coupling device so that maximum power is transferred. Controlled distribution of the signal in space is a requirement of equal importance imposed upon the ECM antenna. Octave bandwidths or greater, medium to high radio-frequency (RF) power-carrying capacity, multiple-polarization response, and a low voltage standing-wave ratio (VSWR) have of course been additional requirements placed upon the ECM-antenna designer.

Antenna Elements

Many different types of antennas have been used in ECM systems. Some of the more common are broadband dipoles, broadband monopoles, spirals, horns and slots, log-periodics, and reflector types. During the past several decades, the RF power-carrying capacity and bandwidth of these basic types have been increased. As an example, the transmitting horns (shown in Fig. 17-10) used in an existing system exhibit a broad frequency bandwidth, high-power capability, high-aspect-ratio beam coverage, and circular-polarization response. Three-to-one-bandwidth, high-aspect-ratio circularly polarized horns have also been developed with satisfactory results.

The basic design problem associated with such horn configurations is the maintenance of a 90° phase differential between two (traveling-wave) quadrature fields in a waveguide structure. Frequency-compensated insertion-phase design techniques

FIG. 17-10 Multiband ECM transmitting horns. *(Courtesy of Transco, Inc.)*

applicable to the problem have been defined. A simplified analysis of the design of an octave-band circularly polarized horn will illustrate the approach applicable to a horn that uses a dielectric slab to obtain the 90° relative phase shift for the quadrature fields. The phase shift in the plane perpendicular to the wide horn aperture (*Y* axis, with the phasing plate) and the phase shift in the orthogonal plane (*X* axis) as illustrated in Fig. 17-11 were calculated; the results are shown in the figure. The calculations are approximate because average horn dimensions were assumed and the lens

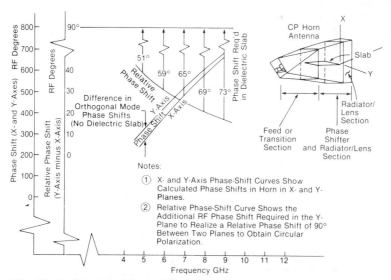

FIG. 17-11 Calculated phase shifts in an asymmetric-beam horn.

portion of the radiator section was assumed to be a waveguide section with solid-dielectric loading. The phase shift in a waveguide section loaded with dielectric is given by

$$\Delta \theta = 30.5X \left(F^2 K_e - \frac{34.9}{a^2} \right)^{1/2} \quad °$$

where X is the length of the waveguide section in inches, F is the frequency in gigahertz, K_e is the relative permittivity of the material within the guide if dielectric material is used, and a is the guide width in inches. ($K_e = 1$ for air-filled guide. The formula is derived from the standard waveguide-wavelength relationship.) An additional phase delay that increases with frequency in a complementary fashion is required in the Y plane to achieve circularly polarized operation (90° relative phase shift) over the complete RF band. The increase in differential-phase-shift requirements at the higher frequencies is compatible with the dielectric-slab phase-shift characteristics.

Extremely broadband monopoles are now available. Using thick (bladelike) configurations and special impedance matching, a 2:1 bandwidth design is relatively straightforward. One company is engaged in research and development to achieve up to a 6:1 bandwidth. Many broadband blades or monopoles have been used in ECM systems.

Fixed-Beam Arrays

An array can be used to create simultaneous fixed beams or to form beams with a high aspect ratio and of a desired shape. One example of a fixed-beam array is presented to illustrate current designs. Twelve horn elements are arranged in the arc of a circle, as illustrated in Fig. 17-12, to form a beam fixed in space that has broad azimuth coverage and a narrower elevation extent. This configuration is used to combine multiple traveling-wave-tube (TWT) powers in space. The azimuth beamwidth, in the plane of the four-horn subgroups, is broad as shown; the elevation beamwidth

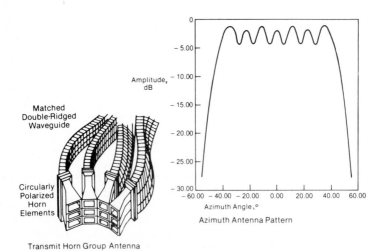

FIG. 17-12   Fixed-beam array for multiple power-amplifier application.

is dependent on the vertical (three-element) array configuration. The individual horn aperture size, both in azimuth and in elevation, must be selected with care. The horn beamwidths are chosen so that approximately one-fourth of the total desired coverage sector is "illuminated" by each horn; the flared section of the horn must be long to reduce aperture phase error, which minimizes phase center spacings. This reduces out-of-phase interference at angles on either side of the in-phase-beam-crossover points. The calculated pattern in Fig. 17-12 shows an approximate 3-dB variation in amplitude; in view of the wide variations in several "factors" affecting jammer-to-radar-signal ratios in a typical ECM application, this pattern variation is quite acceptable. The RF power of 12 TWTs feeding the horn group can be added in space if phase-tracking requirements are satisfied.

Arrays with Beam Control

Combinations of broad- and narrow-beam antenna response are often required. Narrow, or high-gain, beams can be directed to difficult-to-jam high-effective-radiated-power (ERP) pulse radars on a very-short-burst sequential basis, whereas the broad-coverage beam, used most of the time, can be employed against the high-duty-cycle, lower-ERP radars that may be spatially separated. High-quality narrow- and broad-beam patterns have been achieved over an octave bandwidth.

A patented[19] matrix-hybrid array configuration is shown in Fig. 17-13. The feed-system loss depends on the number of radiating elements; amplifiers are inserted near the radiators to negate this feed-loss characteristic. Simultaneous high-ERP beams with fixed positions in space can be created with the design. Beam switching and broad-beam operation with all inputs fed in phase are also possible by using this approach.

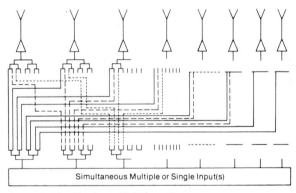

Simultaneous Multiple or Single Input(s)

FIG. 17-13   Matrix multiple-beam array.

Another multibeam, multifeed array configuration now receiving considerable attention for both airborne and shipborne applications is the Rotman lens.[20] RF amplifiers are normally inserted in the assembly between the lens feed and each radiating element. The Rotman lens antenna has several advantages: frequency-independent beam pointing is achieved, printed-microwave-circuit technology can be used to fabricate the lens, and very rapid control of high-ERP beams is possible because beam switching is exercised at the low-RF drive power. By using external-polarization-

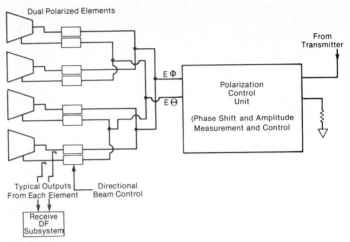

FIG. 17-14 Corporate-feed array.

adjustment techniques, a circularly polarized version of this antenna has been designed for a 3:1 bandwidth.

A corporate-feed array is illustrated in Fig. 17-14. This array offers some significant advantages: it is compatible with a single-output, high-power jammer; a zoom, or broad-beam pattern response, is available by adjusting the phase shifters in the array to yield a circular phase front; frequency-independent beam pointing is available if nondispersive phase shifters are used; and continuously pointed beams in any direction are available with fine phase-shifter adjustments. For applications in which single-aperture transmit-receive operation is needed, directional couplers can be added near the radiation elements for direction finding and beam control. A polarization-control unit is shown on the right side of the diagram. By using hybrids and phase shifters in the polarization section, any polarization response can be attained.

A concept for very-high-power amplifier paralleling and antenna switching is shown in Fig. 17-15. With the input switch in the position shown, the signals passing through the amplifiers will exhibit a phase front as illustrated; all the power is thus directed to the bottom horn antenna. Low-drive-power switching at the input can direct very large amounts of power to high-gain horns pointed at different angles at the outputs. Without switching, the back-to-back lens configuration can be used as a high-power broadband switch and/or power combiner.

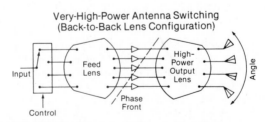

FIG. 17-15 Parallel-amplifier configurations.

17-6 ANTENNA ISOLATION IN JAMMING SYSTEMS

Isolation between transmitting and receiving antennas is an important consideration in an ECM system. Analyses and expressions for the calculation of field-strength propagation beyond obstacles have been published. Propagation around a curved surface (cylinders) and beyond knife edges (diffraction) is documented in Ref. 21.

Another antenna-coupling-analysis and measurements program[22] addressed the problem of defining the isolation between antennas separated by two conducting planes forming a corner with a variable included angle (see Fig. 17-16). The expressions for antenna coupling, determined by physical-optics analysis and by experiment, are

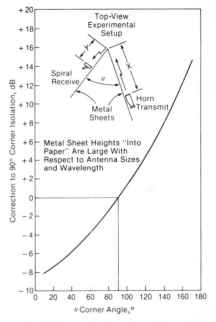

$$\text{Isolation} = \frac{K_1 G_T G_R \lambda^3}{x^2 Y} \text{ for } x > Y$$

$$= \frac{K_1 G_T G_R \lambda^3}{x Y^2} \text{ for } Y > x$$

K_1 is -36.5 dB for circularly polarized horns and spiral antennas and -39.5 dB for monopole antennas. x and Y are defined in Fig. 17-16. G_T and G_R are the transmit and receive antenna gains, each antenna being pointed at the common corner edge. The gain values are in power units with the gain values associated with the polarization perpendicular to the edge surface. These formulas yield reasonably accurate answers for antennas

FIG. 17-16 Variation of isolation with corner angle for antenna coupling beyond a corner obstacle.

moved 10 to 20° (measured to the corner) away from the ground planes. Measurements beyond 20° were not performed. The investigation was carried out in the 2- to 4-GHz range. Measured isolations for several different corner angles and antenna types correspond to the calculated isolations to within 2 dB for large variations in x and Y.

17-7 ADVANCED ARRAY SYSTEMS

Solid-state arrays offer hope for significant improvements in ECM systems.[23] By using a building-block approach, significant ERPs can be obtained because the gain of the array increases with the number of elements; also the total power increases with the number of elements (one amplifier per element). Moreover, transmission-line losses will be reduced because the power amplifiers are located near the radiators. An all-solid-state array configuration, using distributed field-effect-transistor (FET) amplifiers, with potential for full-polarization-diversity capability is shown in Fig. 17-17.

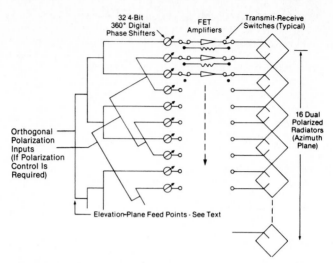

FIG. 17-17 30-kW-EIRP, 8 × 16 solid-state array with full-polarization-diversity capability.

An 8- by 16-element array is represented; only one layer of 16 elements in the azimuth plane is indicated. The orthogonal-polarization inputs near the left side of the diagram would, in turn, be fed by an elevation-plane feed network (not shown). Eight layers of antennas (one of which is shown), each layer including its own drive amplifier, are needed to achieve the 30-kW effective isotropic radiated power (EIRP) indicated in the figure. Appropriate feed-subsystem losses have been assumed. The eight layers in elevation would be fed by using a constant circular phase front so that beam control in azimuth only and direction finding in azimuth only would be required. When cost, element power output, and FET efficiencies are improved, this concept will be used in many applications.

REFERENCES

1 J. D. Kraus, *Antennas,* McGraw-Hill Book Company, New York, 1950, pp. 217–229.

2 H. Jasik, *Antenna Engineering Handbook,* 1st ed., McGraw-Hill Book Company, New York, 1961, pp. 3-10–3-13.

3 L. Young, L. A. Robinson, and C. A. Hacking, "Meander-Line Polarizer," *IEEE Trans. Antennas Propagat.,* vol. AP-21, no. 3, May 1973, pp. 376–378.

4 N. E. Lindenblad, "Antennas and Transmission Lines in the Empire State Television Station," *Communications,* vol. 21, no. 4, April 1941, pp. 13–14.

5 J. J. Epis and F. E. Robles, "Partial Radial-Line Antennas," U.S. Patent 3,831,176, 1974.

6 K. L. Walton and V. C. Sundberg, "Broadband Ridged Horn Design," *Microwave J.,* March 1964, pp. 96–101.

7 T. Sexon, "Quadruply Ridged Horn," Rep. EDL-M1160, Cont. DAAB07-67-C-0181, par. 2.1, 1968.

8 D. K. Barton, *Radar Systems Analysis,* Prentice-Hall, Inc., Englewood Cliffs, N.J., 1964, pp. 51–57, 275–310, 317–347.

9   M. I. Skolnik, *Introduction to Radar Systems,* McGraw-Hill Book Company, New York, 1962, pp. 476–482.

10   L. G. Bullock, G. R. Oeh, and J. J. Sparagna, "Precision Broadband Direction Finding Techniques," *Antennas Propagat. Int. Symp. Proc.,* University of Michigan, Ann Arbor, Oct. 17–19, 1967, pp. 224–232.

11   L. G. Bullock et al., "An Analysis of Wide-Band Microwave Monopulse Direction Finding Techniques," *IEEE Trans. Aerosp. Electron. Syst.,* vol. AES-7, January 1971, pp. 188–203.

12   N. M. Blachman, "The Effect of Noise on Bearings Obtained by Amplitude Comparison," *IEEE Trans. Aerosp. Electron. Syst.,* September 1971, pp. 1007–1009.

13   S. Silver, *Microwave Theory and Design,* McGraw-Hill Book Company, New York, 1949, sec. 13.8.

14   A. S. Dunbar, "Calculations of Doubly Curved Reflectors for Shaped Beams," *IRE Proc.: Waves and Electron. Secs.,* October 1948, pp. 1289–1296.

15   A. Brunner, "Possibilities of Dimensioning Doubly Curved Reflectors for Azimuth Search Radar Antenna," *IEEE Trans. Antennas Propagat.,* vol. AP-19, no. 1, January 1971, pp. 52–57.

16   Skolnik, op. cit, pp. 164–189.

17   K. L. Walton and V. C. Sundberg, "Constant Beamwidth Antenna Development," *IEEE Trans. Antennas Propagat.,* vol. AP-16, September 1968, pp. 510–513.

18   G. A. Deschamps and J. D. Dyson, "The Logarithmic Spiral in a Single-Aperture Multimode Antenna System," *IEEE Trans. Antennas Propagat.,* vol. AP-19, no. 1, January 1971, pp. 90–96.

19   E. Kadak, "Conformal Array Beamforming Network," U.S. Patent 3,968,695.

20   W. Rotman and R. F. Turner, "Wide-Angle Microwave Lens for Line-Source Applications," *IEEE Trans. Antennas Propagat.,* vol. AP-11, November 1963.

21   J. L. Bogdner, M. D. Siegel, and G. L. Weinstock, "Air Force ATACAP Program: Intra-Vehicle EM Capability Analysis," AFAL-TR-71-155, July 1971.

22   D. F. Yaw and J. G. McKinley, "Broadband Antenna Isolation in the Presence of Obstructions," Westinghouse Tech. Mem. EVTM-75-124, December 1975.

23   D. F. Yaw, "Electronic Warfare Antenna Systems—Past and Present," *Microwave J.,* September 1981, pp. 22–29.

24   R. H. Duhamel and G. G. Chadwick, "Frequency-Independent Antennas," in R. C. Johnson and H. Jasik (eds.), *Antenna Engineering Handbook 2/e,* McGraw-Hill Book Company, New York, 1984, Chap. 14.

Chapter 18

Radio-Telescope Antennas

John D. Kraus

The Ohio State University

18-1   RADIO TELESCOPES: DEFINITION

A radio telescope consists of an antenna for collecting celestial radio signals and a receiver for detecting and recording them. The antenna is analogous to the objective lens or mirror of an optical telescope, while the receiver-recorder is analogous to the eye-brain combination or a photographic plate. By analogy the entire antenna-receiver-recorder system may be referred to as a *radio telescope,* although it may bear little resemblance to its optical counterpart. Radio telescopes are used in much the same manner as optical telescopes for the observation and study of celestial objects. However, the appearance of the sky at radio wavelengths is very different from the way in which it looks optically. Thus, the sun is much less bright. On the other hand, the Milky Way radiates with tremendous strength, and the rest of the sky is dotted with radio sources almost entirely unrelated to any objects visible to the unaided eye (see Fig. 18-1).

The earth's atmosphere and ionosphere are opaque to electromagnetic waves with two principal exceptions, a band or window in the optical region and a wider window in the radio part of the spectrum. This radio window extends from a few millimeters to tens of meters in wavelength, being limited on the short-wavelength side by molecular absorption and on the long-wavelength side by ionospheric reflection. Because of this wide range of wavelengths many forms of antennas are used.

18-2   POSITION AND COORDINATES

The accurate position of a radio source is necessary to distinguish the source from others and to assist in its identification with optical objects when possible. The position is conveniently expressed in *celestial equatorial coordinates: right ascension* α and *declination* δ. The poles of this coordinate system occur at the two points where the earth's axis, extended, intersects the celestial sphere. Midway between these poles is the *celestial equator,* coinciding with the earth's equator, expanded.

The declination of an object is expressed in degrees and is the angle included between the object and the celestial equator. It is designated as a positive angle if the object is north of the equator and negative if south. For example, at the earth's equator a point directly overhead (the zenith) has a declination of $0°$, while at a north latitude of $40°$ the declination of the zenith is $+40°$.

The *meridian* is a great circle passing through the poles and a point directly overhead (the zenith). The *hour circle* of an object is the great circle passing through the object and the poles. The *hour angle* of the object then is the arc of the celestial equator included between the meridian and the object's hour circle. This angle is usually measured in hours.

A reference point has been chosen on the celestial equator. It is called the *vernal equinox.* The arc of the celestial equator included between the vernal equinox and the object's hour circle is termed the *right ascension* of the object. It is measured eastward from the vernal equinox and is usually expressed in hours, minutes, and seconds of time.

The right ascension and declination of an object define its position in the sky, independent of the earth's diurnal rotation. However, because of the earth's preces-

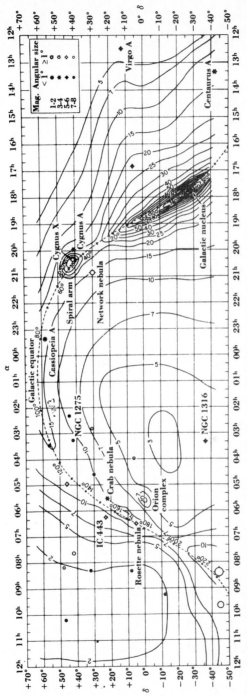

FIG. 18-1 Radio sky at 250 MHz as observed with the Ohio State University 96-helix radio telescope. (*After Ko and Kraus.[1]*)

sion, there is a gradual change in these coordinates for a fixed object in the sky, the change completing one cycle in 26,000 years. Thus, the right ascension and declination of an object will again be the same as they are now in 26,000 years. To be explicit, it is necessary to specify the date to which the right ascension and declination refer. This date is called the *epoch*. At present the epoch 1950.0 is commonly used (that is, the right ascension and declination are those of January 1, 1950), but epoch 2000.0 is being used increasingly.

Sometimes the positions of celestial objects are given in *galactic coordinates,* which are based on the geometry of our galaxy. These coordinates are independent of the earth and hence require no date or epoch. However, there are two systems, *old* (used before 1960) and *new* (used after 1960).[2] Their poles differ by about 1.5°.

By placing an antenna on an *equatorial* or *polar mounting,* that is, on axes one of which is parallel to the earth's axis and the other perpendicular to it, a source can be tracked as it moves across the sky by motion in only one coordinate (right ascension). If the antenna is mounted on vertical and horizontal axes, tracking requires motion in two coordinates. The coordinates in this case are *altitude* (or elevation) and *azimuth* (or horizontal angle around the horizon), and hence this type of mounting is called an *altazimuth* mounting.

18-3   BRIGHTNESS AND FLUX DENSITY

Radiation over an extended area of the sky is conveniently specified in terms of its brightness, that is, the power per unit area per unit bandwidth per unit solid angle of sky. Thus,

$$B = \text{brightness (Wm}^{-2}\text{Hz}^{-1}\text{sr}^{-1})$$

Solid angle may be expressed in steradians (sr) or in square degrees (deg²):

$$1 \text{ sr} = 57.3^2 \text{ deg}^2 = 3283 \text{ deg}^2$$

The integral of the brightness B over a given solid angle of sky yields the power per unit area per unit bandwidth received from that solid angle. This quantity is called the power *flux density, S.* Thus

$$S = \int \int B \, d\Omega = \text{power flux density (Wm}^{-2} \text{ Hz}^{-1})$$

where B = brightness, $\text{Wm}^{-2} \text{ Hz}^{-1} \text{ sr}^{-1}$

$d\Omega$ = element of solid angle, sr

Integrating the power flux density from a radio source with respect to frequency yields the *total power flux density* S_T in the frequency band over which the integration is made. Thus,

$$S_T = \int S \, df = \text{total power flux density (Wm}^{-2})$$

where S = power flux density, $\text{Wm}^{-2} \text{ Hz}^{-1}$

df = element of bandwidth, Hz

18-4   TEMPERATURE AND NOISE

A resistor of resistance R and temperature T matched to a receiver by means of a lossless transmission line as in Fig 18-2a delivers a power to the receiver given by

$$P = kT \, \Delta f \qquad\qquad (18\text{-}1)$$

where P = power received, W
  k = Boltzmann's constant = 1.38×10^{-23} J·K^{-1}
  T = absolute temperature of resistor, K
  Δf = bandwidth, Hz

If the resistor is replaced by an antenna of radiation resistance R, this equation also applies. However, the radiation resistance is not at the temperature of the antenna structure but at the effective temperature T of that part of the sky toward which the antenna is directed, as in Fig. 18-2b. In effect, the radiation resistance is distributed over that part of the sky included within the antenna acceptance pattern. From this point of view the antenna and receiver of a radio telescope may be regarded as a *bolometer* (or heat-measuring device) for determining the effective temperature of distant regions of space coupled to the system through the radiation resistance of the antenna.

The power received by a radio telescope from a celestial source is

$$P = SA_e \, \Delta f \qquad\qquad (18\text{-}2)$$

where P = power received, W
  S = source flux density, W m^{-2} Hz^{-1}
  A_e = effective aperture of telescope antenna, m^2
  Δf = bandwidth, Hz

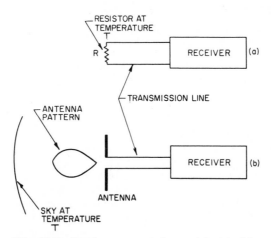

FIG. 18-2   Receiver connected to matched resistor and to antenna.

If this power equals that of a matched (calibration) resistor at temperature T connected to the receiver in place of the antenna, we have by equating Eqs. (18-1) and (18-2)

$$S = \frac{kT}{A_e} \qquad (18\text{-}3)$$

However, if the source is unpolarized (as is true of most celestial sources), only half of its power will be received by any antenna, so the flux density of the source is twice the above value, or

$$S = \frac{2kT}{A_e} \qquad (18\text{-}4)$$

where T = temperature of celestial object, K.

It should be noted that the celestial object is not necessarily at the thermal temperature T. Rather, T is the temperature that a blackbody radiator would need to have to emit radiation equal to that actually emitted by the object at the wavelength of observation. Hence, T is called the *equivalent temperature* of the object. This temperature is the same as the thermal temperature that the reference or calibration resistor must have to give the same response.

Sensitivity is a function of aperture size and also a system temperature and other factors. Thus, the sensitivity, or *minimum detectable (power) flux density*, is given by

$$S_{\min} = \frac{2k}{A_e} T_{\min} \qquad (18\text{-}5)$$

where k = Boltzmann's constant = 1.38×10^{-23} J·K^{-1}
$\quad T_{\min}$ = minimum detectable temperature, K
$\quad A_e$ = effective aperture = $\epsilon_{ap} A_p$, m^2
where ϵ_{ap} = aperture efficiency, dimensionless ($0 \leq \epsilon_{ap} \leq 1$)
$\quad A_p$ = physical aperture = $\pi (D^2/4)$ for circular aperture, m^2
where D = diameter of aperture.
The *minimum detectable temperature*

$$T_{\min} = \frac{K_s}{\sqrt{\Delta f t}} T_{\text{sys}} \qquad (18\text{-}6)$$

where K_s = receiver sensitivity constant, dimensionless ($1 \leq K_s \leq 2.8$)
$\quad \Delta f$ = intermediate-frequency bandwidth, Hz
$\quad t$ = output time constant, s
$\quad T_{\text{sys}}$ = system temperature, K
The *system temperature* depends on the noise temperature of the sky, the temperature of the ground and environs, the antenna pattern, the antenna thermal efficiency, the receiver noise temperature, and the efficiency of the transmission line or waveguide between the antenna and the receiver. The system temperature at the antenna terminals is given by

$$T_{\text{sys}} = T_A + T_{AP}\left(\frac{1}{\epsilon_1} - 1\right) + T_{LP}\left(\frac{1}{\epsilon_2} - 1\right) + \frac{1}{\epsilon_2} T_R \qquad (18\text{-}7)$$

where T_A = antenna noise temperature, K

$\quad T_{AP}$ = physical temperature of antenna, K

$\quad \epsilon_1$ = antenna (thermal) efficiency, dimensionless ($0 \le \epsilon_1 \le 1$)

$\quad T_{LP}$ = physical temperature of transmission line or waveguide between antenna and receiver, K

$\quad \epsilon_2$ = line or guide efficiency, dimensionless ($0 \le \epsilon_2 \le 1$)

$\quad T_R$ = receiver noise temperature, K

The *antenna temperature* is given by

$$T_A = \frac{1}{\Omega_A} \int \int T(\theta,\phi) P_n(\theta,\phi) \, d\Omega \qquad (18\text{-}8)$$

where Ω_A = antenna-pattern solid angle or antenna beam area, deg^2 or sr

$\quad T(\theta,\phi)$ = noise temperature of sky and environs as a function of position angle at wavelength of operation, K

$\quad P_n(\theta,\phi)$ = normalized antenna power pattern, dimensionless

$\quad d\Omega$ = elemental solid angle, deg^2 or sr

The *elemental solid angle* is given by

$$d\Omega = \sin \theta \, d\theta \, d\phi \qquad (18\text{-}9)$$

where θ = angle from zenith (assuming antenna is pointed at zenith), deg or rad

$\quad \phi$ = azimuthal angle, deg or rad

The *antenna beam area* is given by

$$\Omega_A = \int \int P_n(\theta,\phi) \, d\Omega \qquad (18\text{-}10)$$

The *receiver noise temperature* is given by

$$T_R = T_1 + \frac{T_2}{G_1} + \frac{T_3}{G_1 G_2} + \cdots \qquad (18\text{-}11)$$

where T_1 = noise temperature of first stage of receiver (usually a low-noise preamplifier), K

$\quad T_2$ = noise temperature of second stage, K

$\quad T_3$ = noise temperature of third stage, K

$\quad G_1$ = gain of first stage

$\quad G_2$ = gain of second stage

Additional terms may be required if the temperature is sufficiently high and the gain sufficiently low on additional stages.

Radio telescopes are often used at one end of a communication circuit, and for this application the *signal-to-noise ratio* is of prime importance. When a radio telescope is the receiving station, we have from the Friis transmission formula that

$$P_r = \frac{P_t A_{et} A_e}{r^2 \lambda^2} \qquad (18\text{-}12)$$

where P_r = power received, W

$\quad A_{et}$ = effective aperture of transmitting antenna, m^2

$\quad A_{er}$ = effective aperture of receiving or radio-telescope antenna, m^2

$\quad r$ = distance between transmitting and receiving station, m

$\quad \lambda$ = wavelength of operation, m

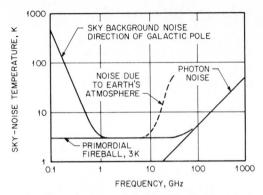

FIG. 18-3 Antenna sky noise temperature as a function of frequency. It is assumed that the antenna is pointed at the zenith, that the beam angle (HPBW) is less than a few degrees, and that the beam efficiency is 100 percent. (*After Kraus and Ko[3] below 1 GHz; after Penzias and Wilson[4] above 1 GHz.*)

TABLE 18-1   Radio Astronomy Units

Dimension or quantity	Symbol	Description	Unit
Brightness	B	$\dfrac{\text{Power flux density}}{\text{Solid angle}}$	$\text{Wm}^{-2}\text{Hz}^{-1}\text{sr}^{-1}$
Power flux density	S	$= \displaystyle\int\int B\, d\Omega$	$\text{Wm}^{-2}\text{Hz}^{-1}$
Total power flux density	S_T	$= \displaystyle\int S\, df$	Wm^{-2}
Power (in terms of temperature)	P	$= kT\,\Delta f$	W
Brightness temperature	T	$= \dfrac{B\lambda^2}{2k}$	K
Power flux density (in terms of temperature)	S	$= \dfrac{2kT}{A_e}$	$\text{Wm}^{-2}\text{Hz}^{-1}$
Minimum detectable flux density	S_{mim}	$= \dfrac{2kT_{\text{min}}}{A_e}$	$\text{Wm}^{-2}\text{Hz}^{-1}$
Minimum detectable temperature	T_{min}	$= \dfrac{K_s}{\sqrt{\Delta ft}}$	K

NOTE: W = watts; m = meters; Hz = hertz; sr = steradians; K = kelvins = degrees celsius (°C) above absolute zero; df = infinitesimal bandwidth; Δf = finite bandwidth = $f_2 - f_1$; λ = wavelength; J = joules; K_s = receiver constant.

By dividing this received power by the minimum detectable power of the radio telescope we obtain the signal-to-noise ratio as

$$\frac{S}{N} = \frac{P_r}{S_{\min}A_{er}\Delta_f} = \frac{P_t A_{et} A_{er}}{r^2 \lambda^2 k T_{\text{sys}}} \frac{\sqrt{t}}{\sqrt{\Delta f}} \qquad (18\text{-}13)$$

where P_t = transmitter power, W
   A_{et} = effective aperture of transmitting antenna, m^2
   A_{er} = effective aperture of receiving antenna, m^2
   t = output time constant, s
   r = distance between transmitting and receiving stations, m
   λ = wavelength of operation, m
   k = Boltzmann's constant = 1.38×10^{-23} J·K^{-1}
   T_{sys} = system temperature of radio telescope, K
   Δf = intermediate-frequency bandwidth, Hz

It is assumed that the receiver is a total-power type ($K_s = 1$) and that the radio telescope is matched to both the polarization and the bandwidth of the transmitted signal.

The antenna temperature T_A (which contributes to the system temperature T_{sys}, as discussed above) includes the contribution of the sky background and the temperature of the ground to the entire antenna pattern (main lobe and backlobes). In no case can T_A be less than 3 K, the limiting temperature of the sky resulting from the primordial fireball. The sky background temperature versus frequency is presented in Fig. 18-3.

Radio-astronomy units are summarized in Table 18-1.

18-5 RESOLUTION

The limiting resolution of an optical device is usually given by *Rayleigh's criterion.* According to this criterion, two identical point sources can just be resolved if the maximum of the diffraction pattern of source 1 coincides with the first minimum of the pattern of source 2.

Assuming a symmetrical antenna pattern as in Fig. 18-2a, Rayleigh's criterion applied to antennas states that the resolution of the antenna is equal to one-half of the beamwidth between first nulls; that is,

$$R = \frac{\text{BWFN}}{2} \qquad (18\text{-}14)$$

where R = Rayleigh resolution or Rayleigh angle
   BWFN = beamwidth between first nulls

An antenna pattern for a single-point source is shown in Fig. 18-4a. The pattern for two identical point sources separated by the Rayleigh angle is given by the solid curve in Fig. 18-4b, with the pattern for each source when observed individually shown by the dashed curves. It is to be noted that the two sources will be resolved provided that the half-power beamwidth (HPBW) is less than one-half of the beamwidth between first nulls, as is usually the case.

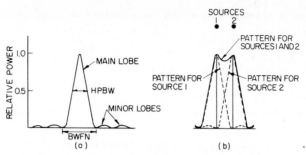

FIG. 18-4 Antenna power pattern (*a*) and power patterns for two identical point sources (*b*) separated by the Rayleigh angle as observed individually (dashed) and together (solid).

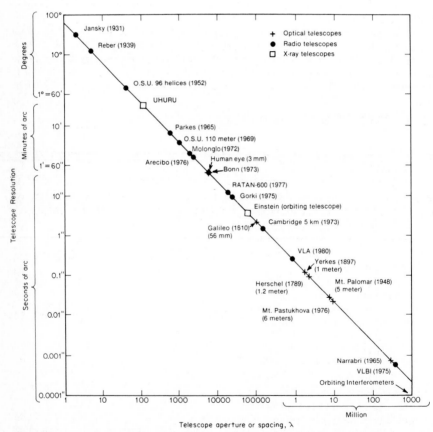

FIG. 18-5 Resolution of radio, optical, and x-ray telescopes as a function of the telescope aperture or interferometer spacing. *(From J. D. Kraus.[6])*

TABLE 18-2

Uniform Apertures	Rayleigh resolution (BWFN/2)	HPBW
Rectangular (length L_λ wavelengths)	$57.3°/L_\lambda$	$50.3°/L_\lambda$
Circular (diameter D_λ wavelengths)	$70°/D_\lambda$	$59°/D_\lambda$

It may be shown[5] that the beamwidth between first nulls for a broadside antenna with a uniform aperture many wavelengths long is given by

$$\text{BWFN} = \frac{114.6}{L_\lambda} °\qquad(18\text{-}15)$$

These results are summarized in Table 18-2, which also gives the beamwidths for uniform circular apertures. The HPBWs are some 15 percent less than the Rayleigh resolution angles for both rectangular and circular apertures.

For two-element interferometers the resolution is given by $57.3°/S_\lambda$, where S_λ is the spacing in wavelengths between the elements. For large spacings, extremely fine resolution can be obtained as indicated in Fig. 18-5, which presents the resolution of radio, optical, and x-ray telescopes.

18-6 PATTERN SMOOTHING

A unique pattern (or far-field pattern) of an antenna is obtained when the radiator is a point source situated at a sufficient distance from the antenna. The distance is sufficient if an increase produces no detectable change in the pattern. (See Sec. 18-8 for a more detailed discussion.) Let this pattern of a receiving antenna be as shown in Fig. 18-6a. If the point source is replaced by an extended source at the same distance, the observed pattern is modified as suggested in Fig. 18-6b. The extended source results in a broadened pattern with reduced minor lobes.

The patterns in Fig. 18-6 are in polar coordinates. In rectangular coordinates, the antenna pattern is as shown in Fig. 18-7a. The extended source distribution is shown in Fig. 18-7b. The observed pattern is then as shown in Fig. 18-7c. These three patterns are superposed in Fig. 18-7d to facilitate intercomparison.

It is to be noted that the observed pattern is only an approximation of the actual source pattern or distribution, the sharp shoulders of the source distribution

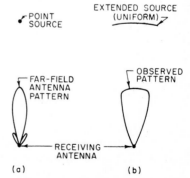

FIG. 18-6 Antenna patterns in polar coordinates for a point source and for an extended source.

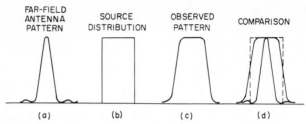

FIG. 18-7 Pattern in rectangular coordinates for antenna (*a*) and for uniform source distribution (*b*). The resultant observed pattern is at *c*. The three patterns are superposed for comparison in *d*.

being rounded off. It is said that the source distribution has been smoothed or blurred by the observing technique. The broader the antenna beamwidth compared with the source pattern, the greater the smoothing effect. On the other hand, the narrower the antenna beamwidth compared with the source pattern, the more nearly is the observed pattern an exact reproduction of the source distribution. However, there is always some smoothing, and one of the important problems of radio astronomy is to reconstruct, insofar as possible, the true source distribution from the observed pattern. It turns out that it is not possible to reconstruct the true source distribution since certain of the finer source details have no effect on the observed pattern and are irretrievably lost. However, partial reconstruction is usually possible, so that a source distribution that is more nearly like the exact source distribution than the observed pattern can be obtained. Pattern smoothing and related topics are discussed in detail in Kraus.[2]

18-7 CELESTIAL RADIO SOURCES FOR PATTERN, SQUINT, AND EFFICIENCY MEASUREMENTS OF ANTENNAS

Celestial radio sources are often useful for the measurement of far-field patterns, squint, and aperture efficiency of antennas, especially large-aperture antennas operating at short wavelengths when the distance to the far field is very large (10s or 100s of kilometers). (See Sec. 18-8 concerning the magnitude of this distance.)

For pattern measurements, the radio source should have a small angular extent (considerably less than the antenna HPBW), be relatively strong, and be well isolated from nearby sources. For aperture-efficiency measurements, accurate flux densities are required over a wide frequency range, and generally the source should be essentially unpolarized (less than 1 or 2 percent). For squint measurements, the position of the sources should be accurately known.

The upper part of Table 18-3 lists a few selected radio sources which meet most or all of the above requirements, that is, they are relatively strong, ½ to 1° from the nearest neighboring sources, of small angular extent, and essentially unpolarized and have accurate positions and also accurate flux densities over a wide frequency range. The lower part of the table lists three sources which do not meet all the above requirements but nevertheless may be useful for some purposes.

TABLE 18-3 Radio Sources for Pattern, Squint, and Efficiency Measurements

Source	Position (epoch 1950.0) Right ascension h	min	s	s error	Declination °	Arc min	Arc s	Arc s error	Size, arc min	Isolation, °	Flux density, (Jy)† 38 MHz	178 MHz	750 MHz	1400 MHz	2695 MHz	5000 MHz	Polarization, %	Distance, light-years
3C20	00	40	19.6	0.4	51	47	09	2	1	1	112	43	17	11.3	6.4	4.2	1.8	8×10^9 (z = 0.9)
3C196	08	09	59.4	0.2	48	22	07	2	0.25	1	166	68	23	13.9	7.7	4.4	1.2	3×10^9 (z = 0.16)
3C273	12	26	32.9	0.1	02	19	39	2	0.35	1	140	63	45	40	42	45	2	7.5×10^9 (z = 0.85)
3C295	14	09	33.6	0.3	52	26	14	2	0.3	0.5	94	83.5	35	22.4	12	6.5	0.4	3×10^9 (z = 0.15)
3C348	16	48	40.1	0.6	05	04	28	5	2	1	1,690	351	84	44.5	22.4	11.9	1.5	7×10^9 (z = 0.7)
3C380	18	28	13.4	0.2	48	42	40	2	0.25	0.5	211	59	22	14.4	10	7.5	1.2	
Cygnus A	19	57	44.5	0.5	40	35	02	2	1.6	...*	22,000	8,700	2,980	1,590	785	371	0.2	10^9
3C123	04	33	55.2	0.2	29	34	14	2	0.1	...	57⁻	189	72	45.8	27	16	1.2	
3C286	13	28	49.7	0.2	30	46	02	2	0.04	0.5	32	24	18	14.6	10	7.5	9	
Cas A	23	21	07		58	33	48		5	...*	37,200	11,600	3,880	2,410	1,470	910	...	11,000

*In complex galactic plane region.
†1 Jy = 1 jansky = 10^{-26} W m^{-2} Hz^{-1}.

18-13

The source designation refers to the third Cambridge (3C) catalog. Distances are given in light-years. The red shift z is also given. The red shift is the physical quantity measured from which a distance is inferred. Cygnus A is an exploded galaxy at a distance of 1 billion (10^9) light-years. Cassiopeia A (Cas A) is a supernova remnant (plasma cloud from an exploded star) at a distance of 11,000 light-years. Cas A is the strongest source in the sky except for the sun. Positions are from Bridle, Davis, Fomalont, and Lequeux,[7] and flux densities are from Kellermann, Pauliny-Toth, and Williams.[8] Flux densities at frequencies between those given in the table can be interpolated from the values given.

The source isolation (proximity of neighboring sources) was determined by inspection of the radio maps of the Ohio Sky Survey.[9] This survey is in seven parts with two supplements.

From the measurement of the antenna temperature difference (ΔT in °C) produced by the observed radio source, the effective aperture (A_e) of the antenna is given by

$$A_e = \frac{2k\Delta T}{S} \qquad (18\text{-}16)$$

where S = flux density of source at frequency of measurement, $\text{Wm}^{-2}\,\text{Hz}^{-1}$
$\qquad k = 1.38 \times 10^{-23}\,\text{J}\cdot\text{K}^{-1}$
The aperture efficiency is then

$$\epsilon_{ap} = \frac{A_e}{A_p} \qquad (18\text{-}17)$$

where A_e = effective aperture, m^2
$\qquad A_p$ = physical aperture m^2
For a detailed discussion of the measurement of celestial radio sources see Kraus.[2]

18-8 NEAR-FIELD–FAR-FIELD CONSIDERATIONS; RADEP

As has been mentioned, the far-field pattern of an antenna is obtained when measurements are made at a sufficiently great distance that an increase in distance produces no detectable change in the pattern. At shorter distances the pattern is different, its shape becoming a function of the distance.

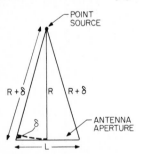

FIG. 18-8 Geometric relations for distance requirement.

The transition to this near-field pattern or patterns from the far-field pattern occurs when the source is sufficiently close that the phase of the incident wavefront across the antenna aperture departs significantly from a constant phase. Thus, if the distance δ in Fig. 18-8 is one-half wavelength, the incident field will be in opposite phase at the edges of the aper-

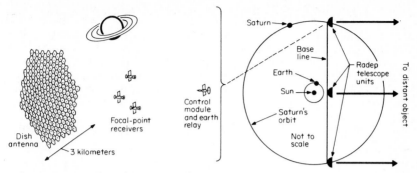

FIG. 18-9 The radep interferometer telescope proposed by a group of Soviet scientists could measure the distance of objects out to the full radius of the universe (15 billion light-years). Three units, each like the one shown at the left, would be deployed in space, two near the orbit of Saturn and one near the Earth.

ture as compared with the phase at the center. This could cause a very significant change in the observed antenna pattern. Thus, for accurate far-field patterns δ should be considerably less, and a value of one-sixteenth wavelength is commonly regarded as desirable. However, if we take δ = one-eighth wavelength, the relation for the required or critical distance becomes very simple; that is,

$$r \geq \frac{L^2}{\lambda} \qquad (18\text{-}18)$$

where r = required or critical distance, m
   L = aperture dimension, m
   λ = wavelength, m
For a more detailed discussion see Kraus.[5]

It has been suggested[10] that the transition between the near-field and far-field conditions be utilized to measure the distance of remote astronomical objects. This passive radio technique for measuring distance, called *radep* for *ra*dio *dep*th, requires radio telescopes in space to avoid atmospheric effects and also to permit the construction of telescopes of sufficient size.

It has been proposed that an interferometer radep telescope of three units be constructed, two units being deployed near the orbit of Saturn with the third near the Earth.[10] Operating at centimeter wavelengths, this radep array (see Fig. 18-9) has the potential, in principle, of measuring the distance of celestial objects out to the full radius of the universe (15 billion light-years).

REFERENCES

1 H. C. Ko and J. D. Kraus, "A Radio Map of the Sky at 1.2 Meters," *Sky Telesc.,* vol. 16, February 1957, p. 160.

2 J. D. Kraus, *Radio Astronomy,* McGraw-Hill Book Company, New York, 1966; Cygnus-Quasar Books, Powell, Ohio, 1982.

3 J. D. Kraus and H. C. Ko, "Celestial Radio Radiation," Ohio State Univ. Radio Observ. Rep. 7, May 1957.

4 A. A. Penzias and R. W. Wilson, "A Measurement of Excess Antenna Temperature at 4080 MHz," *Astrophys. J.,* vol. 142, 1965, p. 419.

5 J. D. Kraus, *Antennas,* McGraw-Hill Book Company, New York, 1950.

6 J. D. Kraus, *Our Cosmic Universe,* Cygnus-Quasar Books, Powell, Ohio, 1980.

7 A. H. Bridle, M. M. Davis, E. B. Fomalont, and J. Lequeux, "Flux Densities, Positions and Structures for a Complete Sample of Intense Radio Sources at 1400 MHz," *Astronom J.,* vol. 77, August 1972, p. 405.

8 K. I. Kellerman, I. I. K. Pauliny-Toth, and P. J. S. Williams, "The Spectra of Radio Sources in the Revised 3C Catalogue," *Astrophys. J.,* vol. 157, July 1969, p. 1.

9 J. D. Kraus, "The Ohio Sky Survey and Other Radio Surveys," *Vistas Astron.,* vol. 20, 1977, p. 445. This article gives a summary of the Ohio Sky Survey and includes a master map and complete references to all installments and supplements.

10 N. Kardashev, J. Shklovsky, and others, *Academy of Sciences USSR Report PR-373,* Space Research Institute, Moscow, 1977.

Index

About the Editors

Richard C. Johnson, Ph.D. has been actively engaged in applied research and development at the Georgia Tech Research Institute since 1956. A fellow of the Institute of Electrical and Electronics Engineers, Dr. Johnson is a former president of the IEEE Antennas and Propagation Society and former editor of the society's newsletter. He is the author of more than 70 technical papers and reports. He received his Ph.D. degree from the Georgia Institute of Technology.

Henry Jasik (deceased) was formerly vice president of the AIL Division of Eaton Corporation and director of its Antenna Systems Division. A well-known authority on antenna engineering and a fellow of the Institute of Electrical and Electronics Engineers, Dr. Jasik was cited for his contributions to the theory and design of VHF and microwave elements. He organized and edited the first edition of *Antenna Engineering Handbook*.